Structural Dynamic Systems Computational Techniques and Optimization

Reliability and Damage Tolerance

Gordon and Breach International Series in Engineering, Technology and Applied Science

Volumes 7–15

Edited by Cornelius T. Leondes

Books on **Structural Dynamic Systems Computational Techniques and Optimization**

Volume 7 *Computer-Aided Design and Engineering*

Volume 8 *Finite Element Analysis (FEA) Techniques*

Volume 9 *Optimization Techniques*

Volume 10 *Reliability and Damage Tolerance*

Volume 11 *Techniques in Buildings and Bridges*

Volume 12 *Seismic Techniques*

Volume 13 *Parameters*

Volume 14 *Dynamic Analysis and Control Techniques*

Volume 15 *Nonlinear Techniques*

Previously published in this series were volumes 1–6 on **Medical Imaging Systems Techniques and Applications**

Forthcoming in the ***Gordon and Breach International Series in Engineering, Technology and Applied Science***

Mechatronics Systems Techniques and Applications

Biomechanical Systems Techniques and Applications

Computer-Aided and Integrated Manufacturing Systems Techniques and Applications

Expert Systems Techniques and Applications

This book is part of a series. The publisher will accept continuation orders which may be cancelled at any time and which provide for automatic billing and shipping of each title in the series upon publication. Please write for details.

Structural Dynamic Systems Computational Techniques and Optimization

Reliability and Damage Tolerance

Edited by

Cornelius T. Leondes

Professor Emeritus
University of California
at Los Angeles

Gordon and Breach Science Publishers

Australia • Canada • China • France • Germany • India •
Japan • Luxembourg • Malaysia • The Netherlands •
Russia • Singapore • Switzerland

Amsteldijk 166
1st Floor
1079 LH Amsterdam
The Netherlands

British Library Cataloguing in Publication Data

Structural dynamic systems computational techniques and
optimization : reliability and damage tolerance. – (Gordon
and Breach international series in engineering, technology
and applied science ; v. 10 – ISSN 1026-0277)
1. Structural dynamics 2. Tolerance (Engineering)
I. Leondes, Cornelius T.
624.1'76

ISBN 90-5699-652-5

CONTENTS

SERIES DESCRIPTION AND MOTIVATION

Many aspects of explosively growing technology are difficult or essentially impossible for one author to treat in an adequately comprehensive manner. Spectacular technological growth is made stunningly manifest by any number of examples, but, just to note one here, the Intel 486 IBM-compatible PC was first introduced in late 1989. At that time the price of this PC was in the $10,000 range and it was thought to be much too powerful for widespread use. By early 1992, a little more than two years later, the price had dropped to $1,000 and it was felt that much more power was needed, leading directly to the Pentium IBM-compatible PC. A similar price reduction pattern has already been projected for the Pentium computer, and, in view of the recent history of the 486, it is difficult to suggest that the same "power hungry" pattern will not occur again in a similar time span. The Pentium is presently planned as a 1,000-MHz processor to be called the Flagstaff in the year 2000. The CD-ROM is presently evolving to the DVD (Digital Versatile Disk) with data storage capability of a greater order of magnitude. A DVD-ROM can hold a database of all the phone numbers and addresses in the United States, which would normally require multiple CD-ROMs. And the DVD format has room to grow. In any event, these examples and their clear implications with respect to the many application-oriented issues in diverse fields of engineering, technology and applied science and their continuing advances make it obvious that this series will fill an essential role in numerous ways for individuals and organizations.

Areas of major significance will be defined and world-class co-authors identified as contributors for essential volumes in respective areas. These areas will be determined by criteria including:

1. Will volumes fill important textbook voids in respective areas?
2. In some cases, a "time void" for an important area will clearly suggest the need for a volume. For example, the important area of Expert Systems might have a textbook void of several years that "requires" an important new volume.
3. Are these technology areas that simply cannot sensibly be treated comprehensively by a single author or even several co-authors?

Examples of areas requiring important volumes will be carefully defined and structured and might include, as the case arises, volumes in:

1. Medical imaging
2. Mechatronics
3. Computer network techniques and applications
4. Multimedia techniques and applications
5. CAD/CAE (Computer Aided Design/Computer Aided Engineering)
6. FEA (Finite Element Analysis) techniques and applications
7. Computational techniques in structural dynamic systems
8. Neural networks (as might possibly be suggested by a significant time void in the textbook literature)
9. Expert Systems (again, depending on a possible significant time void).

One of the most important aspects of this series will be that, despite rapid advances in technology, respective volumes will be defined and structured to constitute works of indefinite or "lasting" reference interest.

SERIES PREFACE

The first industrial revolution, with its roots in James Watt's steam engine and its various applications to modes of transportation, manufacturing and other areas, introduced to mankind novel ways of working and living, thus becoming one of the chief determinants of our present way of life.

The second industrial revolution, with its roots in modern computer technology and integrated electronics technology — particularly VLSI (Very Large Scale Integrated) electronics technology, has also resulted in advances of enormous significance in all areas of modern activity, with great economic impact as well.

Some of the areas of modern activity created by this revolution are: medical imaging, structural dynamic systems, mechatronics, biomechanics, computer-aided and integrated manufacturing systems, applications of expert and knowledge-based systems, and so on. Documentation of these areas well exceeds the capabilities of any one or even several individuals, and it is quite evident that single-volume treatments — whose intent would be to provide practitioners with useful reference sources — while useful, would generally be rather limited.

It is the intent of this series to provide comprehensive multi-volume treatments of areas of significant importance, both the above-mentioned and others. In all cases, contributors to these volumes will be individuals who have made notable contributions in their respective fields. Every attempt will be made to make each book self-contained, thus enhancing its usefulness to practitioners in a specific area or related areas. Each multi-volume treatment will constitute a well-integrated but distinctly titled set of volumes. In summary, it is the goal of the respective sets of volumes in this series to provide an essential service to the many individuals on the international scene who are deeply involved in contributing to significant advances in the second industrial revolution.

PREFACE

Structural Dynamic Systems Computational Techniques and Optimization

Reliability and Damage Tolerance

The principal causes of failure in structural dynamic systems are the exceeding of maximum system design limits and structural fatigue. These causes are analyzed in this volume and concern the use of stochastic processes interacting with finite element analysis along with other techniques. Once these aspects are understood and applied, methods for enhancing the reliability and damage tolerance of structural systems can be utilized in specific instances. Numerous illustrative examples are included.

This is the fourth volume in the set of 9 volumes on structural dynamic systems. Subjects treated are:

1. Computer-Aided Design and Engineering
2. Finite Element Analysis (FEA) Techniques
3. Optimization Techniques
4. Reliability and Damage Tolerance
5. Techniques in Buildings and Bridges
6. Seismic Techniques
7. Parameters
8. Dynamic Analysis and Control Techniques
9. Nonlinear Techniques.

In the first chapter of this volume, Casciati and Colombi tell us failure of a structural element due to fatigue is a complex phenomenon affected by a considerable number of factors. Finite or boundary element techniques allow evaluation of stress and strain fields in a cracked structural component. Damage accumulation rules or fracture-mechanics-based fatigue-crack-propagation laws provide tools to evaluate fatigue lifetime, i.e., time (or number of duty cycles) to failure. Nevertheless, most design quantities, such as fatigue resistance, external loading and initial crack size, are inherently random in nature. For a rational design, therefore, these quantities cannot be treated as deterministic and safety of the structural element against fatigue failure should be judged in a probabilistic sense. This chapter is an in-depth treatment of structural fatigue crack propagation under environmental conditions, and illustrative examples are included — as, indeed, they are throughout the book.

According to Parton and Zobnin (chapter 2) despite many significant advances in the field of fracture mechanics and its numerous applications, formulation and solution of the dynamic problems of this theory remained unknown, until recently, because of their complicated nature. Only the latest elegant analytic solutions of certain model problems and development of new effective numerical methods have helped in surmounting the obstacles involved. In order to understand growing interest toward investigations in dynamic fracture mechanics, it is necessary to grasp the essence of the subject and its interaction with quasi-static fracture mechanics. Indeed, the process of fracture is characterized (at least in its final stage) by a rapid propagation of the arterial crack or a set of branched cracks and is therefore essentially a dynamic process. Results and developments obtained from fracture mechanics can be utilized to enhance or ensure the strength, reliability and extended life of structures.

In chapter 3, Aliabadi describes fundamental concepts of boundary integral formulations in elastodynamics. Computational algorithms required to implement these formulations in practical engineering analysis also are presented. Dual boundary element methods are given as an efficient method for the solution of crack problems in elastodynamics. The method is combined with a time domain, Laplace transform formulation or the dual reciprocity method. The quarter-point elements or the J integral are used to evaluate the dynamic stress intensity factors. Indirect boundary element formulations, known as fictitious stress method and displacement discontinuity method, are covered; these formulations are presented in the Laplace transform space. Several examples of crack problems subjected to dynamic loadings are addressed for both two- and three-dimensional problems.

Damage from earthquakes can be extensive, and destroy buildings, bridges or urban highways, and other structures, we are told by Saadatmanesh in chapter 4. Because this damage or destruction can have serious economic consequences, which, in turn, can have negative impact on economies of major countries on the international scene, it is essential to address these issues in the current initial construction of bridges and highways for major urban arteries. In actual fact, many such structures are already in place in major urban areas throughtout the world; therefore, it is important to address effective methods for enhancing the reliability and damage tolerance of such essential constructions.

The final chapter in this book, by Jian-Jun Chen and Bao-Yan Duan, presents a comprehensive treatment of reliability aspects in structural dynamics for both exceeding structural maximum stress design limit as well as structural reliability failure analysis for the case of structural fatigue. Techniques for optimization of structural reliability for both maximum design stress exceedance loads as well as structural loading under fatigue conditions are covered.

This volume on reliability and damage tolerance in structural dynamic systems clearly reveals the effectiveness and essential significance of techniques available and, with further development, the essential role they will play in the future. The authors are all to be highly commended for their splendid contributions; these papers will provide a significant and unique reference source for students, research workers, practitioners, computer scientists and others on the international scene for years to come.

1 FATIGUE CRACK PROPAGATION UNDER ENVIRONMENTAL ACTIONS

F. CASCIATI[1] and P. COLOMBI[2]

[1]*Department of Structural Mechanics, University of Pavia,*
Via Abbiategrasso 211, I27100, Pavia, Italy
[2]*Department of Structural Engineering, Polytechnic of Milan,*
Piazza L. Da Vinci 32, I20133, Milan, Italy

1.1. INTRODUCTION

The failure of a structural element by fatigue is a very highly complex phenomenon affected by a considerable number of factors. Finite or boundary element techniques allows one to evaluate the stress and strain fields in a cracked structural component.[1] Either damage accumulation rules or fracture-mechanics-based fatigue-crack- propagation laws provides then the tool to evaluate the fatigue lifetime, i.e. the time (or number of duty cycles) to failure. Nevertheless, most of the design quantities, as the fatigue resistance, the external loading and the initial crack size, are inherently random in nature. For a rational design, therefore, these quantities cannot be treated as deterministic and the safety of the structural element against fatigue failure should be evaluated in a probabilistic sense. The fatigue process is composed by three different phases:

1. crack nucleation;
2. propagation of a macroscopic crack;
3. failure of the structural element;

The fatigue crack nucleation process cannot yet be regarded as fully understood. Under cyclic loading conditions there is a migration of dislocations that result in localized plastic deformations. Moreover stress concentrations always takes place at a microscopic level. Due to a detoriation of the material structure, microscopic cracks are created. They grow and join together to produce major cracks. Depending upon the material properties and the applied loading, the nucleation phase can be of different importance in estimating the fatigue life. As indicated by experimental observations, at low load levels the crack initiation period may consume a significant percentage of the usable fatigue life, whereas, at high load amplitude, fatigue cracks just develop in the early cycles. In problems where defects are practically unavoidable due to the fabrication process, crack propagation may also begin at the first load application. Since the propagation of a dominat crack will be addressed in this chapter, the nucleation phase is not taken into account. Indeed a macrocrack is supposed to be present at the beginning of the operation period of the structural element.

For the fatigue crack growth process the randomness of the external loading, of the fatigue resistance, of the initial shape and position of the crack, of the geometry of the crack and of the structural element cannot be neglected. This suggests the development of fatigue crack growth reliability model to evaluate the lifetime distribution. This chapter studies the stochastic fatigue crack propagation of a dominant crack up to the final failure: Section 1.4 explicitly deals with this subject, after the necessary background relations have been built in Sections 1.2 and 1.3. As well as in a deterministic context two alternatives approaches can be followed. The first one makes use of fracture mechanics concepts: the fatigue crack growth rate depends on the applied load, on the crack size and on the material parameters. The second approach adopts the Wöhler (or $S - N$) curves. Under variable amplitude loading, damage accumulation laws can then be used to predict the fatigue damage of the structural element. The remainder of this introduction gives some fracture mechanics concepts, cumulative damage accumulation laws and information on the sequence effect. It is the necessary background for the fatigue crack growth models discussed later.

The fatigue failure of the structural element is eventually discussed in Section 1.5.[24]

1.1.1. Fracture Mechanics Concepts

The estimation of the fatigue lifetime of a structural component requires the use of fracture mechanics principles in order to describe the crack extension and the stress distribution near the crack tip. If the applied

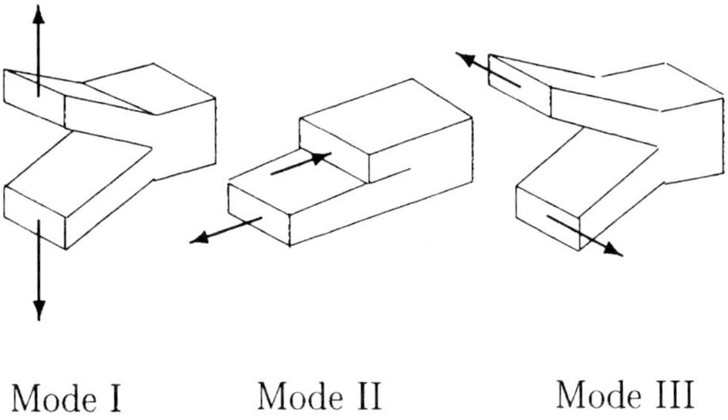

Figure 1. Crack extension modes.

loading generates sufficiently low stress levels (not exceeding the limit of elasticity) the stress distribution in a cracked element can be calculated by linear elastic theory[9,10,40] (i.e. assuming a linear elastic behaviour in the material of the structural element under consideration). The stress field near the crack tip can be divided into three basic types, each of them associated with a different mode of crack surface displacement (see Figure 1).

These three modes are:

1. mode I: opening or tensile mode;
2. mode II: sliding or shearing mode;
3. mode III: tearing or antiplane mode.

In fatigue analysis, mode I is the most common and has received the greatest attention. Using the methods of linear elastic theory, the stresses at any point in the element can, in principle, be characterized. In general they cannot be given in explicit form. But for the mode I crack opening of Figure 2 the stresses in the neighborhood of the crack tip are given as:

$$\begin{Bmatrix} \sigma_x \\ \sigma_y \\ \sigma_{xy} \end{Bmatrix} = \frac{K_I}{\sqrt{2\pi \cdot r}} \cdot \cos\frac{\theta}{2} \cdot \begin{Bmatrix} 1 - \sin\frac{\theta}{2} \cdot \sin\frac{3\theta}{2} \\ 1 + \sin\frac{\theta}{2} \cdot \sin\frac{3\theta}{2} \\ \sin\frac{\theta}{2} \cdot \sin\frac{3\theta}{2} \end{Bmatrix} \tag{1}$$

where r and θ are cylindrical coordinates with origin in the crack tip. Similar expressions are also available for modes II and III.[40] The factor K_I in Equation (1) is the stress intensity factor. This factor defines the magnitude of the local stresses around the crack tip. It depends on loading, crack size, crack shape and geometry of the specimen.

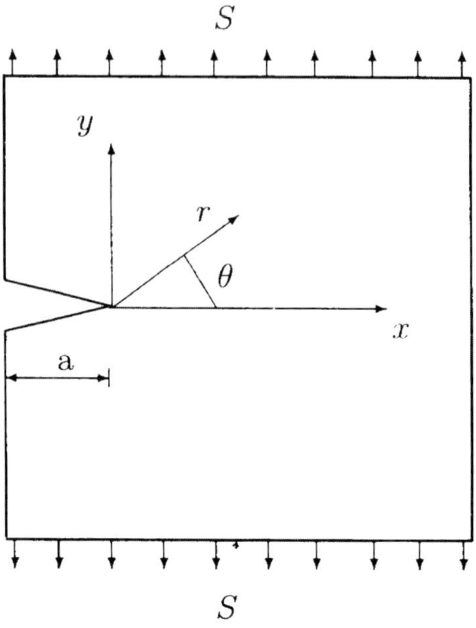

Figure 2. Mode I: axis orientation with respect to the crack tip.

In general, the stress intensity factor has the form[9,63,73]:

$$K_I = Y(a) \cdot S \cdot \sqrt{\pi \cdot a} \tag{2}$$

where a is the crack length, S represents the far-field stress resulting from the applied load and $Y(a)$ is a factor that accounts for the shape of the specimen and the crack geometry. It is seen that the stress field has a singularity at $r = 0$. Since infinite stresses cannot exist in a physical material, the elastic solution has to be modified in order to allow for the crack tip plasticization. Thus a plastic zone around the crack tip is usually introduced. However if the plastic zone is sufficiently small, in comparison with the crack length and the specimen geometry, the stress field solution given by Equation (1) can still be applied outside the plastic zone. The size and shape of the plastic zone can be estimated in terms of the stress intensity factor and of the yield stress σ_y. For the case of a purely plastic zone embedded in an elastic matrix, the extension r_p of the roughly circular plastic zone is:

$$r_p = \frac{K_I^2}{2\pi\sigma_y^2} \tag{3}$$

for plane stress[9,10], and:

$$r_p = \frac{K_I^2}{18\pi\sigma_y^2} \tag{4}$$

for plane strain.[9,10] Note that the extension of the plastic zone is greater for plane stress than for plane strain. Equations (3) and (4) underestimate the extension of the plastic zone due to the stress redistribution effect. For this reason the previous relations should be replaced by:

$$r_p = \frac{K_I^2}{\zeta^2\pi\sigma_y^2} \tag{5}$$

where ζ is a coefficient which accounts for the stress field distribution around the crack tip. Putting Equation (2) into Equation (5) one has:

$$r_p = \frac{S^2 \cdot Y^2(a) \cdot a}{\zeta^2 \cdot \sigma_y^2} \tag{6}$$

The critical value K_{Ic} of K_I is refered to as fracture thoughness.[9,10]

1.1.2. Cumulative Damage Laws

A systematic analysis of fatigue damage accumulation is due to Wöhler. He introduced the so-called Wöhler curves or $S - N$ diagrams where the stress amplitude ΔS is reported as a function of the number of cycles to failure:

$$N \cdot \Delta S^b = k \tag{7}$$

or:

$$N = k \cdot \Delta S^{-b} \tag{8}$$

where k and b are material parameters of a random nature due to the randomness which characterizes the fatigue crack growth process. Experimental evidence shows that the structural element is not damaged if the stress amplitude is smaller than a threshold value ΔS_0 (i.e. the lifetime is unbounded in this case):

$$N = \begin{cases} k \cdot S^{-b} & S > S_0 \\ \infty & S \leq S_0 \end{cases} \tag{9}$$

The Wöhler diagrams, although used to design structural elements subjected to fatigue, do not represent the fatigue processes accurately from a physical point of view. Moreover the constant amplitude idealization of the stress process required by the $S - N$ curves is not realistic. Palmgren and Miner studied the fatigue damage accumulation due to time varying loading using a linear accumulation model. The Palmgren-Miner damage accumulation scheme postulates that the damage fraction D_i associated with stress amplitude ΔS_i is proportional to the ratio between the number of cycles n_i with stress amplitude ΔS_i and the number of cycles to failure N_i obtained from the Wöhler curves:

$$D_i = \frac{n_i}{N_i} = \frac{n_i}{k \cdot S_i^{-b}} \qquad n_i \leq N_i \tag{10}$$

The total damage is then computed as:

$$D = \sum_i D_i = \sum_i \frac{n_i}{N_i} = \sum_i \frac{n_i}{k \cdot S_i^{-b}} \tag{11}$$

and failure takes place when $D \geq 1$. Note that the fracture resistance is modelled here as the number of cycles to failure N_i. Moreover the Palmgren-Miner law cannot model the sequence effect which arises under variable amplitude load cycles.

1.1.3. Sequence Effect in Crack Growth

A rational analysis of crack growth under environmental actions requires the modeling of the external loads as a random process. In this chapter they are treated as a stationary Gaussian random process and the sequence effect will be taken into account.[27,29,64,77,82] Load sequence effect are observed in fatigue experiments under variable amplitude loading: the lifetime, in fact, strongly depends on the arrangement of the sequence of load maxima.[59] The lifetime appears to be greater for a sequence of cycles with decreasing load maxima than for a sequence with increasing arrangement of maxima. Several experiments show rapid changes of the fatigue crack propagation rate after a load cycle with maximum of greater amplitude (overload).[53,61] This phenomenon is called the retardation of the crack growth. The duration and the intensity of the retardation phase depend on many factors including specimen geometry, environmental effect, material properties, magnitude of the overload and of subsequent extremes. The physical nature of this phenomenon has not yet been completely explained. The following mechanisms have been suggested to rationalise it:

- crack tip blunting[71];
- crack tip strain hardening and the formation of a favourable residual stress field ahead the crack tip[71];
- plasticity induced fatigue crack closure[31];
- crack branching and micro-roughness of fractures surface.[72]

Though all these mechanisms are present and observed in fatigue experiments to affect the post-overload fatigue crack growth, the plasticity-induced fatigue crack closure is generally considered as a dominant cause of the retardation in Mode I of fatigue crack growth.[61] Most of the models that were proposed in the literature to predict the fatigue crack growth with regard to the load sequence effect refer to the overload induced plastic zone and a decrease of the effective stress intensity factor range after an overload.[60,78] This scheme can be qualitatively described as follows.[31,71] The predicted elastic stress at a crack tip is infinite. Indeed, there is always a plastic zone at the crack tip. A tensile load opens the crack and plastically deforms the material at the tip. As the load is released, the material at the crack tip is compressed and the crack actually closes. In fact the material at the crack tip is plastically deformed in tension while the bulk of the material remains elastic. When the rest of the material seeks its original shape upon unloading, it acts as clamp on the crack tip. This produce a compressive stress at the crack tip that depends on the past loadings. A large tensile load produces a large plastic zone and a large residual compressive stress, which does not allow the crack to reopen until a sufficient tensile load has counteracted the compression at the crack tip. The part of the stress range that is necessary to overcome the residual compression does not contribute to crack growth. The effective stress range is the difference between the maximum stress and the stress at which the crack tip opens. Such a model is also used in the present chapter to assess some probabilistic characteristics of the structural lifetime given a critical crack length.[29,77,82]

1.1.4. Scope and Organization

As already stated, this chapter investigates the fatigue crack growth under random loading allowing for the load sequence effect. In Section 1.2 deterministic fatigue crack growth models are introduced. Both constant and variable amplitude load cycles are considered. The Veers model is used as a standard crack growth scheme: it includes the load sequence effect.

Random load models, which idealize the loading in a form suitable for the adopted crack growth model, are outlined in Section 1.3. The models range from a simple random variables description, that is useful in sequenceless applications, to random process simulations that include the effect of the actual load sequence.

Section 1.4 introduces the probabilistic crack growth model. First a Markov chain approximation of the extreme sequence is illustrated and discussed. Based on this assumption the fatigue crack growth phenomenon is described as a Markov diffusion process. A simulation technique is then proposed to evaluate the relevant drift and diffusion coefficients.

Section 1.5 addresses the problem of fatigue crack growth reliability under random loading. The model proposed in the previous section is adopted to describe sequence effect under random loading. Randomness in the material fatigue resistance and initial crack size is introduced. The resultant fatigue lifetime distribution is computed by suitable reliability methods.

1.2. FATIGUE CRACK GROWTH MODELS

Results of laboratory experiments and observations of in-service structures clearly indicate that the fatigue life N_f (defined as the number of cycles to grow to a critical value of the crack length) is affected by a variety of factors, e.g. the state of stress, the material properties, the temperature and other environmental effects. The fatigue crack growth rate is given as a function of classical fracture mechanics parameters such as the stress intensity factor. Different expressions are available for fatigue crack under constant and variable amplitude load. In the latter case, in fact, the sequence effect play an important role: without it one underestimates the actual fatigue lifetime. In this section fatigue crack propagation laws under constant and variable amplitude load cycles are introduced and discussed.

1.2.1. Constant Amplitude Fatigue Crack Growth

A wide class of fatigue crack growth laws can be written in the general form:

$$\frac{da}{dN} = F(a, \theta) \tag{12}$$

where N indicates the number of cycles corresponding to the crack length a and θ represents the factors (stress range, material properties and environmental conditions) which affect the fatigue crack growth rate. Investigations of fatigue crack growth in elastic materials have shown that the most suitable quantity for characterizing the fatigue crack growth rate is the stress intensity factor K or more specifically the stress intensity factor range $\Delta K = K^+ - K^-$:

$$\frac{da}{dN} = F(a, \Delta S) = F(\Delta K) \tag{13}$$

The Paris-Erdogan law is the most widely used model within this class of fatigue crack growth equations:

$$\frac{da}{dN} = C \cdot (\Delta K)^m = C \cdot (Y(a) \cdot \Delta S \cdot \sqrt{\pi a})^m \tag{14}$$

where C and m are material constants. Note that the Paris-Erdogan equation can be reformulated in terms of a classical damage accumulation rule, the Palmgren-Miner rule. Integrating Equation (14) the parameter of the fatigue damage accumulation rule b and k (see Equation 7) can be evaluated:

$$\begin{aligned} b &= m \\ k &= \frac{1}{C} \cdot \int_{a_0}^{a_c} \frac{da}{(Y(a) \cdot \sqrt{\pi a})^m} \end{aligned} \tag{15}$$

The parameters C ed m are not constant as in the Paris law but depend on environmental conditions as the temperature and the stress rate R:

$$R = \frac{S^-}{S^+} \tag{16}$$

To model the effect of the stress rate on crack propagation, Equation (13) is rewritten in the form:

$$\frac{da}{dN} = F(\Delta K, R) \tag{17}$$

Different expressions are available in the literature for the function $F(\cdot)$ in Equation (17). For example[9,10]:

$$\frac{da}{dN} = C \cdot \left(\frac{1}{1-R}\right)^2 \cdot (\Delta K)^3 \tag{18}$$

Another possibility is[9,10]:

$$\frac{da}{dN} = C \cdot (1 + \varsigma \cdot R)^q \cdot (\Delta K)^q \tag{19}$$

where ς and q are material constants to be estimated from experiments results. The most flexible fatigue crack growth equation, capable of modelling the stress ratio effect, is the Forman law:

$$\frac{da}{dN} = \frac{C \cdot (\Delta K)^m}{(1-R) \cdot K_{Ic} - \Delta K} \tag{20}$$

where K_{Ic} is the fracture thoughness of the material.

1.2.2. Variable Amplitude Fatigue Crack Growth

Different models have been proposed in the literature to predict fatigue crack growth under variable amplitude load. Since the stress intensity factor range ΔK plays a key role in the case of constant amplitude loading, it is usually believed that this quantity should also be a main factor in dealing with fatigue crack growth under variable amplitude loading. One introduces the effective stress intensity factor range ΔK_{eff} to model the fatigue crack growth rate under variable amplitude load:

$$\Delta K_{\text{eff}} = \Delta S_{\text{eff}} \cdot Y(a) \cdot \sqrt{\pi a} \tag{21}$$

In this chapter the Veers model[77,82] will be used to define the term ΔS_{eff}. Other models are available in the literature[21,64] as for example the Elber model[31], the equivalent-stress-intensity- factor-range model[64], the Wheeler model[79], the Nelson model[51] and the Willemborg model.[80] The Veers model is adopted since it is capable of predicting accurately the fatigue crack growth. Moreover it is also capable of reflecting the relative changes in fatigue life due to different loadings and material choices when a minimum of data are available. When the loading is random and probabilistic analyses or numerical simulations are required, a simple model, as the Veers model, is very valuable. Its simplicity will reduce, in fact, the computation time involved in repeated calculations of crack growth. In the Veers model the effective stress cycle amplitude ΔS_{eff} is given by:

$$\Delta S_{\text{eff}} = S^{+} - S_{op} \tag{22}$$

where S_{op} indicates the opening stress. Experimental evidence shows that the opening stress under variable amplitude loading temporally increases after the application of an overload. As a consequence this decreases the effective stress amplitude, ΔS_{eff}, and eventually the fatigue crack growth rate. In some retardation models[79,80], the retarded fatigue crack growth after an overload is assumed to continue as long as the current plastic zones, due to the maxima following the overload, are contained in the plastic zone produced by the overload (see Figure 3).

Since Elber[31] noticed the crack closure phenomenon and pointed out its significance for fatigue crack growth, the effective stress cycle amplitude (see Equation (22)), instead of ΔS, is usually considered in fatigue crack equations when S_{op} is greater than the current stress minimum S^{-}. The stress at the crack openeing S_{op} is constant under constant amplitude but varies under variable amplitude and depends on the last previous stress extremes. In the literature there is no universal formula describing such a relation. Most of the proposals are based on experimental data.[11] The bilinear form (see Figure 4):

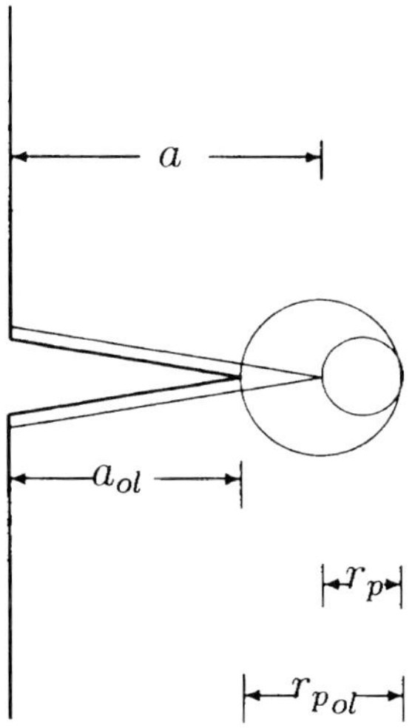

Figure 3. Plastic zone produced by the overload and plastic zone produced by the current load cycle.

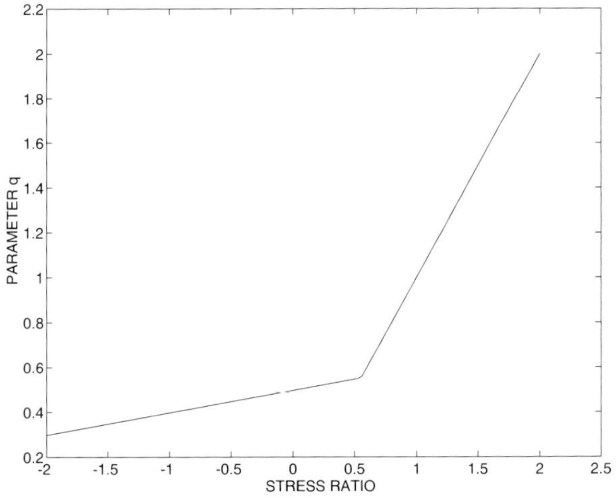

Figure 4. Parameter q vs. the stress ratio R as given by Veers ($q_0 = 0.496$ and $R_0 = -5$).

$$q = \frac{S_{ap}}{S_r} = \max\left\{q_0 \cdot \left(1 + \frac{R}{|R_0|}\right), R\right\} \tag{23}$$

proposed by Veers[77] is used in this chapter with $q_0 \in [0.2, 0.5]$ and $R_0 \in [-5, -2]$ as material parameters. The ratio $R = \frac{S^-}{S^+}$ in Equation (23) defines the stress rate (see Equation (16)).

The crack closure effect resulting from crack tip plasticity is usually modelled by referring to the plastic zone that develops at the crack tip due to the stress cycle maximum. The range of the plastic zone can be estimated as (see Equation (6)):

$$r_p(a, S) = \frac{S^2 \cdot Y^2(a) \cdot a}{\zeta^2 \cdot \sigma_y^2} \tag{24}$$

where σ_y is the yielding stress. The coefficient ζ depends on the stress state at the crack tip.[9,10,40] In order to specify the retardation intensity Veers[77] introduced the so-called reset stress S_r. It determines the stress level that is necessary to reset the maximum extend of the plastic zone $r_p(a_{ol}, S_{ol})$, produced by the overload S_{ol} when the crack length is equal to a_{ol}. Applying the equality (see Figure 3):

$$a_{ol} + r_p(a_{ol}, S_{ol}) = a + r_p(a, S_r) \tag{25}$$

the reset stress can be calculated as:

$$S_r - \frac{\zeta \cdot \sigma_y}{Y(a)} \cdot \sqrt{\frac{a_{ol} + r_p(a_{ol}, S_{ol}) - a}{a}} \tag{26}$$

Generalising the concept of crack opening stress $S_{op} = q \cdot S^+$ at constant amplitude loading for variable amplitude loading, where $S_{op} = q \cdot S_r$, the effective stress amplitude ΔS_{eff} can be explicitly written as:

$$\Delta S_{\text{eff}} = \begin{cases} S^+ - q \cdot S^+ & \text{if } S^+ > S_r \text{ and } S^- < q \cdot S^+ \\ S^+ - S^- & \text{if } S^+ > S_r \text{ and } S^- > q \cdot S^+ \\ S^+ - q \cdot S_r & \text{if } S_r > S^+ > q \cdot S_r \text{ and } S^- < q \cdot S_r \\ S^+ - S^- & \text{if } S_r > S^+ > q \cdot S_r \text{ and } S^- \geq q \cdot S_r \\ 0 & \text{if } S^+ \leq q \cdot S_r \end{cases} \tag{27}$$

Introducing the reset stress S_r, which is a variable governing the retardation effect and depending on the current crack length a, on the crack length a_{ol} at the time of application of the last overload S_{ol} and on the overload value itself, a convenient property of the fatigue crack growth equations (see Equation (12)) is lost: the crack length a and the number of cycles N are no longer separable anymore.

1.3. MODELS FOR RANDOM LOADING

Although crack growth rates are generally defined in terms of constant amplitude loadings, service loads on a structural component show generally variable amplitude. Variable amplitude loads can be described by the sequence of maxima and minima or by a limited number of statistics that reflect the load characteristics. One common way to represent variable amplitude loads is to list precise sequences of peaks and valley in a sample block.[49,50] In this way only a single sample of all possible load sequences is represented and, also, artificial sequences effect may be introduced in the sample block. Another way of specifying a random loading is through a frequency-of-exceedence diagram[10], which defines for a single block the number of times that the load exceeds any given level. This is equivalent to a description of the loading peaks as a random variable. For crack growth analysis without sequence effect, random variables models provide an efficient method of calculating crack growth by summing contributions from all the cycles: the need for cycle-by-cycle integration is therefore avoided. In this section first a definition of random loading is given. Simulation techniques are then introduced. The main counting methods are eventually illustrated and compared.

1.3.1. Definition of Random Loading

To retain a complete description of both the load peaks and ranges and the possible load sequence, the loading must be described as a random process. Random loadings are often described by their frequency content through the power spectral density and simulated in the frequency domain.[3,5,25] Estimating the effect of a random loading requires knowledge of both the appropriate loading statistics and analysis techniques that make use of this statistics. In this chapter it is assumed that the random process is both stationary and ergodic.[6,25,45] A stationary process idealizes a steady state condition, i.e. the mean, the variance and the other statistics are constant in time. This also means that the correlation between the process values at any two points in time only depends on the time between the points and not on the absolute time. It is not possible to say that any real process is stationary in a strict sense, but loading processes can often be divided into segments with stationary behavoiur. The ergodic property means that esemble averages are equivalent to time averages: averaging the values of different realizations of the process at a single time is equivalent to averaging a single realization over the time. The basic descriptors of a stationary and ergodic random loading $\underline{x}(t)$ are[6,25,45]: its average $\mu_{\underline{x}}$ and the standard deviation $\sigma_{\underline{x}}$. The mean is the central tendency; the standard deviation measures the variation about the mean. The time varying nature of the loading is described by the correlation between the load values at any two times; it can be put in the form of the autocorrelation function $R_{\underline{x}}(\tau)$.[6,25,45] For a stationary process, it only depends on the time lag τ and is defined as:

$$R_{\underline{x}}(\tau) = E[\underline{x}(t)\underline{x}(t+\tau)] \tag{28}$$

Note that when $\tau = 0$, the autocorrelation is equal to the second moment $E[\underline{x}^2(t)] = \mu_{\underline{x}}^2 + \sigma_{\underline{x}}^2$ of the process. The correlation length of the stationary process $\underline{x}(t)$ is defined as:

$$\tau_{cor} = \frac{\int_0^\infty \tau \cdot R_{\underline{x}}(\tau)}{\int_0^\infty R_{\underline{x}}(\tau)} \tag{29}$$

The two-sided power spectral density $S_{\underline{x}}(\omega)$ is defined as the Fourier transform of $R_{\underline{x}}(\tau)$.[6,25,45]:

$$S_{\underline{x}}(\omega) = \frac{1}{2\pi} \cdot \int_{-\infty}^{\infty} R_{\underline{x}}(\tau) e^{-i\omega\tau} d\tau \tag{30}$$

Since the autocorrelation is a real valued function which is symmetric about $\tau = 0$ the power spectral density is also real and symmetric. To avoid negative frequencies, a one-sided power spectral density is often introduced over the positive frequency axis. It assumes values double of the magnitude of the two-sided spectral density $S_{\underline{x}}(\omega)$. The integral of the spectral density is equal to the variance of the process, $\sigma_{\underline{x}}^2$. Wide band processes are composed of frequency components from a wide band of frequencies. Narrow band processes have most of their energy concentrated in a narrow range of frequencies. Some useful measures of the load bandwidth can be calculated in term of the spectral moments m_k, i.e weighted averages of the power spectral density[45]:

$$m_k = \int_{-\infty}^{\infty} \omega^k \cdot S_{\underline{x}}(\omega) d\omega \tag{31}$$

The irregularity factor α is defined as[45]:

$$\alpha = \frac{m_2}{\sqrt{m_0 \cdot m_4}} \tag{32}$$

and can be regarded as the ratio between the mean-up-crossing rate and the peak rates. The bandwidth parameter ϵ is defined in term of the regularity factor α as[45]:

$$\epsilon = \sqrt{1 - \alpha^2} \tag{33}$$

For a narrow band process $\alpha \simeq 1$ and $\epsilon \simeq 0$.

1.3.2. Simulation of Random Loading

The power spectral density defines the distribution of energy in a random process as a function of frequency. A time domain realization of the random process can be generated by discretizing the power spectral density into frequency bands and summing the resulting sine waves.[3,5,62] The random nature of the process is preserved by making the relative phase of each sine wave an independent, uniformly distributed, random variable on the interval $[0, 2\pi]$. The resulting time domain realization will approach a Gaussian process as the number of frequency components becomes large:

$$S(t) = \sum_{i=1}^{N} [2 \cdot S_{\underline{S}}(-\bar{\omega} + i \cdot \Delta\omega) \cdot \Delta\omega]^{1/2} \cdot \cos[(-\bar{\omega} + i \cdot \Delta\omega) \cdot t + \phi_i] \quad (34)$$

where ϕ is the random phase and $\bar{\omega}$ is the central frequency of each band. In practice actual sine waves are rarely added together to simulate a time series. One of the difficulties in using this algorithm to generate time series is the high resolution required by the need of an accurate definition of peaks and troughs. This error can be reduced by fitting a parabola to the three points nearest each peak and extrapolating to the true peak value.

1.3.3. Counting Methods

Cycle counting is straighforward for a sine wave, but ambiguities arise as soon as the bandwidth is not very narrow.[7,8] Several cycle counting algorithms have been commonly used for wide band load histories. The three most popular algorithms[7] are the all transitions count, the zero-crossing transitions count, and the so-called rainflow count:

1. The all transition count is the most straightforward method (see Figure 5a). Each transition between a local extremum and the following one is counted as half a cycle. Its main drawback is that large bandwidth signals (for instance a signal with a high frequency component superimposed on a lower frequency one) result in a very large number of very small cycles and almost no large ones;
2. With the zero crossing transition count only the highest maximum H^+ between an up-zero-crossing and the next down-zero-crossing is retained. In the same way for minima, only the lowest minimum H^- between a down-zero-crossing and the next up-zero-crossing (see Figure 5b) is retained. This method has the drawback that it requires to know the average load values ("zero"). Moreover, cycles of small amplitude without zero-crossing are lost;

3. Rainflow count[30] is an algorithm that attempts to avoid the drawbacks of the two preceding methods. As a matter of fact, rainflow counting replaces some of the transitions between extremum values with transitions defined in the zero-crossing count. It provides a higher number of small cycles than the first scheme and a higher number of large cycles than the second one. The rainflow algorithm is not easy to implement and its computational complexity may be computer-resource consuming. The algorithm was recently reformulated by Rychlik.[55] Let u be the intensity of the maximum of the load process at time t. Also, let t^+ be the time of the first up-crossing of the level u and t^- the time of the first down-crossing of the level u. One has (see Figure 5c):

$$\begin{aligned} \Delta H^-(t) &= \max_{t^- < \tau < t} \{S(t) - S(\tau)\} \\ \Delta H^+(t) &= \max_{t < \tau < t^+} \{S(t) - S(\tau)\} \end{aligned} \tag{35}$$

The cycle amplitude is then defined as:

$$\Delta S(t) = \min[\Delta H^+(t), \Delta H^-(t)] \tag{36}$$

1.4. FATIGUE CRACK GROWTH UNDER RANDOM LOADING

In this section the fatigue crack propagation model for fatigue crack growth under random loading, recently proposed in[21–23], is outlined. Most of the proposed deterministic model refers to the plastic zone produced by the overload at the crack tip and introduces the concept of effective stress intensity factor range ΔK_{eff}. In this section the model developed by Veers[77], and described previously, is adopted and a diffusion model for the fatigue crack growth process is developed. This technique is often used in literature to solve the fatigue crack propagation problem when randomness is present in the definition of the fatigue crack growth resistance[12,14–17,42,43,65,67–69,74] or in both the definition of the external loading (without sequence effect) and the fatigue crack growth resistance.[28,37,52,75,76] The in-service fatigue life is modelled as a sequence of successive *retardations* and *post-retardation phases*. During the retardation phase the fatigue crack growth is retarded, i.e. the fatigue crack growth rate is reduced. The retardation phase is eventually followed by a post-retardation phase with the actual fatigue crack growth rate. Each couple of retardation and post-retardation phases is called *load block* which is started and closed by an overload. The extreme sequence is modelled as a Markov chain. This makes possible the evaluation of the probability distribution of some load parameters such as maxima, minima,

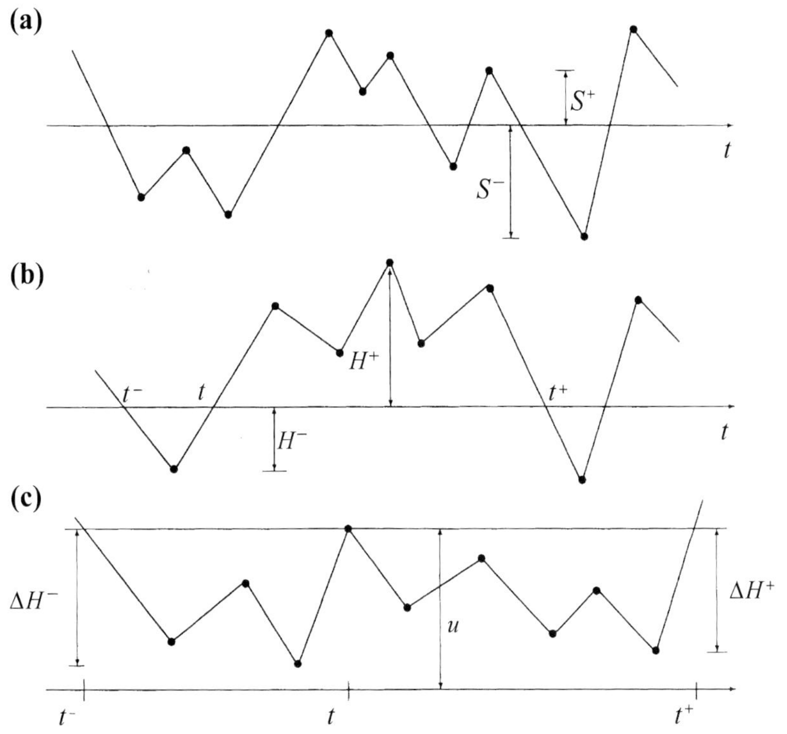

Figure 5. Cycles counting methods.

and cycle amplitudes in a very simple way. The drift and diffusion coefficients of the proposed diffusion model are evaluated by simulation, making use of the Markov chain approximation of the extreme sequence.

1.4.1. Markov Chain Approximation of the Extreme Sequence

The extremes are the only parameters of the external loading which are present in Equation (27). Their probabilistic characterization is the information one needs for the evaluation of the probability distribution of the fatigue life.[41,45] A full probabilistic description of the extreme sequence is not available in a general case. Recently in[46,54,56] it was shown that the extreme sequence for a wide class of stationary Gaussian and non Gaussian processes can be well approximated by a Markov chain. Let $[u_i, u_j]$ be the

l-th stress cycle of the external loading with maximum $u_i = \underline{S}^+(l)$ and minimum $u_j = \underline{S}^-(l)$. By a statistical analysis, one can evaluate the transition matrix $\mathbf{P} = [P_{ij}]$ between a maximum and the consequent minimum and the transition matrix between a minimum and the consequent maximum $\hat{\mathbf{P}} = [\hat{P}_{ij}]$. These transition matrices are then defined as:

$$P_{ij} = \mathrm{Prob}[\underline{S}^-(l+1) = u_j | \underline{S}^+(l) = u_i] \tag{37}$$

and

$$\hat{P}_{ij} = \mathrm{Prob}[\underline{S}^+(l) = u_j | \underline{S}^-(l) = u_i] \tag{38}$$

Of course, by definition, $\mathbf{P}$ is an upper triangular matrix and $\hat{\mathbf{P}}$ is a lower triangular matrix (in a cycle in fact the maximum is always greater than the corresponding minimum). The transition matrix between a maximum and the next one in the loading sequence is denoted as $\mathbf{P}^+ = [P_{ij}^+]$ and is given by (see Figure 6):

$$\mathbf{P}^+ = \mathbf{P} \cdot \hat{\mathbf{P}} \tag{39}$$

The transition matrices defined above allow one to calculate the probability distribution function of several parameters of the load cycles (maximum, minimum, cycle amplitude etc.).

The probability distribution of the maximum, $\mathbf{p}^+ = [p_i^+]$, in fact, is given by (see Figure 6):

$$\begin{gathered} \mathbf{p}^+ = \mathbf{p}^+ \cdot \mathbf{P} \cdot \hat{\mathbf{P}} \\ \sum_i p_i^+ = 1 \end{gathered} \tag{40}$$

In a similar way the probability distribution of the minimum $\mathbf{p}^- = [p_i^-]$ can be evaluated as:

$$\begin{gathered} \mathbf{p}^- = \mathbf{p}^- \cdot \hat{\mathbf{P}} \cdot \mathbf{P} \\ \sum_i p_i^- = 1 \end{gathered} \tag{41}$$

The probability distribution $\mathbf{p}^{\Delta S} = [p_i^{\Delta S}]$ of the cycle amplitude:

$$\Delta\underline{S}(l) = \underline{S}^+(l) - \underline{S}^-(l) \tag{42}$$

is given by:

$$p_i^{\Delta S} = \sum_{u_j} \hat{P}_{j(i+j)} \cdot p^-{}_j \tag{43}$$

Of course the Markov approximation of the extreme sequence is not possible for all the load processes. One needs a criterion to check if the approximation is acceptable from a practical point of view.[36,47,54] Let:

$$N_T(u, v) = \{[x, y] \;\; x > u \geq v > y\} \tag{44}$$

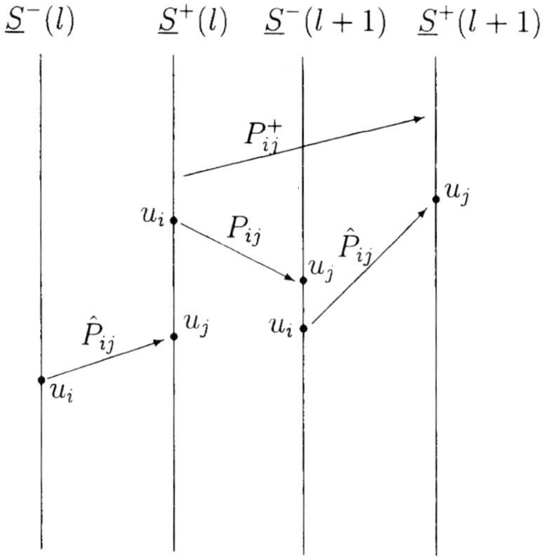

Figure 6. Computational scheme for the evaluation of the transition matrices $\mathbf{P}$, $\hat{\mathbf{P}}$ and $\mathbf{P}^+$ and of extreme probability distribution function.

be the number of cycles in the time interval $[0, T]$ with amplitude greater than the one of the load cycle $[u, v]$. Then $N_T(u, u)$ indicates the number of crossing of the level u in the $[0, T]$. Let $\Omega(u, v)$ the mean value of $N_T(u, v)$:

$$\Omega_T(u, v) = E[\underline{N}_T(u, v)] \tag{45}$$

For a stationary process one has:

$$\Omega_T(u, v) = T \cdot \Omega(u, v) \tag{46}$$

where $\Omega(u, v)$ indicates the value of $\Omega_T(u, v)$ for a unit interval. Let:

$$p_k = \sum_{u_l \geq u_{j+1}} P_{kl} \tag{47}$$

The entries of vector $\mathbf{p}$ are the conditional probabilities that a maximum u_k is followed by a minimum lower than u_j. For a given cycle $[u_i, u_j]$ one has:

$$\Omega_T(u_i, u_j) = c \cdot \boldsymbol{\pi} \cdot \mathbf{d} \tag{48}$$

where:

$$\begin{aligned} \boldsymbol{\pi} &= [p^+{}_1\, p^+{}_2 \ldots p^+{}_{i-1}] \\ \mathbf{d}^T &= [p_1\, p_2 \ldots p_{i-1}] \end{aligned} \tag{49}$$

The constant c in Equation (48) is evaluated by a normalization procedure.[36] A check of the Markov approximation of the extreme sequence is then obtained by comparing $\Omega_T(u, v)$ (Equation (48)) with simulation results. A second check is obtained by comparing $\Omega_T(u, u)$ with simulation results (or the exact solution when available).

1.4.2. Markov Diffusion Approximation of the Fatigue Life

For random loads, any maximum can potentially be an overload. With reference to the fatigue crack growth model introduced above, the maximum $\underline{S}^+(k)$ is an overload if and only if:

$$\underline{S}^+(k-1) \leq \underline{S}^+(k) \quad \text{and} \quad \underline{S}^+(k+1) \leq \underline{S}^+(k) \tag{50}$$

Then if the condition on $\underline{S}^+(k)$ introduced in Equation (50) is satisfied a retardation block starts and continues as long as:

$$\underline{S}^+(l) \leq \underline{S}_r \tag{51}$$

where l is a given load cycle in the retardation block. If the condition introduced in Equation (51) is not satisfied a post-retardation phase starts and continues as long as:

$$\underline{S}^+(l-1) \leq \underline{S}^+(l) \leq \underline{S}^+(l+1) \tag{52}$$

One can extend this scheme to the whole fatigue life which is then idealized as a sequence of blocks. In each block one has a retardation and a post-retardation phase (see Figure 7).

The block concept is the base of the probabilistic analysis of the fatigue crack growth which will be performed in the next subsections. The block length will be indicated hereafter as $\underline{n}_b$. In the next section a diffusion model for the fatigue crack growth process is introduced and discussed in detail.

1.4.3. Probability Distribution of the Fatigue Life

As discussed previously the in-service life is viewed as a sequence of blocks. Each block is given by a retardation and post-retardation phase. In practical applications the correlation length n_{corr} of the extreme sequence is much smaller than the block length mean value $E[\underline{n}_b]$. Moreover the mean value $E[\underline{N}_f]$ of the fatigue life is much greater than $E[\underline{n}_b]$. The fatigue process is idealized as a continous process with time measured by duty cycles since the number of cycles to failure is very large.

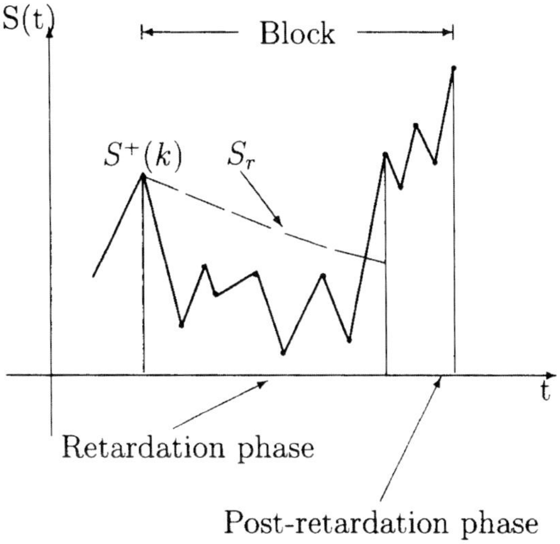

Figure 7. Retardation and post-retardation phase in a given block.

The property $n_{corr} \ll E[\underline{n}_b] \ll E[\underline{N}_f]$ suggests then to introduce a Markov approximation of the fatigue crack growth equation.[2,4,66] For this purpose, consider the Paris law:

$$\frac{d\underline{a}}{dN} = C \cdot (Y(\underline{a}) \cdot \sqrt{\pi \underline{a}})^m \cdot \Delta \underline{S}_{\text{eff}}{}^m \tag{53}$$

By Markov approximation, one substitutes Equation (53) with an Itô stochastic differential equation[66]:

$$\frac{d\underline{a}}{dN} = \eta(a) + d(a) \cdot \underline{\xi}(N) \tag{54}$$

where $\eta(a)$ and $d(a)$ are the drift and diffusion coefficients, respectively, and $\underline{\xi}(t)$ is a Gaussian white noise.[45,66] The stochastic averaging technique[2,4,44,70] gives for the coefficients $\eta(a)$ and $d(a)$:

$$\begin{aligned} d^2(a) &= (C \cdot (Y(a) \cdot \sqrt{\pi a})^m)^2 \cdot \alpha_{\Delta S_{\text{eff}}} \\ \eta(a) &= \bar{\eta}(a) + \frac{1}{4} \cdot \frac{d}{da} d^2(a) \end{aligned} \tag{55}$$

where:

$$\bar{\eta}(a) = C \cdot (Y(a) \cdot \sqrt{\pi a})^m \cdot E[\Delta \underline{S}_{\text{eff}}] \tag{56}$$

In Equation (55) $\alpha_{\Delta S_{\text{eff}}}$ is the equivalent white noise intensity[44,66], i.e. the area under the autocorrelation function $R_{\Delta \underline{S}_{\text{eff}}}(\tau)$ of $\Delta \underline{S}_{\text{eff}}(t)$:

$$\alpha_{\Delta S_{\text{eff}}} = \int_{-\infty}^{\infty} R_{\Delta \underline{S}_{\text{eff}}}(\tau) d\tau \tag{57}$$

The Markov approximation of the fatigue crack growth process results in an inverse Gaussian distribution of the fatigue life.[26] The probability density function $p_{\underline{N}_f}(N_f)$ is given by[64]:

$$p_{\underline{N}_f}(N_f) = \sqrt{\frac{d}{2\pi}} \cdot N_f^{-3/2} \cdot \exp\left[-\frac{d \cdot (N_f - c)^2}{2 \cdot c^2 \cdot N_f}\right] \tag{58}$$

while the probability distribution function is equal to[64]:

$$P_{\underline{N}_f}(N_f) = \Phi\left[\sqrt{\frac{d}{N_f}} \cdot \left(\frac{N_f}{c} - 1\right)\right] + e^{2d/c} \cdot \Phi\left[-\sqrt{\frac{d}{N_f}} \cdot \left(1 + \frac{N_f}{c}\right)\right] \tag{59}$$

The parameters c and d are functions of the mean value and standard deviation of the fatigue life:

$$\begin{aligned} c &= E[\underline{N}_f] \\ d &= \left(\frac{E[\underline{N}_f]}{\sigma_{\underline{N}_f}}\right)^3 \end{aligned} \tag{60}$$

The moments of the fatigue life are evaluated as the solution of the following system of differential equation[64]:

$$\eta(a_0) \cdot \frac{d}{da_0} E[\underline{N}_f{}^n(a_0)] + \frac{1}{2} \cdot d^2(a_0) \cdot \frac{d^2}{da_0^2} E[\underline{N}_f{}^n(a_0)] = -n \cdot E[\underline{N}_f{}^{n-1}(a_0)] \tag{61}$$

where $E[\underline{N}_f{}^0(a_0)] = 1$. The following boundary conditions must be forced to the moments $E[\underline{N}_f{}^n(a_0)]$[64]:

$$\begin{aligned} E[\underline{N}_f^n(a_0)] &= 0 \text{ per } a_0 = a_c \\ \frac{d}{da_0} E[\underline{N}_f^n(a_0)] &= 0 \text{ per } a_0 = 0 \end{aligned} \tag{62}$$

where a_0 and a_c are the initial and critical crack size, respectively. For a Markov process the parameters $\eta(a)$ and $d^2(a)$ are the drift and diffusion coefficients; they can be regarded as the mean value and second moment of the crack size increment in one cycle:

$$\begin{aligned}\bar{\eta}(a) &= E[\underline{a}(i+1) - \underline{a}(i)|\underline{a}(i) = a]\\ d^2(a) &= E[(\underline{a}(i+1) - \underline{a}(i))^2|\underline{a}(i) = a]\end{aligned} \tag{63}$$

The simulation procedure introduced for evaluating the drift and diffusion coefficients[21–23] will be briefly illustrated in the next section.

1.4.4. Characterization of the Markov Diffusion Model

The drift and diffusion coefficients are evaluated by simulating the fatigue crack growth process in a block. Due to the randomness of the external load, the crack length at the beginning of the block a_{ol} and the overload intensity S_{ol} are random (see Figure 8).

The simulation starts by selecting a crack length at the beginning of the block $\underline{a}_{ol} = a_k$ (see Figure 8). The overload intensity is also set equal to $\underline{S}_{ol} = u_k$ (note that the extreme sequences are modelled as a Markov chain and then they can only take discrete values). Crack values larger than a_k are discretized as shown in Figure 8. Let the crack in the generic position a_i of the discretization mesh: the generic load cycle $[u_m, u_n]$ (which does not stop the block) will advance the crack. The new crack position a_j is evaluated by Veers model. The probability that the crack grows from a_i to a_j is then equal to the probability P^l_{nm} that the load cycle $[u_m, u_n]$ does not stop the block. Repeating these steps for all possible load cycles the element T^l_{ij} of the transition matrix $\mathbf{T}^l = [T^l_{ij}]$ is evaluated by adding the probabilities P^l_{nm} for all the cycles started at a_i and closed at a_j. The procedure can then be summarized as follows:

1. discretize the crack tip position a_j at the end of the generic load cycle $[u_m, u_n]$ which does not stop the block started with overload $\underline{S}_{ol} = u_k$ and a crack tip position $\underline{a}_{ol} = a_k$;
2. select the load cycles $[u_m, u_n]$ which do not stop the block;
3. evaluate the crack position a_j at the end of each load cycle $[u_m, u_n]$ by Veers model;
4. compute the probability P^l_{nm} that the cycle $[u_m, u_n]$ does not stop the block;
5. sum the probability P^l_{nm} to the element T^l_{ij} of the transition matrix $\mathbf{P}^l$;

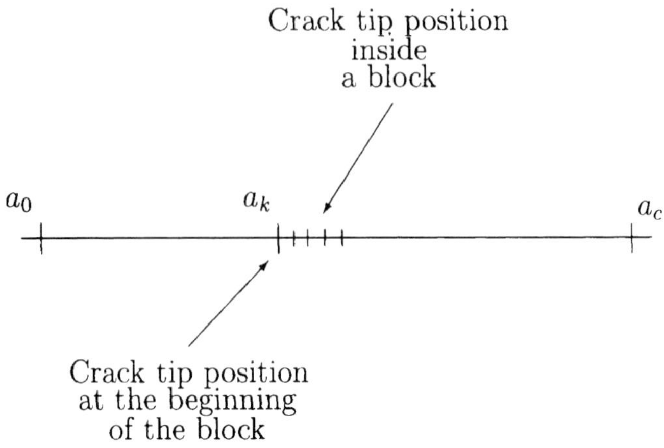

Figure 8. Discretization mesh of the crack tip position in a block.

6. repeat steps from 3 to 5 for all the possible crack tip positions a_i. Each row of the transition matrix $\mathbf{T}^l$ represents the transition probability between the crack tip position a_i and a_j.

The sum of the entries of the i-th row of the transition matrix $\mathbf{T}^l$ is equal to the probability that the block does not close in position a_i. One can then evaluate the joint probability distribution $\mathbf{P}^b(a_k, u_k) = [P^b_{li}(a_k, u_k)]$ of the crack length $\underline{a}_b$ at the end of the block and the number of cycles $\underline{n}_b$ in the block:

$$P^b_{li}(a_k, u_k) = \left(1 - \sum_j T^{l+1}_{ij}\right) \cdot p^l_i \tag{64}$$

where $p^l = [p^l_i]$ is the probability that the crack length is equal to a_i at the end of the l-th load cycle. Note that in principle $\mathbf{P}^b(a_k, u_k)$ is a function of the crack size at the beginning of the block and of the overload intensity which started the block. Repeating the simulation for different overload intensities $\underline{S}_{ol} = u_k$, one has:

$$P^b_{li}(a_k) = \sum_j P^b_{li}(a_k, u_j) \cdot p^r_j / \sum_j p^r_j \tag{65}$$

where p^r_j indicates the probability that the block was started by the overload $\underline{S}_{ol} = u_j$. At this point it is possible to evaluate the probability density

$\mathbf{P}^{n_b}(a_k) = [p_l^{n_b}(a_k)]$ of the number of cycles in a block:

$$p_l^{n_b}(a_k) = \sum_i P_{li}^b(a_k) \tag{66}$$

and the probability density $\mathbf{P}^{a_b}(a_k) = [p_i^{a_b}(a_k)]$ of the crack tip position at the end of the block:

$$p_i^{a_b}(a_k) = \sum_l P_{li}^b(a_k) \tag{67}$$

Numerical results clearly show that the mean value of the crack length at the end of the block $\underline{a}_b$ satisfies the condition:

$$E[\Delta\underline{a}_b] = E[\underline{a}_b] - a_k \ll a_k \tag{68}$$

This justifies the averaging procedure:

$$\begin{aligned} E[\Delta\underline{a}(a_k)] &= \sum_i \sum_l \frac{(a_i - a_k)}{l} \cdot P_{li}^b(a_k) \\ E[(\Delta\underline{a}(a_k))^2] &= \sum_i \sum_l \frac{(a_i - a_k)^2}{l} \cdot P_{li}^b(a_k) \end{aligned} \tag{69}$$

which gives the mean value $E[\Delta\underline{a}(a_k)]$ and the second moment $E[(\Delta\underline{a}(a_k))^2]$ of the crack size increment in a cycle. Provided the property reported in Equation (68) holds, the second moments of the crack increment (see Equation (69)) are eventually considered as the diffusion coefficient $d(a)$. The drift coefficient $\eta(a)$ is evaluated by inserting Equation (69) in Equation (55).

1.4.5. Two Numerical Examples

Two numerical examples are presented in order to show the applicability of the proposed method to both Gaussian and non-Gaussian loads.

1.4.5.1. Fatigue crack growth under Gaussian loading

The first numerical example deals with an infinite plate with a part through crack of initial length $a_0 = 0.127$ mm subjected to Gaussian load. The crack propagates up to a final length of $a_c = 3.810$ mm. The Paris-Erdogan fatigue crack propagation law is used in connection with Veers model to evaluate the crack size increment. Consider for simplicity $Y(a) = 1$ and the following material parameters (the units are MPa and mm/cycle):

$$\begin{aligned} C &= 4.13 \cdot 10^{-13} \\ m &= 3.5 \end{aligned} \tag{70}$$

Further material parameters are required to define the Veers model and to evaluate the effective stress cycle amplitude ΔS_{eff}:

$$\begin{aligned} \sigma_y &= 137.6 \quad \text{MPa} \\ R_0 &= -5 \\ q_0 &= 0.496 \end{aligned} \tag{71}$$

Consider a stationary Gaussian load process $\underline{S}(t)$. The spectral density is assumed constant over the interval $[f_1, f_2]$ and equal to zero otherwise (see Figure 9). The lower value f_1 is kept constant and equal to 10 Hz. Realizations of the stress process are obtained for the following values of the upper limit f_2:

$$f_2 = 11, 13, 20, 40, 50, 100 \quad \text{(Hz.)} \tag{72}$$

The standard deviation $\sigma_{\underline{S}}$ of the stress process is kept constant and equal to 30.76 MPa.

The simulated extreme sequence is discretized in 51 discrete levels. At this point it is possible to evaluate the transition matrices $\mathbf{P}$, $\mathbf{P}^+$ and $\hat{\mathbf{P}}$ by statistical analysis. Before the evaluation of the drift and diffusion coefficients starts, it is necessary to check the applicability of the Markov chain approximation for the extreme sequence. Figure 10 reports the comparison between the iso-level lines of $\Omega(u, v)$ obtained by the Markov approximation and the ones obtained by simulation.

Figure 11 gives the comparison between $\Omega(u, u)$ and the simulation results. As one can see from Figures 10 and 11, the Markov approximation of the load sequence produces results which match very well with the simulated ones.

The simulation procedure introduced in the previous section can be used to simulate the fatigue crack growth process in a block and to evaluate the joint probability density of the crack increment $\Delta\underline{a}_b$ in a block and the number of cycles $\underline{n}_b$ in the block. This result is illustrated in Figure 12 (note that $\Delta\underline{a}_b = \underline{a}_b - a_k$) for $f_2 = 40$ Hz and $\underline{a}_{ol} = a_k = 2.5$ mm.

The probability density of the number of cycles and the crack increment in a block are eventually evaluated by Equations (66) and (67). The relevant plots are drawn in Figures 13 and 14.

The coefficient $\bar{\eta}(a)$ and the diffusion coefficient $d(a)$ are given by (see Equations (55) and (56)):

$$d^2(a) = C^2 \cdot \pi^m \cdot \alpha_{\Delta S_{\text{eff}}} \cdot a^m = {d_0}^2 \cdot a^m \tag{73}$$

and

$$\bar{\eta}(a) = C \cdot \pi^{m/2} \cdot a^{m/2} \cdot E[\Delta\underline{S}_{\text{eff}}] = \eta_0 \cdot a^{m/2} \tag{74}$$

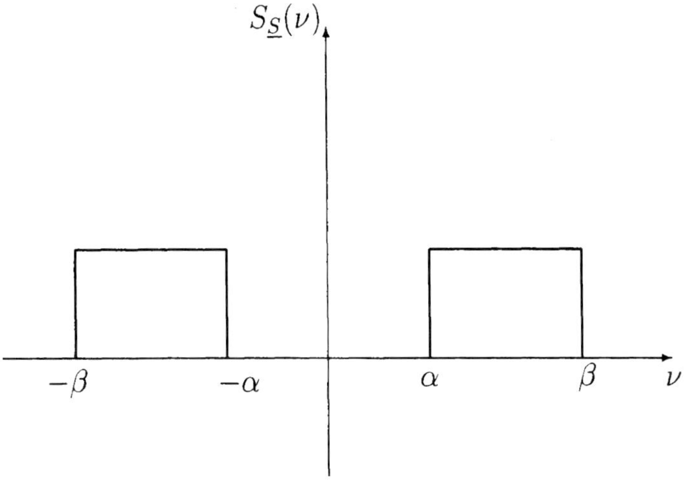

Figure 9. Spectral density function of the load process.

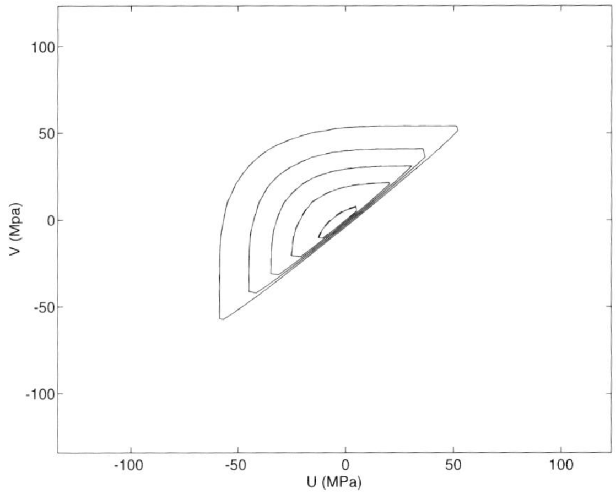

Figure 10. Comparison between the iso-level of $\Omega(u, v)$ ($f_2 = 40$ Hz) given by the Markov approximation (solid line) and the simulation results (dotted line).

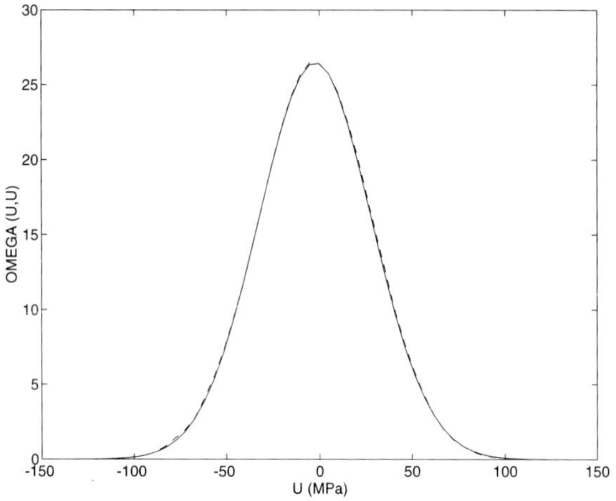

Figure 11. Comparison between $\Omega(u, u)$ ($f_2 = 40$ Hz.) obtained by the Markov approximation (solid line) and the simulation results (dotted line).

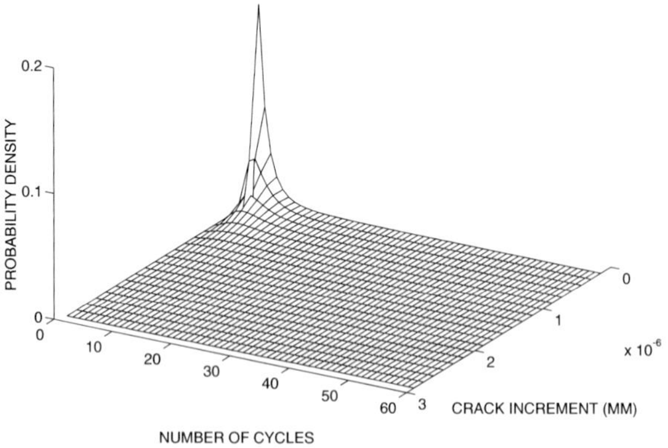

Figure 12. Joint probability density of the crack increment and number of cycles in a block ($f_2 = 40$ Hz and $a_k = 2.54$ mm).

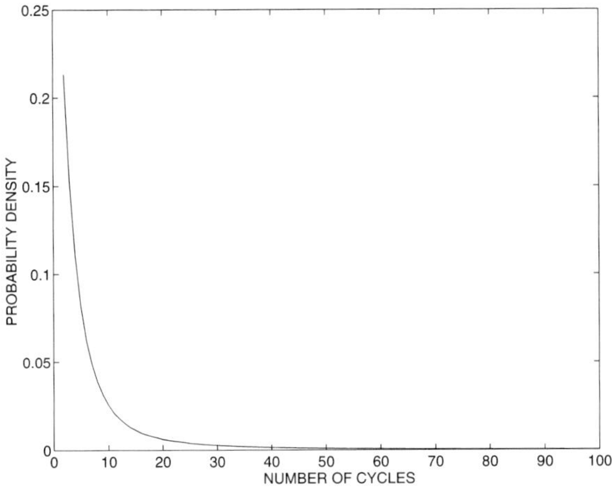

Figure 13. Probability density of the number of cycles in a block ($f_2 = 40$ Hz and $a_k = 2.54$ mm).

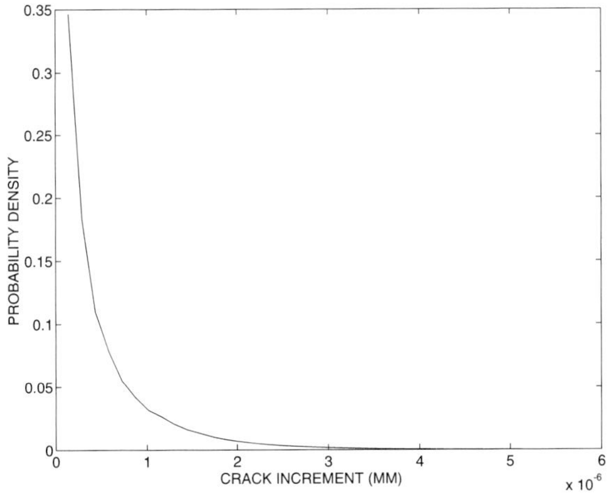

Figure 14. Probability density of the crack increment in a block ($f_2 = 40$ Hz and $a_k = 2.54$ mm).

The drift coefficient $\eta(a)$ is then:

$$\eta(a) = \eta_0 \cdot a^{m/2} \cdot \left(1 + \frac{{d_0}^2}{2 \cdot \eta_0} \cdot \frac{m}{2} \cdot a^{(m-2)/2}\right) \tag{75}$$

The numerical values of the coefficients η_0 and d_0 in Equation (75) are evaluated using Equation (69). Simulation results show that the numerical values of the coefficient η_0 is a few orders of magnitude greater than d_0. The mean value of the fatigue life is then the only significant parameter ("quasi deterministic fatigue crack growth"). This does not mean that the whole fatigue process is a deterministic one. In fact randomness is always present in the definition of the initial and final crack length and on the fatigue crack growth resistance. This aspect will be investigated in the next section. The final expression of the fatigue life is then evaluated by integrating Equation (61):

$$\bar{N}_f = \frac{1}{\eta_0} \cdot {a_c}^{1-b} \cdot \left\{\frac{1}{1-b} \cdot (1 - u^{1-b})\right\} \quad \text{for} \quad b \neq 1 \tag{76}$$

and

$$\bar{N}_f = -\frac{1}{\eta_0} \cdot {a_c}^{b-1} \cdot \ln(u) \quad \text{for} \quad b = 1 \tag{77}$$

In Equations (76) and (77) one has:

$$\begin{aligned} b &= \frac{m}{2} \\ u &= \frac{a_0}{a_c} \end{aligned} \tag{78}$$

The mean value $E[\underline{N}_f]$ of the fatigue life is reported in Figure 15 as a function of the bandwidth parameter ϵ. In the same picture the mean value of the fatigue life for the unretarded fatigue crack growth is also shown. This is done in order to emphasize the effect of the bandwidth parameter on the fatigue life by setting the reset stress $\underline{S}_r$ to zero.

From Figure 15 it is seen that the retardation effect is more significant for narrow band processes. This is due to the cluster effect, i.e. for a narrow band process the overloads are grouped together ("cluster") and are followed by a quite large number of relatively small cycles. By contrast, for a wide band process where the retardation phase is reduced to a small number of cycles after an overload.

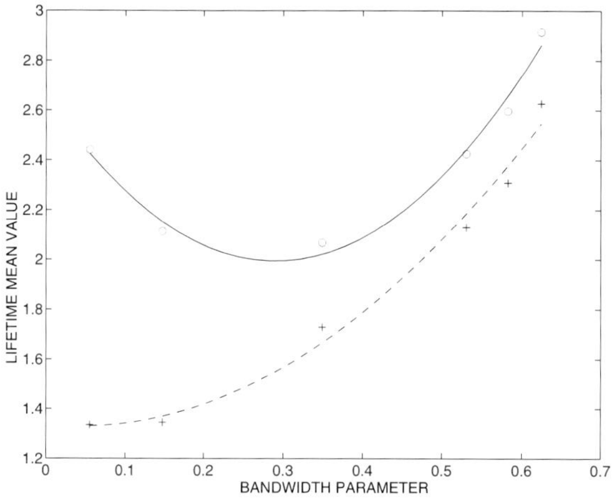

Figure 15. Fatigue life mean value for retarded (solid line) and inretarded (dashed line) fatigue crack growth as a function of the bandwidth parameter ϵ of the external load.

1.4.5.2. Fatigue crack growth under non-Gaussian loading

The second proposed numerical example deals with fatigue crack growth under non-Gaussian loading $\underline{y}(t)$ of the form:

$$\underline{y}(t) = g(\underline{x}(t)) \tag{79}$$

will be used in the numerical analysis. In Equation (79) $\underline{x}(t)$ is a Gaussian process and $g(\cdot)$ is a non linear transformation function. Note that $g(\cdot)$ must be a monotonically increasing odd function in order to obtain non-normal processes that have the same rate of zero up-crossing and peak rates as those of the normal process (i.e. such that $\underline{x}(t)$ and $\underline{y}(t)$ would have the same frequency content). Then if $\underline{x}(t)$ is a narrow band process also $\underline{y}(t)$ is narrow band and, neglecting the effect of the non-Gaussianity, the mean value of the fatigue life $E[\underline{N}_f]$ can be obtained by the Paris-Erdogan fatigue crack growth law as[87]:

$$E[\underline{N}_f] = \frac{k}{(2\sqrt{2}\sigma_{\underline{y}})^m \cdot \Gamma(\frac{m}{2}+1)} \tag{80}$$

where $k = \frac{1}{C}\int_{a_0}^{a_c} \frac{da}{(Y(a)\sqrt{\pi a})^m}$. In the literature the analysis of fatigue crack growth under non-Gaussian load is performed by computing the mean value of the fatigue life by Equation (80) and then applying a correction factor γ. It is defined as:

$$\gamma = \frac{E[\underline{N}_f] \quad \text{Gaussian load}}{E[\underline{N}_f] \quad \text{non-Gaussian load}} \tag{81}$$

Numerical results show that in general one has $\gamma > 1$, i.e. neglecting the non-Gaussianity produce in some cases unconservative results. If the load $\underline{y}(t)$ is not a narrow band process a second coefficient δ must be introduced:

$$\delta = \frac{E[\underline{N}_f] \quad \text{wide band process}}{E[\underline{N}_f] \quad \text{narrow band process}} \tag{82}$$

Neglecting the effect of the bandwidth produces conservative results, i.e. $\delta > 1$. The degree of non normality is measured by the kurtosis of $\underline{y}$:

$$\kappa_{\underline{y}} = \frac{E[(\underline{y} - \mu_{\underline{y}})^4]}{\sigma_{\underline{y}}^4} \tag{83}$$

For normal process it is exactly equal to 3. Any deviation from this value indicates that the process is not normal. In particular unconservative results are obtained for $\kappa_{\underline{y}} > 3$ while the cases $\kappa_{\underline{y}} < 3$ produces conservative estimations of the fatigue life. The final expression proposed in the literature for fatigue crack growth under a wide band non-Gaussian process is then obtained by multiplying $E[\underline{N}_f]$ (see Equation (80)) by $\frac{\delta}{\gamma}$. Note that retardation effect is not taken into account by the coefficients γ and δ.

In the literature several methods are proposed to estimate the coefficients γ and δ. A presentation of the available formulas for δ is given in[8]. They include some complicated expressions involving the irregularity factor α (see Equation (32)), the standard deviation of the load process and the parameters of the fatigue crack propagation law. In particular in this chapter the following schemes are considered for comparison purposes:

1. the Miles formula which adopts a Rayleigh approximation of the stress range probability distribution;
2. the Hancock formula which adopts a theoretical approximation of the stress range probability distribution;
3. the Wirsching formula which adopts a Rayleigh approximation of the stress range probability distribution and empirical rainflow correction factor;

4. the Zhao and Baker formula which adopts a theoretical approximation of the stress range probability distribution;
5. the Dirlik formula which adopts a theoretical approximation of the stress range probability distribution;
6. the Krenk formula which adopts a theoretical approximation of the stress range probability distribution;

Several researchers attempted a theoretical evaluation of the correction factor γ. Jensen[38] proposed a Gram-Charlier expansion of the joint probability density function $p_{\underline{y},\underline{\dot{y}}}$ of $\underline{y}(t)$ and its first time derivative $\underline{\dot{y}}(t)$. This result is then used to estimate the probability distribution of the extremes and of the cycle amplitude. The final result for γ is:

$$\gamma = 1 + \frac{1}{24}(m^2\lambda_{40} - m(4\lambda_{40} - 6\lambda_{22})) \tag{84}$$

where λ_{mn} are functions of the central moments of $p_{\underline{y},\underline{\dot{y}}}$ and m is the exponent of the Paris-Erdogan crack propagation law.

Winterstein[83–85] proposed a Hermite expansion of the process $\underline{y}(t)$. With reference to the approximation level introduced in the estimation of the coefficients of the Hermite expansion, one has two different expressions for the coefficient γ. The first one corresponds to a first order approximation:

$$\gamma = 1 + \frac{1}{24}(m^2\lambda_{40} - m\lambda_{40}) \tag{85}$$

By a second order approximation one has:

$$\gamma = (\frac{\sqrt{\pi}\kappa}{2V_r!})^m \frac{(mV_r)!}{(\frac{m}{2})!} \tag{86}$$

where:

$$V_r{}^2 = \frac{4}{\pi}(1 + h_4 + \bar{h}_4) - 1 \tag{87}$$

In Equation (87) h_4 is the fourth Hermite moment of $\underline{y}(t)$ and $\bar{h}_4$ and κ are the coefficients of the Hermite expansion.[81] Lutes *et al.*[48] evaluated γ by numerical simulation starting from the definition of γ reported in Equation (81). In particular $E[\underline{N}_f]$ for $\underline{y}(t)$ is evaluated by simulation while Equation (80) is used to estimate $E[\underline{N}_f]$ for the process $\underline{x}(t)$.

Kanegaonkar *et al.*[39] proposed to approximate the probability distribution of the stress amplitude by a mixture of a Gaussian and shifted exponential distributions. The corresponding density function is used to estimate γ.

Sarkani *et al.*[57,58] suggested to estimate γ by applying the nonlinear transformation introduced in Equation (79) to the Rayleigh distribution which describes the probability distribution of the cycle amplitude for a narrow band Gaussian process.

Consider the following expression for the function $g(\underline{x}(t))$:

$$g(\underline{x}(t)) = 17 \cdot (\underline{x}(t) + 0.04 \cdot \underline{x}(t)^3) \tag{88}$$

and the following expression for the geometry function $Y(a)$:

$$Y(a) = 2.3 \cdot \exp[-1.0648 \cdot a^{0.44}] + 1 \tag{89}$$

Two different power spectra for the Gaussian process $\underline{x}(t)$ are considered (see Figure 16).

The first one is a unimodal spectrum:

$$S_{\underline{x}}(f) = 0.25 \quad \text{per} \quad 1 \leq f \geq 5 \tag{90}$$

while the second one is the classical bimodal spectrum used in offshore technology[87]:

$$S_{\underline{S}}(f) = \frac{9.9018\text{E+03} \cdot \exp[-5.585\text{E-05}/f^4]}{1.1812\text{E+08} \cdot f^5 \cdot [(1 - 12.25 \cdot f^2)^2 + 0.0206 \cdot f^2]} \tag{91}$$

Both the adopted power spectra were appropriately scaled in order to have the process standard deviations equal to one. The mean value of the number of cycles to failure is given in Table 1 for both the power spectra. The mean value for the unretarded fatigue crack growth is also reported in order to quantify the retardation effect.

In order to make possible comparisons and to show the accuracy of the proposed technique to evaluate the mean value of the fatigue life under random loading, the retardation effect is neglected by setting to zero the reset stress. Moreover the effective stress cycle amplitude is set equal to the full stress amplitude. Under these assumptions it is possible to estimate the coefficients γ and δ as described previously and to compute the mean value of the fatigue life by Equation (80). The relevant numerical results are reported in Tables 2 and 3 for the unimodal and bimodal spectra for $\underline{x}(t)$, respectively.

Table 1. Mean value of the fatigue life for the retarded and unretarded fatigue crack growth

Unimodal spectrum for $\underline{x}(t)$	
Retarded	Unretarded
6.8896E+07	6.0036E+07
Bimodal spectrum for $\underline{x}(t)$	
Retarded	Unretarded
7.6745E+07	6.8962E+07

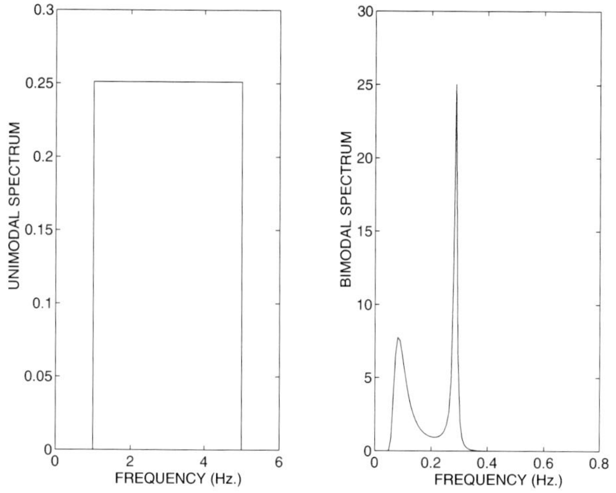

Figure 16. Power spectrum densities adopted for the process $\underline{x}(t)$.

Results from the proposed model are listed in the first row of each table. In the second, third, and fourth column the mean value of the fatigue life as computed from several models is given.

Tables 2 and 3 show that the proposed model overestimates the number of cycle to failure with respect to the scheme which are classified in the literature as very conservative. The simulation performed in the Gaussian case[23] showed that the results are still conservative but close to the simulation ones. Note that better results are obtained for the unimodal spectrum rather than the bimodal one.

1.5. FATIGUE RELIABILITY UNDER RANDOM LOADING

The model illustrated in the previous section only accounts for the randomness in the external loading. It was shown that the coefficient of variation of the fatigue life is very small, i.e. the standard deviation of the number of cycles to failure is negligeble. This does not mean that the fatigue crack growth under stochastic load is a deterministic phenomenon because randomness is present in the definition of the initial and final crack length and in the fatigue crack growth resistance. In this section two techniques are proposed to evaluate

Table 2. Comparison between the mean value of the fatigue life given by the proposed method and several models available in the literature (unimodal spectrum for $\underline{x}(t)$)

Unimodal spectrum for $\underline{x}(t)$			
Proposed model		1678890	
Different models for the coefficient δ	Eq.(84) ($\gamma = 1.3408$)	Eq.(85) ($\gamma = 1.2976$)	Eq.(86) ($\gamma = 1.2674$)
Miles	911081	941413	963760
Hancock	1102073	1138764	1165800
Wirsching	1071781	1107463	1133800
Zaho and Baker	1199557	1239493	1268900
Dirlik	1122366	1159732	1187300
Krenk	1333097	1377478	1410300

Table 3. Comparison between the mean value of the fatigue life given by the proposed method and several models available in the literature (bimodal spectrum for $\underline{x}(t)$)

Bimodal spectrum for $\underline{x}(t)$			
Proposed model		2056620	
Different models for the coefficient δ	Eq.(84) ($\gamma = 1.3408$)	Eq.(85) ($\gamma = 1.2976$)	Eq.(86) ($\gamma = 1.2674$)
Miles	951495	983173	1006600
Hancock	1194003	1233754	1263100
Wirsching	1125760	1163240	1190900
Zaho and Baker	1315951	1359762	1392100
Dirlik	1221880	1262559	1292600
Krenk	1494971	1544742	1581500

the effect of this randomness on the mean value of the lifetime. They couple suitable reliability tools with the model developed in the previous section to evaluate the fatigue lifetime of a structural component: the first is known in literature as advanced mean value first order method (AMVFO), the second adopt a response surface scheme.

1.5.1. First Order Reliability Methods

A very brief review of first order reliability methods (FORM) is given.[20,32] They make use of a simple representation of the design variables by their mean value vector and the covariance matrix. A failure criterion:

$$g(\underline{\mathbf{x}}) = g(\underline{x}_1, \underline{x}_2, ..., \underline{x}_n) \leq 0 \tag{92}$$

must be formulated in terms of the set of basic random variables $\underline{\mathbf{x}} = \underline{x}_1, \underline{x}_2, ..., \underline{x}_n$. The distribution of the basic random variables must be known and any continuous type of distribution is allowed. Equation (92) describes a hypersurface which divides the n-dimensional space of realizations of $\underline{\mathbf{x}}$ in two distinct sets called the failure set $\mathcal{F}$ and the safe set $\mathcal{S}$. The border line between the two set is called the failure surface or the limit state surface. The failure probability p_f is:

$$p_f = \int_{\mathcal{F}} p_{\underline{\mathbf{x}}}(\underline{x}_1, \underline{x}_2, ..., \underline{x}_n) dx_1, ..., dx_n \tag{93}$$

where $p_{\underline{\mathbf{x}}}$ is the joint density function of the basic random variables. The practical problem consists in finding an efficient algorithm to calculate a sufficiently close approximation of p_f. A generalized reliability index β_G can be introduced for this purpose[32]:

$$\beta_G = \Phi^{-1}(1 - p_f) = -\Phi^{-1}(p_f) \tag{94}$$

where $\Phi(\cdot)$ is the standardized normal cumulative distribution function. Good approximation of p_f is easy to compute if the basic variables are mutually independent and standardized normally distributed. The first step is then a one to one transformation Ψ of the set of basic variables into a set $\underline{\mathbf{z}} = \underline{z}_1, \underline{z}_2, ..., \underline{z}_n$ of random variables with these properties:

$$\underline{\mathbf{x}} = \Psi(\underline{\mathbf{z}}) \tag{95}$$

The failure surface in the x space is then transformed into a failure surface in the standardized z space. The failure surface is then replaced with the tangent hyperplane at the point $\mathbf{z}^*$ which is closest to the origin. If the distance from the origin to this plane is denoted by β one has $\beta_G \simeq \beta$[20] and:

$$p_f \simeq \Phi(-\beta) \tag{96}$$

The point $\mathbf{z}^*$ (or the equivalent point $\mathbf{x}^*$ in the x space) is often referred as the design point[20,32] and has coordinates:

$$\mathbf{z}^* = \beta \boldsymbol{\rho} \tag{97}$$

The coordinates of the unit vector $\boldsymbol{\rho}$ are called the sensitivity factors.[20]

1.5.2. The AMVFO Method

In order to reduce the computational time the advanced mean value first order (AMVFO) method was proposed in[86,88–90] Following the advanced mean value concept the probability distribution of $\underline{N}_f(\mathbf{x})$ is evaluated by introducing a first-order Taylor's series expansion of $N_f(\mathbf{x})$ about the mean value $E[\underline{\mathbf{x}}]$ of $\underline{\mathbf{x}}$. This requires evaluation of the lifetime mean value at $E[\underline{\mathbf{x}}]$ and at small perturbations about $E[\underline{\mathbf{x}}]$ in order to evaluate the parameters of the linear function. The function $N_f(\mathbf{x})$ is then replaced by $N'_f(\mathbf{x})$:

$$N'_f(\mathbf{x}) = c_0 + \sum_i c_i \cdot (x_i - E[\underline{x}_i]) \tag{98}$$

where:

$$c_i = \left[\frac{\partial N_f}{\partial x_i}\right]_{E[\underline{\mathbf{x}}]} \tag{99}$$

The probability distribution is approximated at discrete points $N_{f\,i}$. The number of points is arbitrary, but it was shown that three or four points on both side of the mean are adequate to define the probability distribution, when the probability of failure is expected to be a smooth function. A first estimate of the probability distribution is made from the linear form. The limit state function at each $N_{f\,i}$ is:

$$g_i(\underline{\mathbf{x}}) = N'_f(\underline{\mathbf{x}}) - N_{f\,i} \tag{100}$$

The probability distribution $P_{\underline{N}_f}(N_{f\,i})$ of the mean number of cycles to failure is eventually computed as:

$$P_{\underline{N}_f}(N_{f\,i}) = \Phi(-\beta_i) \tag{101}$$

where β_i is the reliability index[20] associated with each $g_i(\underline{\mathbf{x}})$ and $\Phi(\cdot)$ the standard normal probability distribution. This is the mean value first order (MVFO) method: it leads one to estimate the design point $\mathbf{x}_i^*$ for the linear approximation of Equation (98). To improve the estimate of the probability distribution, the actual mean value of the number of cycles to failure $N_f(\mathbf{x}_i^*)$ is evaluated at each design point of the sample space identified. These values are associated with the probability of failure $P_{\underline{N}_f}(N_{f\,i})$ obtained from Equation (101): the corresponding plot is regarded as an improved form of the searched probability distribution. This improvement gives rise to the advanced mean value first order (AMVFO) method. The total number of function evaluations is given by:

$$J = n_v + n_p + 1 \tag{102}$$

where n_v is the total number of random variables and n_p is the number of points used to define the probability distribution. The estimate of $P_{\underline{N}_f}(N_f)$ could be improved by introducing a new linear approximation for N_f at each design point $\mathbf{x}_i^*$. The reliability analysis is then performed for each of the linear functions. As a result the probability estimate at each value of $N_{f\,i}$ is more accurate. Still another improvement could be made by evaluating the real function at each design point. This process can be repeated, but experience suggests that the probability distribution obtained by the first function evaluation (AMVFO) is generally very close to the exact solution.[90]

1.5.3. The Response Surface Method

AMVFO method can deal only with random variables for the description of the randomness in the fatigue crack growth process. In a more realistic situation some parameters (as for example the fatigue resistance) must be modelled more coherently as a random process or random fields. In this case a response surface scheme[20,33–35] must be used to evaluate the fatigue lifetime distribution.[13,18,19,21]

1.5.4. Numerical Example

Consider a structural element with a crack which grows in a region of stress concentration. Again the Paris law is assumed to describe the fatigue crack growth rate. The same geometry function used in the second example of the previous section is adopted (see Equation (89)):

$$Y(a) = 2.3 \cdot \exp[-1.0648 \cdot a^{0.44}] + 1 \tag{103}$$

The initial crack size a_0 and the Paris coefficient C are modelled as random variables. Their probabilistic definition is given in Table 4. The external loading $\underline{S}(t)$ is modelled as a stationary Gaussian process with bimodal power spectrum used in offshore technology[87] (see Figure 17):

$$S_{\underline{S}}(f) = \frac{1.2541\text{E+}07 \cdot \exp[-2.7245\text{E-}05/f^4]}{2.4215\text{E+}08 \cdot f^5 \cdot [(1 - 12.25 \cdot f^2)^2 + 0.0206 \cdot f^2]} \tag{104}$$

Realizations of the process $\underline{S}(t)$ are performed by classical frequency domain simulation techniques.[20]

Table 4. Numerical data for the deterministic and random parameters in the fatigue crack growth equation

Variable	Distribution	Mean Value (MPa units)	Standard Deviation (MPa units)
m	Constant	3.5	–
C	Weibull	4.1E-13	8.2E-14
a_0	Lognormal	0.15	0.03
a_c	Constant	4	–
K_t	Constant	3.3	–
γ	Constant	1.0648	–
δ	Constant	0.44	–
σ_y	Constant	350	–
R_0	Constant	–5	–
q_0	Constant	0.496	–

Consider now the retardation effect and the randomness of the Paris coefficient $\underline{C}$ and the initial crack length $\underline{a}_0$. Following the AMVFO concept N_f is computed at the mean values of the random variables (see Table 4) by the method described in the previous section. Moreover solutions are computed at perturbated values (arbitrally chosen as the median plus 0.10 standard deviations). The resulting first order approximation to N_f is:

$$N_f' = 4.2093E{+}06 - 1.0280E{+}19 \cdot C - 1.1030E{+}07 \cdot a_0 \tag{105}$$

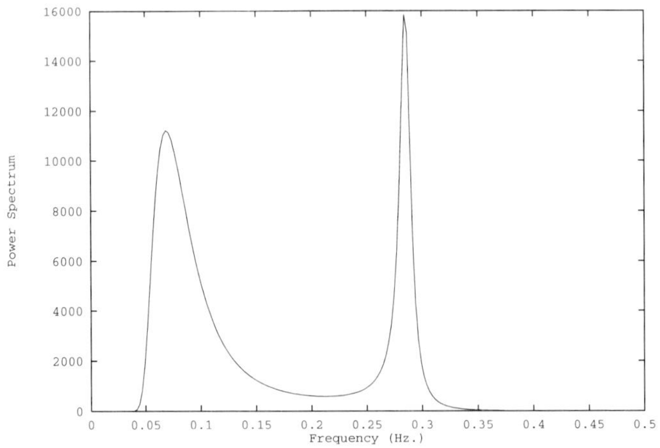

Figure 17. Bimodal power spectrum adopted for the stress process $\underline{S}(t)$.

Table 5. Data used to construct the AMVFO approximation of the probability distribution of $\underline{N}_f$

$N_{f\,i} \times 10^6$	$P_{\underline{N}_{f\,i}}$	C_i^*	$a_{0\,i}^*$ MPa units	$N_{f\,i}^* \times 10^6$ MPa units
2.0	3.219E-3	5.687E-13	2.024E-1	2.6931
2.5	1.980E-2	5.416E-13	1.823E-1	2.9493
3.0	7.960E-2	5.089E-13	1.675E-1	3.2465
3.2	1.239E-1	4.944E-13	1.628E-1	3.3797
3.5	2.154E-1	4.716E-13	1.569E-1	3.5952
3.8	3.324E-1	4.476E-13	1.520E-1	3.8357
4.0	4.190E-1	4.311E-13	1.493E-1	4.0106
4.2	5.074E-1	4.142E-13	1.469E-1	4.2010
4.5	6.339E-1	3.884E-13	1.438E-1	4.5178
4.8	7.436E-1	3.620E-13	1.412E-1	4.8821
5.0	8.044E-1	3.442E-13	1.396E-1	5.1575
5.5	9.115E-1	2.993E-13	1.362E-1	5.9889
6.0	9.666E-1	2.542E-13	1.329E-1	7.1195
6.5	9.899E-1	2.094E-13	1.293E-1	8.7358
7.0	9.977E-1	1.656E-13	1.247E-1	11.203

First the MVFO method is used to compute $P(\underline{N}_f \leq N_i)$ at preselected values of $N_{f\,i}$ (see Table 5).

This solution is shown in Figure 18. The subsequent evaluation at the design points C_i^* and $a_{0\,i}^*$ (see Table 5) of the actual values of $N_{f\,i}^*$, AMVFO method, provides the improved solution: it is shown in Figure 18.

1.6. CONCLUSIONS AND FUTURE WORK

This chapter could be divided in two parts. In the first one a stochastic model is presented for fatigue crack growth under stochastic loads. Sequence effects due to the randomness of the external load are taken into account. The fatigue crack growth is modelled as a diffusion process and drift and diffusion parameters are modelled by simulation. To this end the lifetime is subdivided in a sequence of blocks. A block is given by a retardation phase and the consequent post-retardation one. In the first phase the fatigue crack growth is retarded, i.e the fatigue crack growth rate is reduced, and the Veers model is used to define the effective stress intensity factor range. The concept of reset stress is introduced in order to evaluate the effective stress range. The adopted fatigue crack growth equation is the Paris law but any other expression available in the literature for the fatigue rate could be used. The extreme sequence is modelled by a Markov chain. This makes the

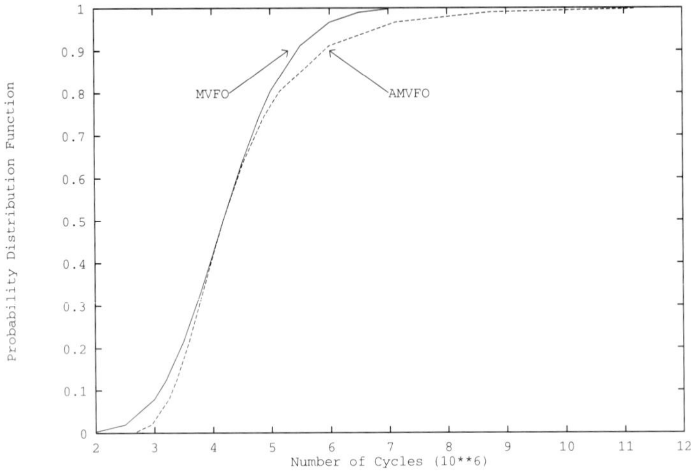

Figure 18. Comparison between the AMVFO solution and the MVFO solution for $P_{\underline{N}_f}(N_f)$.

simulation scheme introduced very efficient to compute the drift and diffusion coefficients. The method is applied to both Gaussian and non-Gaussian load. The bandwidth effect is well captured and the conservativism introduced by similar models available from the literature is decreased. Numerical results show that the coefficient of variation of the lifetime is very small, i.e. the standard deviation of the number of cycles to failure is negligible. This does not mean that the fatigue crack growth process is a deterministic one, because randomness is present in the definition of the initial and final crack size and on material parameters. Models for the evaluation of the actual fatigue reliability of a cracked structural component subjected to fatigue are illustrated in the second part of this chapter. Reliability techniques are briefly recieved with special attention to the so called advanced mean value first order method (AMVFO) and response surface method. The AMVFO is coupled with the stochastic model presented in the first part to evaluate the fatigue lifetime probability distribution.

The directions of future work consider mixed fatigue crack propagation and surface crack growth. Mode I was only considered in this chapter but in a more realistic situation mixed crack propagation will occur, i.e the crack is not only subjected to tensile stress but shear stress is also present. Moreover, more than one crack could be present in a structural component which interacts during propagation and these could eventually join together. In these two cases no closed expressions for the stress intensity factor are available and numerical techniques must be used to solve the structural problem. Surface crack growth of a semi-elliptical crack is also a complicated open problem since propagation occurs along both the semi-axis of the crack. Fatigue sensitive structural elements are subjected to periodic controls in order to identify and eventually repair cracks. The inspection results can be used to update the computed lifetime distribution. This analysis can be applied in principle also to the stochastic fatigue crack growth model illustrated in this chapter.

AKNOWLEDGEMENTS

This research has been supported by funds from the Italian Ministry of University and Scientific and Technological Research (MURST) and from the Italian National Research Council (CNR).

References

1. Aliabadi, M.H. and D.P. Rooke, 1991, *Numerical Fracture Mechanics*, Kluwer Academic Publishers.
2. Arnold, L., 1974, *Stochastic Differential Equation: Theory and Applications*, John Wiley and Sons.
3. Augusti, G., A. Baratta and F. Casciati, 1984, *Probabilistic Methods in Structural Engineering*, Chapman & Hall.
4. Bellomo, N. and R. Riganti, 1987, *Nonlinear Stochastic Systems in Physics and Mechanics*, World Scientific Pubblications.
5. Bendat, J.S. and A.G. Piersol, 1971, *Random Data: Analysis and Measurement Procedures*, Wiley.
6. Benjamin, J.R. and C.A. Cornell, 1970, *Probability, Statistics, and Decision for Civil Engineers*, McGraw-Hill.
7. Bignonnet, A. and M. Olagnon, 1991, Fatigue Life Prediction for Variable Amplitude Loading, *Proc. 10th Offshore Mechanics and Artic Engineering Conference*, Eds. M.M. Salama, H.C. Ree, J.P. Kam, G.S. Booth and J.G. Williams, Vol. **III-B**, ASME, pp. 395–401.
8. Bouyssy, V., S.M. Naboishikov and R. Rackwitz, 1993, Comparison of analytical counting methods for Gaussian processes, *Structural Safety*, **12**, 35–57.
9. Broek, D. 1982, *Elementary Engineering Fracture Mechanics*, Martinus Nijhoff Publishers.
10. Broek, D., 1988, *The Practical Use of Fracture Mechanics*, Kluwer Academic Publishers.
11. Bulloch, J.H., The Influence of Mean Stress or R-ratio on the Fatigue Crack Threshold Characteristics of Steels – A Review, *Int. Jour. Pres. Ves. & Piping*, **47**(3), 263–292.
12. Casciati, F. and P. Colombi, 1993, Load Combination and Related Fatigue Reliability Problems, *Structural Safety*, **13**, 93–111.
13. Casciati, F. and P. Colombi, 1994, Fatigue Lifetime Prediction for Uncertain Systems, *Proc. Iutam Symposium on Probability Structural Mechanics: Advances in Structural Reliability Methods*, Eds. P. Spanos and Y.T. Wu, Springer-Verlag, 87–99.
14. Casciati, F., P. Colombi and L. Faravelli, 1991, Filter Technique for Stochastic Crack Growth, *Proc. Computational Stochastic Mechanics*, Eds. P. Spanos and C.A. Brebbia, Computational Mechanical Publications, pp. 485–496.

15. Casciati, F., P. Colombi and L. Faravelli, 1991, Stochastic Crack Growth by Filter Technique, *Proc. 6th International Conference on Application of Statistics and Probability in Civil Engineering*, Eds. S. Ruiz and L. Estexa, **1**, 71–81.
16. Casciati, F., P. Colombi and L. Faravelli, 1991, Stochastic Crack Growth and Reliability Analysis, *Proc. 10th International Conferences on Offshore Mechanics and Artic Engineering*, Ed. C.G. Soares, C. Ostergaard, M.J. Baker, A. Pittaluga, M. Huter and P. Thof-Christensen, ASME, pp. 107–111.
17. Casciati, F., P. Colombi and L. Faravelli, 1992, Fatigue Crack Size Probability Distribution via Filter Technique, *Fat. & Fract. Engng. Mater. Struct.*, **15**(5), 463–475.
18. Casciati, F., P. Colombi and L. Faravelli, 1993, Lifetime Prediction of Fatigue Sensitive Structural Elements, *Structural Safety*, **12**, 105–111.
19. Casciati, F., P. Colombi and L. Faravelli, 1992, Fatigue Lifetime Evaluation via Response Surface Methodology, *Proc. European Safety and Reliability Conference '92*, Eds. K.E. Petersen and B. Rasmussen, Elsevier, pp. 157–166.
20. Casciati, F. and L. Faravelli, 1991, *Fragility Analysis of Complex Structural System*, Research Studies Press.
21. Colombi, P. 1995, Vita Residua a Fatica di Componenti Strutturali Metallici Soggetti ad Eccitazione Stocastica (in Italian), Ph.D. thesis, University of Pavia-Polytechnic of Milan.
22. Colombi, P. and K. Dolinski, 1993, Markov Approach to Fatigue Crack Growth Under Stochastic Load, *Proc. Localized Damage III - Computer Aided Assessment and Control*, Computational Mechanical Publications, pp. 89–95.
23. Colombi, P. and K. Dolinski, Fatigue Reliability Under Stochastic Load, *Proc. 14th International Conference on Offshore Mechanics and Artic Engineering*, ASME, in press.
24. Collins, J.A., 1981, *Failure of Materials in Mechanical Design: Analysis, Prediction, Prevention*, John Wiley & Sons.
25. Crandall, S.H. and W.D. Mark, 1963, *Random Vibration in Mechanical Systems*, Academic Press.
26. Ditlevsen, O., 1986, Random Fatigue Crack Growth – A First Passage Problem, *Engng. Fract. Mech.*, **23**(2), 467–477.
27. Ditlevsen, O. and K. Sobczyk, Random Fatigue Crack Growth with Retardation, *Engng. Fract. Mech.*, **24**(6), 861–878.
28. Dolinski, K., Stochastic Loading and Material Inhomogeneity in Fatigue Crack Propagation, *Engng. Fract. Mech.*, **25**(5/6), 809–818.
29. Dolinski, K., 1987, Fatigue Crack Growth with Retardation Under Stationary Stochastic Loading, *Engng. Fract. Mech.*, **27**(3), 279–290.
30. Downing, S.D. and D.F Socie, 1982, Semplified Rainflow Counting Algorithms, *Int. Jour. of Fatigue*, **4**(1), 31–40.
31. Elber, W., 1971, The Significance of Fatigue Crack Closure, in *Damage Tolerance in Aircraft Structures*, STP-486, ASTM, pp. 230–242.
32. Faravelli, L., 1988, *Sicurezza Strutturale* (in Italian), Pitagora Editrice.
33. Faravelli, L., 1992, Structural Reliability via Response Surface, *Proc. Iutam Symposium on Nonlinear Stochastic Mechanics*, Eds. N. Bellomo and F. Casciati, Springer-Verlag, pp. 213–223.
34. Faravelli, L., Response Surface Approach for Reliability Analysis, *Jour. of Engng. Mech. (ASCE)*, **115**(12), 2763–2781.
35. Faravelli, L., 1989, Finite Element Analysis of Stochastic Nonlinear Continua, *Computational Mechanics of Probabilistic and Reliability Analysis*, Eds. W.K. Liu and T. Belytschko, Elmepress, pp. 264–280.
36. Frendahl, M. and I. Rychlik, 1992, *Rainflow analysis - Markov method*, Private Communication.
37. Ishikawa, H., A. Tsurui, H. Tanaka and H. Ishikawa, 1993, Reliability Assessment of Structures Based Upon Probabilistic Fracture Mechanics, *Prob. Engng. Mech.*, **8**, 43–56.
38. Jensen, J.J., 1990, Fatigue Damage Due to Non-Gaussian Response, *Jour. of Engng. Mech. (ASCE)*, **116**(1), 240–246.

39. Kanegaonkar, H.B. and A. Haldar, 1987, Non-Gaussian Response of Offshore Platforms: Fatigue, *Jour. of Struct. Engng. (ASCE)*, **113**(9), 1899–1908.
40. Kanninen, M.F. and C.H. Popelar, 1985, *Advanced Fracture Mechanics*, Oxford University Press.
41. Leadbetter, M.R., R.G. Lindgren and H. Rootzen, 1983, *Extremes and Related Properties of Random Sequences and Processes*, Springer-Verlag.
42. Lin, Y.K. and J.N. Yang, 1983, On Statistical Moments of Fatigue Crack Propagation, *Engng. Fract. Mech.*, **18**(2), 243–256.
43. Lin, Y.K. and J.N. Yang, 1985, A Stochastic Theory of Fatigue Crack Propagation, *AIAA Jour.*, **23**(1), 117–124.
44. Lin, Y.K., 1986, Some Observation on the Stochastic Averaging Method, *Prob. Engng. Mech.*, **1**(1), 23–27.
45. Lin, Y.K., 1967, *Probabilistic Theory of Structural Dynamics*, McGraw-Hill.
46. Lindgren, G. and I. Rychlik, 1982, Wave Characteristic Distribution for Gaussian Waves – Wave Length, Amplitude and Steepness, *Oc. Engng.*, **9**, 411–432.
47. Lindgren, G. and I. Rychlik, 1987, Rainflow Cycle Distribution for Fatigue Life Prediction Under Gaussian Load Processes, *Fat. & Fract. of Engng. Mater. and Struct.*, **10**(3), 251–260.
48. Lutes, L.D., M. Corazao, S.J. Hu and J. Zimmerman, 1984, Stochastic Fatigue Damage Accumulation, *Jour. of Struct. Engng. (ASCE)*, **110**(11), 2585–2601.
49. Madsen, H.O., 1989, Stochastic Modeling of Fatigue Crack Growth, Lecture Notes for Advanced Seminar on Structural Reliability, Ispra Course.
50. Madsen, H.O., 1983, *Probabilistic and Deterministic Models for Predicting Damage Accumulation due to Time Varying Loading*, DIALOG 5–82, Danish Engeneering Academy.
51. Nelson, D.V., 1978, *Cumulative Fatigue Damage in Metals*, Ph.D. thesis, Stanford University.
52. Nienstedt, J., A. Tsurui, H. Tanaka and G.I. Schueller, 1990, Time Variant Structural Reliability Analysis Using Bivariate Diffusive Crack Growth Models, *Int. Jour. Fat.*, **12**(2), 83–89.
53. Reynolds, A.P., 1992, Constant Amplitude and Post-Overload Fatigue Crack Growth Behaviour in PM Aluminium Alloy AA 8009, *Fat. & Fract. Engng. Mater. Struct.*, **15**(6), 551–562.
54. Rychlik, I., 1989, Simple Approximations of the Rainflow Cycle Distribution for Discretized Random Loads, *Prob. Engng. Mech.*, **4**, 40–48.
55. Rychlik, I., 1987, A New Definition of the Rainflow Cycle Counting Method, *Int. Jour. of Fat.*, **9**, 119–121.
56. Rychlik, I., 1988, Rainflow Cycle Distribution for Ergodic Load Processes, *SIAM Jour. of App. Math.*, **48**, 662–679.
57. Sarkani, S., 1991, L.D. Lutes, P.J. Hughes and D.P. Kihl, 1991, Sequence Effects on Stochastic Fatigue of Welded Joints, *Jour. of Struct. Engng. (ASCE)*, **117**(6), 1852–1867.
58. Sarkani, S., D.P. Kihl and J.E. Beach, 1994, Fatigue of Welded Joints Under Narrowband Non-Gaussian Loadings, *Prob. Engng. Mech.*, **9**, 179–190.
59. Schijve, J., 1973, Effect of Load Sequence on Crack Propagation Under Random and Program Loading, *Engng. Fract. Mech.*, **5**, 269–280.
60. Schijve, J., 1976, Observation on the Prediction of Fatigue Crack Growth Propagation under Variable Amplitude Loading, *Fatigue Crack Growth Under Spectrum Loads*, SPT-595, ASTM.
61. Shin, C.S. and N.A. Fleck, 1987, Overload Retardation in Structural Steel, *Fat. & Fract. Engng. Mater. Struct.*, **9**(5), 379–393.
62. Shinozuka, M. and C.M. Jan, 1972, Digital Simulation of Random Processes and its Applications, *Jour. of Sound and Vibr.*, **25**(1), 111–128.
63. Sih, G.C., 1973, *Handbook of Stress Intensity factors*, Lehigh University.
64. Sobczyk, K. and B.F. Spencer Jr., 1992, *Random Fatigue: From Data to Theory*, Academic Press Inc.
65. Sobczyk, K., 1986, Modelling of Random Fatigue Crack Growth, *Engng. Fract. Mech.*, **24**(4), 609–623.

66. Sobczyk, K., 1991, *Stochastic Differential Equation with Application to Physics and Engineering*, Kluwer Academic Publishers.
67. Solomos, G.P., 1989, First-Passage Solutions in Fatigue Crack Propagation, *Prob. Engng. Mech.*, **4**(1), 33–39.
68. Solomos, G.P. and A.C. Lucia, 1990, Markov Approximation to Fatigue Crack Size Distribution, *Fat. & Fract. Engng. Mater. Struct.*, **13**(5), 457–471.
69. Spencer Jr., B.F and J. Tang, 1988, Markov Process Model for Fatigue Crack Growth, *Jour. Engng. Mech. (ASCE)*, **114**(12), 2134–2157.
70. Stratonovich, R.L., 1967, *Topics in the Theory of Random Noise*, Vol. **2**, translated by R.A. Silverman, Gordon and Breach.
71. Suresh, S., 1991, *Fatigue of Materials*, Cambridge University Press.
72. Suresh, S., 1983, Micromechanism of Fatigue Crack Growth Retardation Following Overloads, *Engng. Fract. Mech.*, **18**(3), 577–593.
73. Tada, H., P. Paris and G. Irwin, 1973, *The Stress Analysis of Cracks Handbook*, Del Research Corporation.
74. Tang, J. and B.F. Spencer Jr., 1989, Reliability Solution for the Stochastic Fatigue Crack Growth Problem, *Engng. Fract. Mech.*, **34**(2), 419–433.
75. Tsurui, A. and H. Ishikawa, 1986, Application of the Fokker-Planck Equation to a Stochastic Fatigue Crack Propagation, *Structural Safety*, **4**, 15–29.
76. Tsurui, A., J. Nienstedt, G.I. Schueller and H. Tanaka, 1989, Time Variant Structural Reliability Analysis Using Diffusive Crack Growth Models, *Engng. Fract. Mech.*, **34**(1), 153–167.
77. Veers, P.J., 1987, *Fatigue Crack Growth Due to Random Loading*, SAND87-2037, Sandia National Laboratories.
78. Wanhill, R.J. and J. Schijve, 1988, *Proc. Fatigue Crack Growth Under Variable Amplitude Loading*, Eds. J. Petit *et al.*, Elsevier.
79. Wheeler, O.E., 1972, Spectrum Loading and Crack Growth, *Jour. Basic Engng.*, **94**, 181–186.
80. Willenborg, J., R.M. Engle and H.A. Wood, 1971, *A Crack Growth Retardation Model Using an Effective Stress Concept*, Report TM-71-1-FBR, Wrigth Patterson Air Force Base.
81. Winterstein, S.R. and O.B. Ness, 1989, Hermite Moment Analysis of Nonlinear Random Vibration, *Computational Mechanics of Probability and Reliability Analysis*, Eds. W.K. Liu and T. Belytschko, Elmepress, pp. 452–478.
82. Winterstein, S.R. and P.S. Veers, 1989, Diffusion Models of Fatigue Crack Growth with Sequence Effects due to Stationary Random Loads, *Proc. 5th International Conference on Structural Safety and Reliability*, pp. 1523–1530.
83. Winterstein, S.R., 1985, Non-Gaussian Response and Fatigue Damage, *Jour. of Engng. Mech. (ASCE)*, **111**(10), 1291–1295.
84. Winterstein, S.R., 1988, Nonlinear Vibration Models for Estremes and Fatigue, *Jour. of Engng. Mech. (ASCE)*, **114**(10), 1772–1790.
85. Winterstein, S.R., 1989, Duscussion of the paper: Non-Gaussian Response of Offshore Platforms: Fatigue, *Jour. of Struct. Engng. (ASCE)*, **115**(3), 749–752.
86. Wirshing, P.H., K. Ortiz and Y.N. Chen, 1987, Fracture Mechanics Fatigue Model in a Reliability Format, *Proc. 6th International Conference on Offshore Mechanics and Artic Engng.*, ASME, pp. 331–337.
87. Wirshing, P.H. and M.C. Light, 1982, Fatigue Under Wide Band Random Stresses, *Jour. of Struct. Div. (ASCE)*, **106**, 1593–1607.
88. Wirshing, P.H., T.Y. Torng and W.S. Martin, 1991, Advanced Fatigue Reliability Analysis, *Int. Jour. Fatigue*, **13**(5), 389–394.
89. Wu, Y.T., O.H. Burnside and J. Dominquez, 1987, Efficient Probabilistic Fracture Mechanics Analysis, *Proc. 4th International Conference on Numerical Methods in Fracture Mechanics*, Pineridge Press, pp. 85–100.
90. Wu, Y.T., H.R. Millwater and T.A. Cruse, 1989, An Advanced Probability Structural Analysis Method for Implicit Performance Functions, *Proc. 30th Structures, Structural Dynamics and Materials Conference*.

2 TECHNIQUES IN DYNAMIC FRACTURE MECHANICS

V.Z. PARTON and A.I. ZOBNIN

Department of Mathematics, Moscow State Academy of Chemical Engineering, 21/4. Staraya Basmannaya ul., 107884, Moscow, Russia

Over the last few decades, fracture mechanics has come to be recognized as a separate branch of mechanics of deformable solids. The results obtained are used to ensure the strength, reliability, and long-life of structures, and to work out effective means of nondestructive testing in order to prevent accidents that may have serious economic and social repercussions.

The successful practical application of fracture mechanics can primarily be attributed to the mechanics of quasi-static cracks. In this case, methods have been worked out and standardized to answer questions concerning the stability of an existing arterial crack under quasi-stationary loads.

As regards dynamic fracture mechanics, which analyzes the stability of stationary cracks subjected to dynamic loading and processes of crack propagation, the theoretical investigations cannot be backed up by practical recommendations, for the time being. This is due to the extremely complicated behavior of fracture mechanics, and also to the existing disproportionality between the development of theoretical and experimental methods in dynamic fracture mechanics. For many years, progress in this field was associated with the solution (by analytical and numerical methods) of simulation problems in idealized situations. This left open the question of a correspondence between the idealized situation and the real conditions of dynamic fracture, as well as the experimental confirmation of theoretical results.

In the present chapter we have endeavored to describe the difference between dynamic fracture mechanics and quasi-static mechanics, and also discuss their principal aims, basic conceptions and techniques. The chapter contains primarily the results of investigations carried out by the authors in this field.

2.1. INTRODUCTION

In spite of all the brilliant achievements in the field of fracture mechanics and its numerous applications, the formulation and solution of the dynamic problems of this theory remained unknown, until recently, on account of their extremely complicated nature. Only the latest elegant analytic solutions of certain model problems and the development of new effective numerical methods have helped in surmounting this obstacle. In order to understand the growing interest towards investigations in dynamic fracture mechanics, it is necessary to grasp the essence of the subject and its interaction with quasi-static fracture mechanics. Indeed, the process of fracture is characterized (at least in its final stage) by a rapid propagation of the arterial crack or a set of branched cracks and is therefore essentially a dynamic process.

A large number of problems still remain unsolved in the description of the process on microscopic and macroscopic levels. Hence, when we state that fracture mechanics is an essential tool for computing the strength of bodies and structures, we mean the quasi-static fracture mechanics which determines whether or not an arterial crack is stable. Indeed, the quasi-static mechanics of brittle fracture, which is based on the idealized model of a sharp arterial crack and the concept of the stress intensity factor (SIF, at its tip, has been developed quite extensively; however, it provides only the first approximation to the description of fracture and can simply indicate whether or not a catastrophic growth of the crack sets in.

The chapter begins with the construction of analytical and numerical solutions (finite element method), then it is followed by the determination of SIF in different problems for bodies with cracks subjected to steady-state vibrations (Sections 2.1 and 2.2).

Section 2.3 is dedicated to the determination of dynamic SIF in limited and unlimited bodies in the case of impact loading. Under investigation is the influence of microcracks on the propagation of arterial cracks.

Section 2.4 compares theoretical and experimental results in dynamic fracture mechanics that lead to the conclusion of the necessity of construction of new models for an adequate description of essentially dynamic processes. In this case, it is necessary to take into consideration a microstructure of material which reacts in the first place to the propagation of waves passing ahead of the fracture.

Finally, Section 2.5 treats the problem of thermal shock in a medium with a crack.

Unlike the static fracture theory, the dynamic theory studies wave propagation. If the system under consideration contains a stationary or propagating defect, the wave field becomes very complicated and this has to be taken into account. Thus the temporal behavior of SIF for a specimen under shock loading with due regard for the reflected waves is characterized by strong oscillations. One more example – the tips of branching cracks themselves became the sources of new propagating waves. Even microdefects formed ahead of the tip of the main crack generate waves and interact with the main crack, which can by no means be neglected.

Thus at the present stage of its development, the theory of dynamic fracture mechanics is rather inconsistent. Today, the efforts of scientists working in fracture mechanics should be focused in this direction. But contradictions and inconsistencies in science always gave an impetus for new investigations and therefore the accumulated knowledge on fracture dynamics should inevitably lead to the formulation of a new rigorous and consistent theory. It is also worth noting that the recent decade has been characterized by a drastic increase of the number of investigations in the field. These include the creation of new models of fracture, analytical and numerical solutions of the problems of dynamic elasticity and plasticity theories for bodies with stationary or propagating cracks, and the development of new experimental methods.

2.2. STEADY-STATE VIBRATIONS — ANALYTICAL METHODS FOR DETERMINING STRESS INTENSITY FACTORS

Two types of problems are considered in the investigation of elastic stress fields in the vicinity of a crack tip being in a harmonic field. In the first type, harmonic loads are applied to crack faces, while in the second type a wave arrives from infinity and the singular field results from the diffraction of waves by the crack. The complete solution of the problem in the second case is presented as a sum of the incident wave (regular solution) and the scattered wave (singular solution). Obviously, the stress intensity factors are determined only by the solution for the scattered wave. In order to obtain this solution, say, for a stress-free opening mode crack in a symmetry plane, we most solve the boundary value problem for a half-plane with the following boundary conditions: on the crack, the tensile stress is equal to the stress due to the incident wave, but with the opposite sign; the normal displacement outside the crack is equal to zero; the shearing stress is also equal to zero everywhere on the boundary of the half-plane. It can be shown that, in the general case, the diffraction problems (in which the stress intensity factors are determined) can be reduced to the first type of problem. Consequently, we shall not accentuate the difference between diffraction problems and the problems of steady-state vibrations in the subsequent discussions.

It should be noted that a large number of problems have been solved to date for bodies with stationary cracks with the help of analytical methods. Solutions have been obtained for the problem of the steady-state vibrations of a plane with a crack, a strip with a crack, a space with a crack 'tunnelled' by antiplane shear, a plane with a period system of cracks caused by normal tensile and shearing loads applied to their faces, a space with a semi-infinite crack produced by concentrated harmonic loads applied to their faces, etc. The most important solutions are given in[1,2]. A special feature of the results

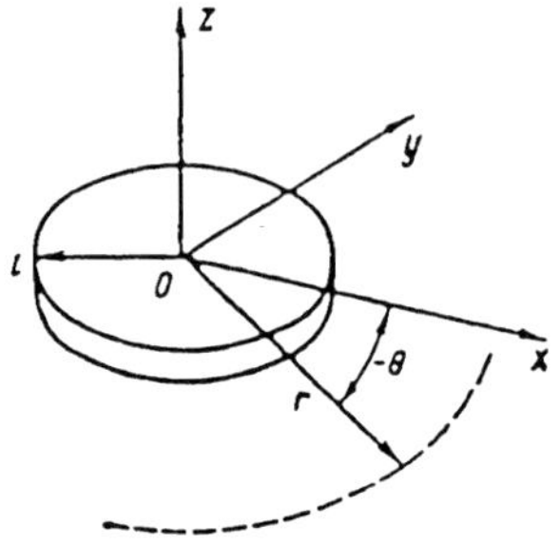

Figure 1. A penny-shaped crack in space.

obtained for semi-infinite cracks is that the influence of inertia effects on the stress intensity factors cannot be estimated in this case, since these problems do not have a static analog (the solution for zero frequency does not exist). However, the solution of dynamic problems for finite cracks at zero frequency must coincide with the static solution for the corresponding load. This allows us to obtain qualitative and quantitative estimates for the effect of dynamic loading on the increased risk of brittle fracture. It is also interesting to study the influence of inertia effect in three-dimensional case of a penny-shaped crack, and this problem will be solved and investigated below.

Let us study the vibrations of a medium containing a plane (circular) penny-shaped crack (Figure 1). We shall first consider the axisymmetric vibrations corresponding to the waves of axial compression-dilatation or radial shear. Then, we shall take up the case of torsional vibrations. We introduce cylindrical polar coordinates in such a way that the crack lies in the plane $z = 0$ and the origin coincides with the center of the crack. In the axisymmetrical case, we have two nonzero components of the displacement vector, viz., u_z and u_ϑ $(u_\vartheta = 0)$. Moreover, all derivatives with respect to ϑ are equal to zero.

If a wave defined by the potentials $\varphi^{(i)}$ and $\psi^{(i)}$ is incident on the crack, the complete solution of the problem is equal to the superposition of the incident and scattered waves

$$\varphi(r, z, t) = \varphi^{(i)}(r, z, t) + \varphi^{(s)}(r, z, t)$$
$$\psi(r, z, t) = \psi^{(i)}(r, z, t) + \psi^{(s)}(r, z, t)$$

The singular nature of stresses is due to the scattered waves. Since the functions φ and ψ are assumed to vary harmonically in time, the equations of motion for the potentials can be reduced to the form

$$(\nabla_1^2 + \alpha_1^2)\varphi = 0 \tag{1}$$

$$(\nabla_2^2 + \alpha_2^2)\psi = 1 \tag{2}$$

where

$$\nabla_j^2 = \frac{1}{r}\frac{\partial}{\partial r} r \frac{\partial}{\partial r} - \frac{(j-1)^2}{r} + \frac{\partial^2}{\partial z^2} \quad (j = 1, 2)$$

Without any loss of generality, the potentials of incident waves can be presented in the form

$$\begin{aligned} \varphi^{(i)} &= w_1(r)\cos(\alpha_1 z)\exp(-i\omega t) \\ \psi^{(i)} &= w_2(r)\cos(\alpha_2 z)\exp(-i\omega t) \end{aligned} \tag{3}$$

where the functions and w_1 and w_2 satisfy ordinary differential equations

$$\frac{d^2 w_j}{dr^2} + \frac{1}{r}\frac{dw_j}{dr} - \frac{(j-1)^2}{r^2} w_j = 1 \qquad (j = 1, 2)$$

The solutions bounded at $r = 0$ are $w_1 = \varphi_0$ and $w_2 = \psi(r/l)$ where φ_0 and ψ_0 are constant amplitudes of the incident wave. Relations (3) then assume the form

$$\begin{aligned} \varphi^{(i)} &= \varphi_0 \cos(\alpha_i z)\exp(-i\omega t) \\ \psi^{(i)} &= \psi_0(r/l)\cos(\alpha_i z)\exp(-i\omega t) \end{aligned}$$

In the investigation of axial dilatation-compression waves, we put $\psi^{(i)} = 0$. Then the stresses created by the incident wave having a potential $\varphi^{(i)}$ are equal to

$$\begin{aligned} \sigma_{rr}^{(i)} &= q^{(1)}(1 - 2n_*^4)\cos(\alpha_1 z)\exp(-i\omega t),\, q^{(1)} = -\mu\alpha_2^2\varphi_0 \\ \sigma_{\vartheta\vartheta}^{(i)} &= \sigma_{rr}^{(i)},\, \sigma_{zz}^{(i)} = q^{(1)}\cos(\alpha_1 z)\exp(-i\omega t) \\ \sigma_{rz}^{(i)} &= \sigma_{r\vartheta}^{(i)} = \sigma_{\vartheta z}^{(i)},\, n_*^2 = C_2/C_1 \end{aligned}$$

Since the crack is free from stresses and the symmetry conditions are satisfied for $Z = 0$, we get

$$\begin{aligned} &\sigma_{zz}^{(s)}(r, 0, t) + \sigma_{zz}^{(i)}(r, 0, t) = 0, \quad 0 \leqslant r < l \\ &u_z^{(s)}(r, 0, t) = 0, \quad r > l \\ &\sigma_{zz}^{(s)}(r, 0, t) = 0, \quad r \geqslant o \end{aligned} \tag{4}$$

Applying Hankel transforms to Equations (1) and (2), we obtain ($\beta_j^2 = s^2 - \alpha_j^2$)

$$\varphi^{(s)}(r,z,t) = \int_0^\infty B_1^{(1)}(s) J_0(rs) \exp[-(\beta_1 z + i\omega t)]dz, \quad z \geqslant 0$$
$$\psi^{(s)}(r,z,t) = \int_0^\infty B_1^{(2)}(s) J_1(rs) \exp[-(\beta_2 z + i\omega t)]dz, \quad z \geqslant 0 \tag{5}$$

where J_0 and J_1 are the zero- and first-order Bessel functions of the first kind, and $B_1^{(1)}$, $B_1^{(2)}$ are functions that have to be determined. Taking into account (4) and (5), we arrive at the dual integral equations

$$\int_0^\infty B_1(s) J_0(rs) = 0, \quad r \geqslant l$$
$$\int_0^\infty s f_1(s) B_1(s) J_0(rs) ds = -\frac{q^{(1)}}{2\mu(1-n_*^4)}, \quad 0 \leqslant r < l \tag{6}$$

where

$$f_1(s) = \frac{(2s^2 - \alpha_2^2)^2 - 4s^2\sqrt{s^2-\alpha_1^2}\sqrt{s^2-\alpha_2^2}}{2\alpha_2^2(1-n_*^4)s\sqrt{s^2-\alpha_1^2}}$$

and the unknown function B_1 is connected with $B^{(1)}$ and $B^{(2)}$ through thc relations

$$B_1^{(1)}(s) = \frac{2}{\alpha_2^2\sqrt{s^2-\alpha_1^2}}\left(s^2 - \frac{1}{2}\alpha_2^2\right) B_1(s)$$
$$B_1^{(2)}(s) = \frac{2s}{\alpha_2^2} B_1(s)$$

The system of integral equations obtained here can be solved by Copson's method.[2] The solution of (6) can be written in the form

$$B_1(s) = \left(\frac{2}{\pi}\right)^{1/2} \int_0^l d_1(t)\sin(st)dt$$

where

$$d_1(\xi) = \frac{1}{(2\pi)^{1/2}} \frac{q^{(1)} l}{\mu(1-n_*^4)} D_1(\xi)$$

and $D_1(\xi)$ is the solution of the Fredholm integral equation

$$D_1(\xi) - \int_0^l K_1(\xi,\eta)D_1(\eta)d\eta = \xi \tag{7}$$

having a kernel

$$K_1(\xi,\eta) = \frac{2}{\pi}\int_0^\infty \left[f_1\left(\frac{s}{l}\right)+1\right]\sin(s\xi)\sin(s\eta)ds$$

Since $D_1(\xi)$ and $K_1(\xi,\eta)$ are complex quantities, Equation (7) can be decomposed into two equations. Near a crack edge, the angular distribution of stresses in the plane of the crack is the same as in the plane problem on an opening mode crack. The stress intensity factor is equal to

$$K_I = \frac{2}{\pi}q^{(1)}\sqrt{\pi l}D_1(1)\exp[-i\omega(t-\xi_1)]$$

where

$$\xi = \frac{1}{\omega}\arctan\left[\frac{\mathrm{Im}D_1(i)}{\mathrm{Re}D_1(1)}\right]$$

The shear stress intensity factors are equal to zero and the solution for $\omega = 0$ coincides with the static solution

$$K_{Is} = (2/\pi)q^{(1)}\sqrt{\pi l}.$$

The results of numerical computations are shown in Figure 2. Let us now consider radial shearing vibrations. We put the potential $\varphi^{(i)} = 0$ and describe the stresses of the incident wave in the form

$$\sigma_{rr}^{(i)} = \frac{2q^{(2)}}{\alpha_2 l}\sin(\alpha_2 z)\exp(-i\omega t), \quad q^{(2)} = -\mu\alpha_2^2\psi_0$$
$$\sigma_{\vartheta\vartheta}^{(i)} = -\sigma_{rr}^{(i)}, \quad \sigma_{zz}^{(i)} = \frac{4q^{(2)}}{\alpha_2 l}\sin(\alpha_2 z)\exp(-i\omega t)$$
$$\sigma_{rz}^{(i)} = -q^{(2)}\left(\frac{r}{l}\right)\cos(\alpha_2 z)\exp(-i\omega t), \quad \sigma_{r\vartheta}^{(i)} = \sigma_{\vartheta z}^{(i)} = 0$$

The boundary conditions and the symmetry conditions are

$$\sigma_{rz}^{(s)}(r,0,t) + \sigma_{rz}^{(i)}(r,0,t) = 0, \quad o \leqslant r < l$$
$$u_r^{(s)}(r,0,t) = 0, \quad r \geqslant l$$

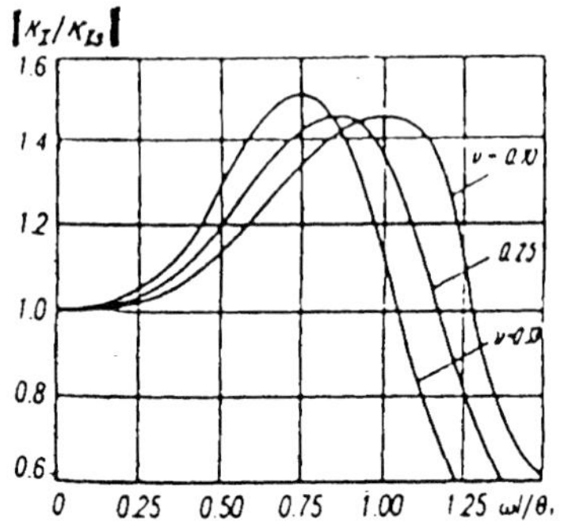

Figure 2. Opening stress intensity factor for the incidence of a dilatational wave on a penny-shaped crack.

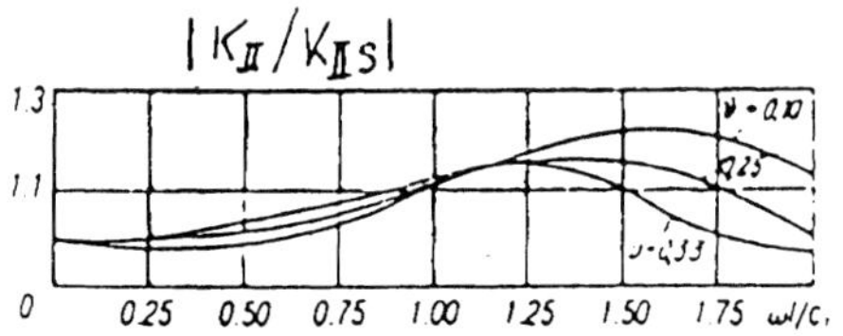

Figure 3. Inplane shear stress intensity factor for the case of incidence of a radial shear wave on a penny-shaped crack.

The resulting stress field is antisymmetric with respect to the plane $z = 0$; the angular distribution of stresses in this plane is the same as in the plane problem on an inplane shear crack.

The results of numerical computations of the quantities K_{II} as functions of the wave number $\omega l/c_1$ are presented in Figure 3.

Finally, let us consider a penny-shaped crack under the action of torsion waves. In this case, $u_r = u_z = 0$, while the remaining nonzero components of displacement and stresses are given by

$$u_\theta = u_\vartheta(r, z, t), \quad \sigma_{r\vartheta} = m\left(\frac{\partial ur}{\partial r} - \frac{u\vartheta}{r}\right), \quad \sigma_{\vartheta z} = \mu\frac{\partial u\vartheta}{\partial z}$$

The displacement u_ϑ is determined from the equation

$$\frac{\partial^2 u_\vartheta}{\partial r^2} + \frac{1}{r}\frac{\partial u_\vartheta}{\partial r} - \frac{u_\vartheta}{r^2} + \frac{\partial^2 u_\vartheta}{\partial z^2} = \frac{1}{c_2^2}\frac{\partial^2 u_\vartheta}{\partial t^2}$$

If u_φ is harmonically dependent on time this equation assumes the form

$$\frac{\partial^2 u_\vartheta}{\partial r^2} + \frac{1}{r}\frac{\partial u_\vartheta}{\partial r} - \frac{u_\vartheta}{r^2} + \frac{\partial^2 u_\vartheta}{\partial z^2} + \alpha_2^2 u_\vartheta = 0$$

Let the incident wave be defined by

$$u_\vartheta(r, z, t) = w_3(r)\sin(\alpha_2 z)\exp(-i\omega t)$$

where ω_3 is the bounded solution of the following equation for $r = 0$:

$$\frac{d^2 w_3}{dr^2} + \frac{1}{r}\frac{dw_3}{dr} - \frac{1}{r^2}w_3 = 0$$

and is equal to

$$w_3(r) = w_0 r/l$$

We then obtain

$$u_\vartheta^{(i)}(r, z, t) = \frac{w_0 r}{l}\sin(\alpha_2 z)\exp(-i\omega t)$$

$$\sigma_{\vartheta_z}^{(i)}(r, z, t) = \frac{q^{(3)} r}{l}\cos(\alpha_2 z)\exp(-i\omega t), \quad q^{(3)} = \mu\alpha_2 w_0$$

The torsional moment of the incident wave is equal to

$$T^{(i)} = 2\pi \int_0^l r^2 \left(\lim_{z\to o} \sigma_{\vartheta_z}^{(i)}\right) dr^{(3)} \frac{J}{l}\exp(-i\omega t)$$

where J is the polar moment of inertia of a circle of radius r. In the plane $z = 0$, we have

$$\sigma_{\vartheta_z}^{(i)} = \frac{T^{(i)} r}{l}$$

We shall seek the displacement $u_\vartheta^{(s)}$ for the scattered wave in the form

$$u_\vartheta^{(s)}(r, z, t) = \int_0^\infty B_3(s) J_1(rs)\exp[-(\beta_2 z + i\omega t)]ds$$

where the unknown function $B_3(s)$ is determined from the boundary condition

$$\sigma_{\vartheta_z}^{(s)}(r, 0, t) + \sigma_{\vartheta_z}^{(i)}(r, 0, t) = 0, \quad 0 \leqslant r < l$$

and from the condition for $z = 0$:

$$u_\vartheta^{(s)}(r, 0, t) = 0, \quad r \geqslant l$$

This leads to the following system of dual integral equations:

$$\int_0^\infty B_3(s)J_1(rs)ds = 0, \quad r \geqslant l$$

$$\int_0^\infty \beta_2/B_3(s)J_1(rs)ds = \frac{\tau_2}{\mu}\frac{r}{l}, \quad o \leqslant r < l$$

The required function $B_3(s)$ is expressed by

$$B_3(s) = s^{1/2}\int_o^l \left(\frac{2t}{\pi}\right)^{1/2}\frac{2\tau_2 l}{3\mu}D_3\left(\frac{r}{l}\right)J_{3/2}(st)dt$$

where $D_3(\xi)$ is the solution of the Fredholm integral equation of the second kind

$$D_3(\xi) + \int_0^l K_3(\xi,\eta)D_3(\eta)d\eta = \xi^2$$

having a symmetric kernel

$$K_3(\xi,\eta) = (\xi\eta)^{1/2}\int_0^\infty [(s^2 - \alpha_2^2 l^2)^{1/2} - s]J_{3/2}(s\xi)J_{3/2}(s\eta)ds$$

It can be shown that the angular distribution of the stress field in the plane $z = 0$ in the vicinity of the crack tip is the same as in the plane problem of a longitudinal shear crack. The results of the numerical computation of the stress intensity factor for the case of longitudinal shear, i.e.,

$$K_{III} = \frac{4}{3\pi}q^{(3)}\sqrt{\pi l}D_3(l)\exp[-i\omega(t - \xi_3)]$$

are shown in Figure 4. Here,

$$\xi_3 = \frac{1}{\omega}\arctan\left[\frac{\mathrm{Im}D_3(1)}{\mathrm{Re}D_3(1)}\right]$$

2.2. STEADY-STATE VIBRATIONS — APPLICATION OF THE FINITE-ELEMENT METHOD FOR CALCULATING STRESS INTENSITY FACTORS

We shall describe a method for calculating the dynamic stress intensity factors in plates with cracks subjected to steady-state vibrations.[1] This method is based on the presentation of these factors as a superposition of 'nominal' stress intensity factors corresponding to the normalized forms of free vibrations with certain weight factors. Another method for determining the stress intensity factors under harmonic loading is based on a direct step-by-step solution of the system of differential equations of motion.

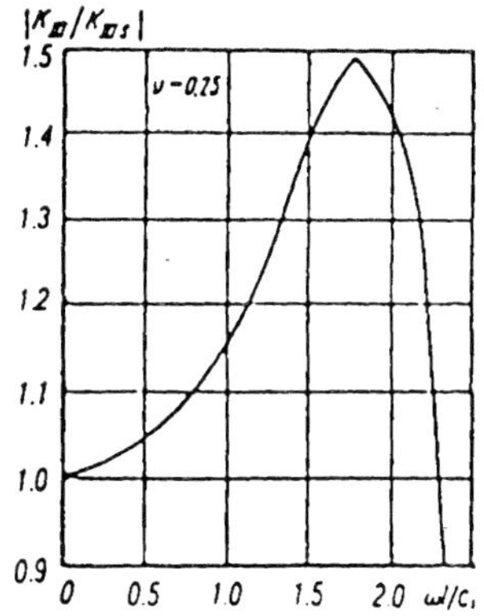

Figure 4. Antiplane shear stress intensity factor for the case of incidence of a torsion wave on a penny-shaped crack.

However, the solution in this case is quite cumbersome and not suitable for analysis since the parameters of motion and stress intensity factors in finite bodies depend on the ratio $(\omega/\omega_i)^2$, where ω_i are the free-vibration frequencies and ω is the loading frequency.

The method of calculating stress intensity factors based on the superposition of free vibration modes is more convenient to use.

The importance of investigations of the role of cracks on the frequencies and shapes of free vibrations in plates and shells has been emphasized before. These investigations are necessary not only for determining the admissible levels of loads and frequencies, but also for diagnosing the size of defects in various structural elements.

It should be noted that the analytical frequency dependences of the stress intensity factors for infinite media cannot be directly applied to finite-sized bodies, as is often done in the case of static stress intensity factors and dynamic stress intensity factors for impact loading (for the initial interval of time up to the instant when a wave scattered at the boundary arrives at the crack tip).

Unlike analytic methods, the method of finite elements is applicable to finite-sized bodies. A special procedure was proposed for estimating the accuracy of the computational solution.

The finite-element equations of motion of an elastic body in the absence of attenuation under harmonic loading have the form

$$[M]\{x\}^{\cdot\cdot} + [K]\{x\} = \{f\}\exp(i\omega t) \tag{8}$$

Here $[M]$ is the mass matrix, $[K]$ is the stiffness matrix, $\{x\}$ is the displacement vector, and $\{f\}$ is the load vector. For $\omega = 0$, we obtain the equilibrium equation

$$[K]\{x\} = \{f\} \tag{9}$$

Denoting by ω_i the eigenvalues in the ascending order, and by $\{x^{(i)}\}$ the eigenvectors from the generalized eigenvalue problem

$$[K]\{x\} = \omega^2[M]\{x\}$$
$$\{x^{(i)}\}^T[M]\{x^{(j)}\} = \delta_{ij} \tag{10}$$

(the superscript 'T' denotes transposition), we can write the particular solution of (8) with the frequency of the perturbing force in the form

$$\{x(t)\} = \sum_i \{x^{(i)}\}\frac{(\{x^{(i)}\}^T\{f\})}{\omega_i^2 - \omega^2}\exp(i\omega t) \tag{11}$$

Consequently, the static solution can be presented as a superposition of free vibration forms:

$$\{x^{(s)}\} = \sum_i \frac{\{x^{(i)}\}^T\{f\}}{\omega_i^2}x^{(i)} \tag{12}$$

Let us denote by K_s the static stress intensity factor corresponding to $\{x^{(s)}\}$, by $K^{(i)}$ the intensity factors corresponding to $\{x^{(i)}\}$, and by $K(t)$ the dynamic stress intensity factor. The dimensionality of $K^{(i)}$ is determined by taking into account the normalization.[10] Since the displacement vector together with a linear functional uniquely determines the stress intensity factor, we obtain

$$K(t) = \sum_i K^{(i)}\frac{(\{x^{(i)}\}^T\{f\})}{\omega_i^2 - \omega^2}\exp(i\omega t) \tag{13}$$

$$K_s = \sum_i K^{(i)}\frac{(\{x^{(i)}\}^T\{f\})}{\omega_i^2} \tag{14}$$

Introducing the dimensionless coefficients

$$z_i = \frac{K^{(i)}(\{x^{(i)}\}^T\{f\})}{K_s\omega^2}$$

we obtain from (12) and (13)

$$K(t) = K_s\sum_i z_i\frac{\omega_i^2}{\omega_i^2 - \omega^2}\exp(i\omega t) \tag{15}$$

$$\sum_i z_i = 1 \tag{16}$$

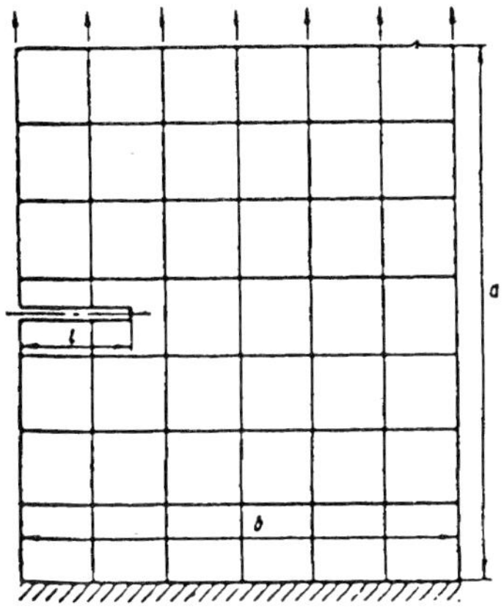

Figure 5. Division of a plate with an edge crack into finite elements.

The last equality serves as the criterion for the accuracy with which the dynamic stress intensity factor has been calculated, if the static stress intensity factor appearing in the expression for z_i is taken not in accordance with (14), but directly from the static system (9) of equilibrium equations. Moreover, the required number of vibration modes in (15) is determined from Equation (16) in the frequency interval $0 \leqslant \omega < \omega_1$.

Thus, the error in the determination of the dynamic stress intensity factors is estimated from the difference $|\sum z_i - 1|$. In other words, the error in the determination of $K(t)$ is estimated by comparing the dynamic stress intensity factor at zero frequency with the static stress intensity factor determined from the equilibrium equation. The dimensionless coefficients z_i used for determining $K(t)$ do not depend in the two-dimensional case of the size of the plate, its thickness, or the Young's modulus, but they do depend on the ratio of the edges, configuration, and the relative crack-length, as well as on the Poisson's ratio.

The method of calculating the dynamic stress factors described above was applied to the case of a strip fixed at one end with an edge crack (Figure 5), a free plate with a central horizontal crack, and a plate with an oblique central crack. The problem of calculating the dynamic stress intensity factors in a plate having an edge crack and fastened at one end may be encountered, for example in the investigations of the strength of turbine blades and wings of airplanes.

Suppose that a plate (see Figure 5) is subjected to a harmonic extension-compression at the edge parallel to the line of fastening. The ratio of sides

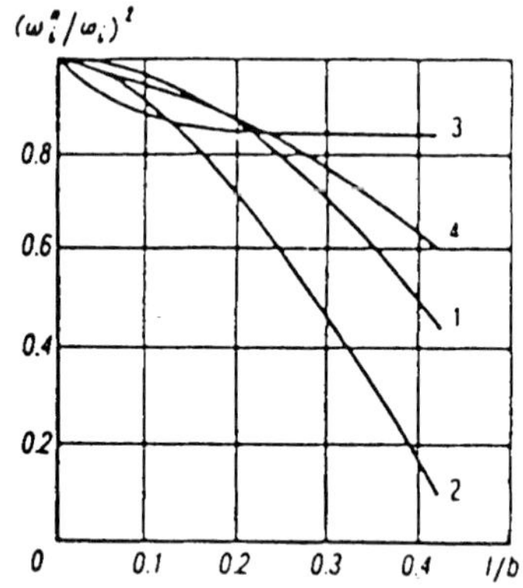

Figure 6. Dependence of the square of vibration frequency on crack length.

of the plate is $a/b = 1.17$. To determine the frequencies and the free longitudinal vibration modes for the plate, the latter is divided into a regular mesh of elements. The number of nodes was varied from 58 to 66 depending on the length of the crack. The special singular element was used. It was shown that any further increase in the number of degrees of freedom does not significantly affect the first eight frequencies and modes.

The results of computations of frequencies were identical to those obtained with the help of regular elements only in similar meshes. (In the case of regular elements, obviously, a considerable decrease in the element size in the vicinity of the crack tip is essential for calculating the stress intensity factors.) The accuracy of computation of eigenvectors was 10^{-4}.

The square of the ratio of vibration frequence is ω_i^* of a plate with a crack to the vibration frequence ω_i of a plate without a crack is plotted in Figure 6 as a function of the relative crack length for the first four frequencies. The dimensionless values of the square of the vibration frequency for a defect-free plate, computed for the ratio of sides equal to 1.17 and Poisson's ration $\nu = 0.3$, are 0.33, 2.15, 2.81 and 9.34. The transition to dimensional frequencies is made using the relation

$$\omega_i^2 = \bar{\omega}_i^2 \frac{E}{\delta_0 a d}$$

where $\bar{\omega}_i$ are dimensionless frequencies, E is Young's modulus, and δ_0 is the density.

It can be seen from Figure 6 that the appearance and propagation of a crack considerably lowers the vibration frequencies. Consequently, even if the range of loading frequencies does not exceed the fundamental (lowest) frequency of free vibrations of a structure, the growth of the crack may lead to resonance. Resonance is most dangerous for high-frequency loading, since the free vibration frequencies in real structural elements are of the order $\sqrt{E/\delta_0}$, i.e., are rather high.

The obtained results show that the crack growth leads to a more rapid attenuation of the longitudinal vibration frequencies as compared to transverse vibration frequencies. For example, the first four longitudinal vibration frequencies in the case of a relative crack length of 0.33 decreased in comparison with the frequencies of a continuous plate by 19.4%, 40.0%, 7.8% and 14.0%.

The results of computation of vibration frequencies of plates with cracks can be used for calculating the admissible loading frequencies as well as for determining the defect size in structural elements from variations in the natural vibration spectrum. While calculating the maximum admissible loading frequencies by taking into account the growth of defects, we should determine the free vibration frequencies corresponding to the largest crack size. Sometimes the corollaries of the Courant-Fisher theorem[3] are used to draw conclusions concerning the magnitudes of frequencies for small cracks. Suppose that a plate with a crack has vibration frequencies $\omega_1 \ll \omega_2 \ll \dots$, while the frequencies of vibrations of a plate with a smaller crack, which can be treated upon finite-element discretization as the first plate with r constraints imposed on it, are given by $\omega_2' \leqslant \omega_2' \leqslant \dots$. Then the following inequality holds:

$$\omega' \leqslant \omega_2' \leqslant \omega_i + \tau$$

In general, it can be stated on the basis of the corollaries of the Courant-Fisher theorem that the frequencies of free vibrations of a structure can be increased not only by decreasing the crack length, but also by fastening parts of the boundary, decreasing the plastic zones, reducing the mass, and increasing the stiffness. On the other hand, these factors increase the risk of brittle fracture in some cases. Moreover, the inertia effects may cause the stress intensity factors to reach critical values before the onset of resonance. Hence, the calculations of admissible frequencies must also be carried out on the basis of an analysis of the dynamic stress intensity factors.

It should be observed that the use of stiffening ribs may serve as an effective means of reducing the risk of brittle fracture. This is so because, first, they increase the vibration frequency of plates and, second, they decrease the value of the stress intensity factors.

A comparison of the vibration modes for a cracked plate with the corresponding modes for defect-free plate showed that the presence of a crack considerably changes the vibration modes. The fundamental vibration modes for a cracked plate are characterized by a considerable opening of the crack and, to a lesser extent, by a relative displacement of the crack faces. This is reflected in the stress intensity factors corresponding to normalized modes of vibrations: for the first four modes, all the ratios $K_{II}^{(i)}/K_I^{(i)}$ except the last one do not exceed unity and are equal to 0.24, 0.17, 0.77, and 1.35. Unlike transverse vibrations, the relative discontinuities in free vibration modes upon passing through the crack are quite large.

The effect of loading frequency on the stress intensity factors K_I for opening mode and K_{II} for inplane shear was investigated for the frequency interval $0 \leqslant \omega < \omega_1$, which is most important from the point of view of applications (the nonzero value of K_{II} during extension is due to pinching of the plate edge). These investigations showed that the opening mode stress intensity factor increases monotonically with loading frequency and ultimately exceeds the static value (for $\omega = 0$) on account of the structure of z_i in[15]. For $l/b = 0.167$, these dimensionless coefficients are equal to 0.104, 1.102, −0.234, 0.060, 0, −0.043, 0.030, and 0.001. It can be seen that for $0 < \omega < \omega_1$, the amplitude ratio $K_I^{(t)}/K_{I_s}$ increases owing to quite large positive values of z_1 and z_2. The condition (16) is satisfied in this case with an accuracy of 98%. Similarly, the coefficients for $l/b = 0.25$ are 0.220, 0.845, −0.061, 0.019, −0.003, 0.020, 0, and 0. The sum of these coefficients is 1.040, which means that the error in this case is 4%. For long cracks, the error increases up to 6%. The dependences of $K_I^{(t)}/K_{I_s}$ on dimensionless frequency are shown in Figure 7. The magnitude of the dynamic stress intensity factor for a given amplitude of the applied stress can be calculated with the help of these curves and the static intensity factor K_{I_s} presented in the form $K_{I_s} = q^{(1)}\sqrt{\pi l}F(l/b, a/b)$. If calculations are carried out by taking the ratio of the plate sides $a/b = 1.17$ and the values of relative crack length given above, the function $F(l/b, a/b)$ is found to be equal to 1.40, 1.65, and 2.53, respectively.

The accuracy with which condition (16) is satisfied for the stress intensity factors K_{II} is somewhat lower (the error in this case is between 6 and 14%). However, this degree of accuracy is sufficient for estimation and qualitative analysis of the stress intensity factor. It is found that the amplitude of the ratio $K_{II}^{(t)}/K_{II_S}$ first increases with increasing frequency from 1 (for $\omega = 0$) to about 1.25, after which it decreases and reverses its sign (Figure 8).

Calculations show that the ratio of the stress intensity factors of first and second kinds in the frequency range $0 \leqslant \omega \leqslant \omega_1$ is small. The modulus $|K_{II}/K_I|$ attains its highest value at $\omega = \omega_1$ and is equal to 26.6% and 19.0%, respectively, for $l/b = 0.25$ and 0.417. For $\omega^2 \leqslant 0.8\omega_1^2$,

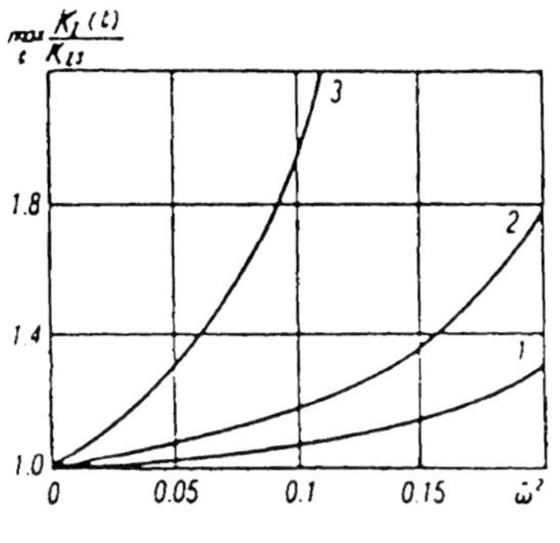

Figure 7. Dependence of the peak value of the ratio of dynamic and static opening stress intensity factors on the square of nondimensionalized vibration frequency (curves 1–3 correspond to relative check lengths 0.167, 0.250, 0.417).

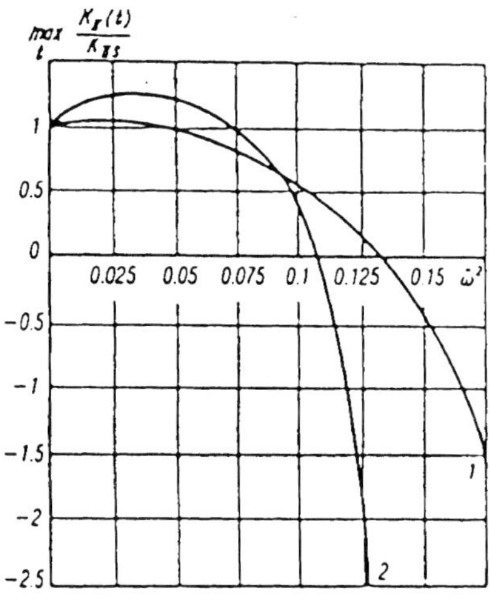

Figure 8. Dependence of the peak value of the ratio of dynamic (inplane shear) and static stress intensity factors on the square of nondimensionalized vibration frequency (curves 1 and 2 correspond to relative lengths 0.250 and 0.417, and the function $F(l/b, a/b)$ in the relation $K_{II_s} = q^{(1)}\sqrt{\pi l}F(l/b, a/b)$ is equal to 0.55 and 0.103, respectively, for these two cases).

the modulus of this ratio does not exceed 7.5%, while its highest value for $\omega^2 \leqslant 0.9\omega_1^2$ is 13.7%. Consequently, the stress intensity factor for fracture is the most significant parameter in the range of operating frequencies. Both stress intensity factors increase indefinitely in absolute value as the loading frequency approaches the fundamental vibration frequency.

The obtained results confirm that as the vibration frequency increases, the magnitude of the fracture load decreases, i.e., the risk of brittle fracture grows.

It is well known that the interaction of turbine blades with a gas jet excites vibrations as a result of which the stresses in the turbine elements become unstationary. The frequencies of these vibrations may become indefinitely close to the resonance frequencies, since their upper limit is equal to the product of the rotational frequency and the number of blades[4], while, according to the results obtained, the fundamental frequencies of free vibrations of plates are of the order of $\sqrt{E/\rho_0}$, which lies in the same range. This, as well as the above conclusions about the increased risk in brittle fracture upon an increase in the loading frequency, and about the decrease in the frequency of free vibrations due to crack growth, can be used to explain a number of cases described in literature on brittle fracture of turboengine elements resulting from operation at rotational frequencies exceeding the rated values[5] (including the fracture during the first few months of operation when fatigue cracks cannot attain significant values).

Dynamic stress intensity factors in a square plate with an oblique central crack (Figure 9) under harmonic extension-compression conditions were also investigated. The crack was inclined at an angle of 45° to the base of the plate and a load of unit intensity was applied to the horizontal edges.

Figure 10 shows the plots of the peak values of these stress intensity factors as functions of $\bar{\omega}^2$, the square of the dimensionless loading frequency in the interval $0 \leqslant \omega \leqslant \omega_1$. It can be seen that the stress intensity factors monotonically increase from their static values at $\omega = 0$ and tend to infinity as the fundamental vibration frequency is approached.

2.3. IMPACT LOADING — THE METHODS OF DETERMINING STRESS INTENSITY FACTORS

In this section we shall consider the problems of cracked bodies subjected to impact loading. As in the case of harmonic loading, the stress intensity factors increase in this case in comparison with their static values. This fact must be taken into consideration while designing machines and structures involving the application of the methods of fracture mechanics. Under the action of impact loading, the behavior of time-dependent dynamic stress intensity factors is more complicated than for harmonic loading. Thus, for example, the increase in the dynamic stress intensity factors for finite cracks takes place until the arrival at the crack tip of a wave scattered from the opposite tip of the crack.[6] In the case of semi-infinite cracks whose faces are subjected to a uniformly distributed tearing impact load, the stress intensity factor

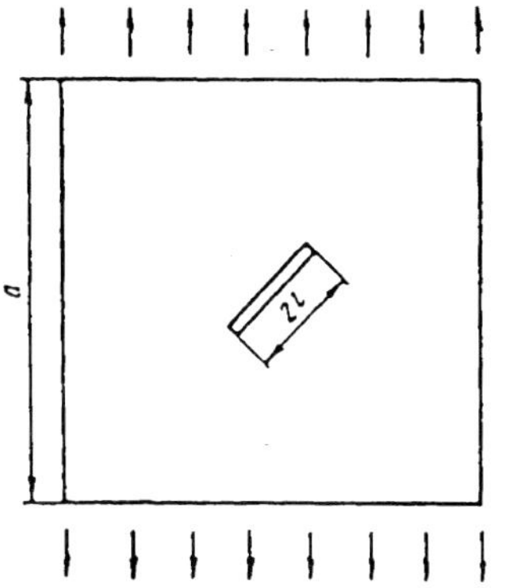

Figure 9. A square plate with an oblique crack ($2l = 4\sqrt{2} \times 10^3$ mm, $a = 22 \times 10^3$ mm, $E = 1$ Pa, $\delta_0 = 0.1$ kg/m^3.

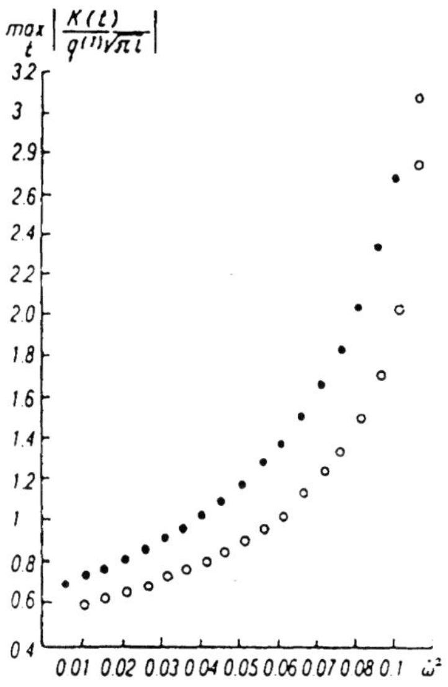

Figure 10. Dependence of stress intensity factors due to opening (light circles) and shear (dark circles) on the square of nondimensional loading frequency.

increases with time in proportion to $t^{1/2}$ and becomes infinite as $t \to \infty$.[7] Another interesting fact worth mentioning here[8] is that the stress intensity factor for a plate with a semi-infinite crack whose faces are subjected to concentrated tearing impact forces assumes a constant (static) value after the passage of a certain time. As in the case of harmonic loading, the problem of impact loading of a body with a crack can be solved analytically to the end only for a few idealized formulations because of the complex nature of the mathematical analysis. Usually, such problems are considered for infinite media. The problem of the stress state at the tip of a semi-infinite crack whose faces are subjected to uniformly distributed tearing impact loads was first solved by Maue.[9] After this Baker[10] obtained the solution of a two-dimensional problem in which a semi-infinite crack (appearing in a uniform tearing field at the instant $t = 0$) propagates with a constant velocity ϑ; in this case, we obtain for $\vartheta = 0$ the solution for a stationary crack. In a certain sense, this is 'calibration' problem, since the analytic and numerical solutions of the problem of impact opening of finite cracks for the initial interval of time (from the zero instant to the moment of arrival at the crack tip of the waves scattered from the boundary of the body or from the other tip of the crack) must coincide with the solution for a semi-infinite crack.

The problems of the behavior of finite cracks under impact loading were considered also in[1,2,11]. The problem is reduced to the numerical solution of Fredholm integral equations for variables in the Laplace transform domain, while the inverse transform was carried out only for the main part of local stresses at the crack tip. A characteristic feature of this approach is that the solution for a finite crack remains finite as $t \to \infty$, and that after the attainment of the peak value at the instant when a wave emitted by the opposite tip of the crack arrives at the crack tip, the stress intensity factor oscillates about the static value with a decreasing amplitude. It should be emphasized once again that up to this instant, the solution for a finite crack coincides with that for a semi-infinite crack.

The mentioned approach does not lead to an exact determination of the points of discontinuity for the time derivative of the stress intensity factor, since the results are known to be smoothed out in the numerical inversion of the Laplace transform. Nevertheless, this method can be used to determine the maximum dynamic stress intensity factor and other qualitative characteristics of its variation with time.

Let us consider the problem for an infinite strip with a crack (Figure 11) and for a medium with a penny-shaped crack (Figure 1). In the case of tearing mode, the boundary conditions have the following form:

$$\sigma_{yy}(x, 0, t) = -q^{(1)}H(t), \quad |x| < l$$
$$u_y(x, 0, t) = 0, \quad |x| \geqslant l$$
$$\sigma_{xy}(x, 0, t) = 0, \quad \sigma_{xy}(x, \pm L, t) = 0, \quad \sigma_{yy}(x, \pm L, t) = 0 - \infty < x < \infty$$

The variation of the stress intensity factor is shown in Figure 12. It is clear that the oscillations observed for small values of t in the case of a narrow

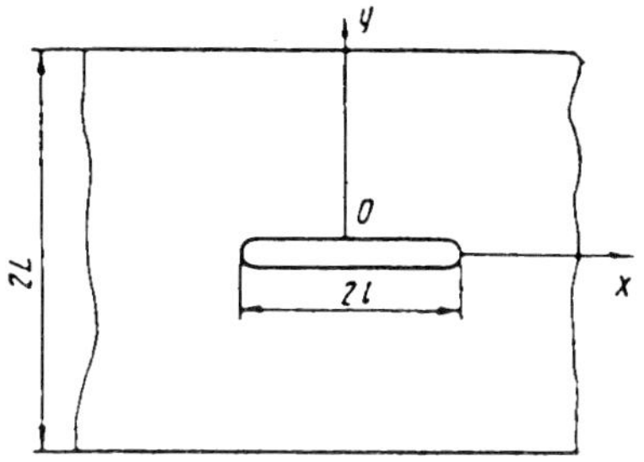

Figure 11. A strip with a crack.

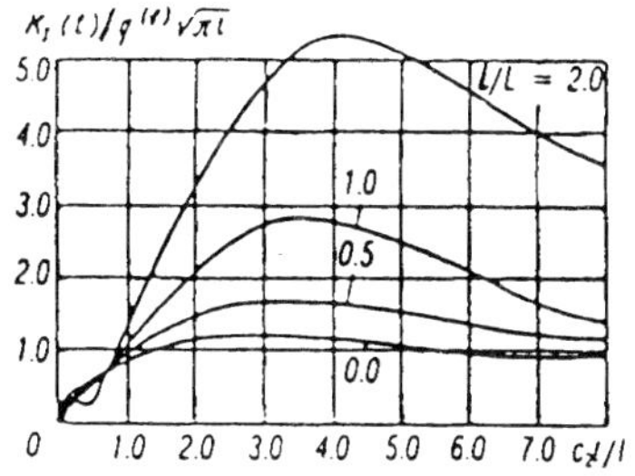

Figure 12. Stress intensity factors in a strip with a crack.

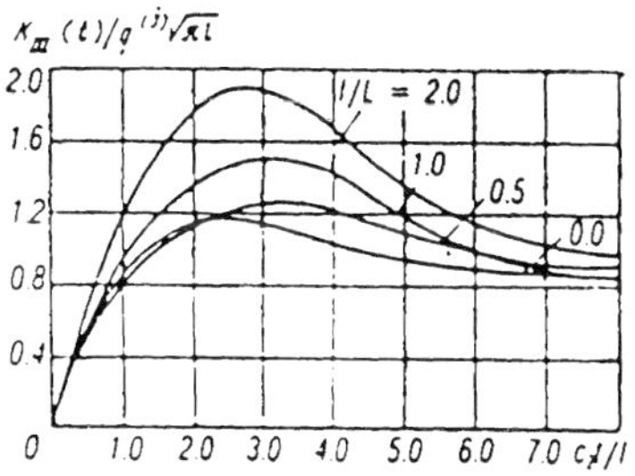

Figure 13. Stress intensity factors in a strip with a crack due to antiplane shear.

strip ($l/L = 2$) are due to the arrival of a wave scattered from the strip edge at the crack tip.

The results for an antiplane shear crack are shown in Figure 13.

The solution of the problem of penny-shaped cracks due to inplane and antiplane shear is also presented. The results of calculations normalized with the help of the static values of the stress intensity factors are shown in Figure 14.

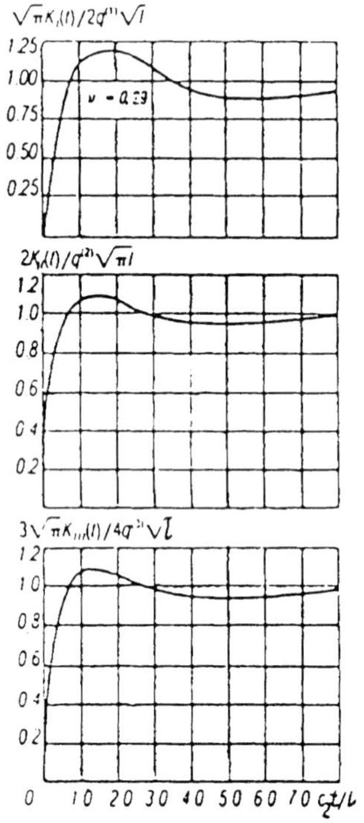

Figure 14. Time dependence of stress intensity factors (of three types) under impact loading of a penny-shaped crack.

The obtained results show that the stress intensity factors attain their peak values after a brief interval of time has elapsed following the application of the load. These values then oscillate with a decreasing amplitude in the vicinity of the static values. Unlike in the case of a plane crack, the initial wave has a toroidal form and the stress waves are generated at each point of the crack face. Consequently, the maximum values of the stress intensity factor are attained much more rapidly. For large intervals of time, when c_2l is considerably larger than the crack radius, the wave front gradually becomes spherical.

Here we also present the results of the numerical investigation of wave propagation in the cracked bodies obtained with the help of our program code which utilizes the finite element method together with singular elements and the implicit method of the integration on time.

The first group of calculations concerns to the modelling of wave interaction between the macrocrack and microcrack, and the tips of branched crack. Such calculations were stimulated by the experimental results[12,13] which have shown the necessity to account the complicated wave interaction between the cracks, specimen boundaries, other cracks and microcracks.

The second group of calculations has been made for the investigation of wave propagation in massive elements of metallurgical equipment undergoing impulse loads, namely in the rolls of rolling mills and no-anvil hammers. Particularly the influence of the hammer shape on stress intensity factor dependencies was analysed.

Our analysis of the stress intensity factors was based on the finite element method. The simulation of stress singularities was provided by means of the singular finite element.[11] The finite element solution of wave problems is reduced to the solution of the system of the second-order differential equations. We have chosen for our purposes the unconditionally stable Θ-method which has been recommended in[15]. For this method the step estimates based on the spectral analysis are known[14], but our numerical experiments have shown that for the problem under consideration another estimates are useful. If one is interested in the wave propagation processes in details the step sometimes less than the time of wave propagation through the smallest element must be chosen. This step was usually less than the step recommended in[14]. But if only the 'inertial effect' (in other words, the SIF amplification), is under investigation one may increase the step essentially.

We have investigated the plate with the central symmetrically branched crack (Figure 15). The length of the branched part projection is equal to 0.2 l, where l is the half of the main crack length. The half of the branch angle is equal to $\pi/4$.

The calculated SIF K_I and K_{II} dependencies on time in the case of the tearing loads suddenly applied to the edges are presented in Figure 16 ($2b$ is the height of the plate).

Corresponding static values of SIF are: $K_{I_s}/\sigma_0\sqrt{\pi l_*} = 1.47, K_{II_s}/\sigma_0\sqrt{\pi l_*} = 0.77$. Here l_* is equal to the main crack half length plus the branch projection length. The total field in the plate is the superposition of the waves reflected from the boundaries and singular points, and all three types of waves (rarefaction, shear and Rayleigh) are essential. The oscillations of the SIF dependencies can be connected with the moments of waves arrival to the crack tip. The maximum ratio of dynamic SIF to static ones is about 2.

The distinguishing feature of the calculated dependencies is that they are identical to each other (when dynamic values are normalized by means of static ones, as it is shown in Figure 17).

The next calculations refer to the case of the plate with macrocrack and two 'microcracks' behind its tips (Figure 18). The 'microcrack' length is 1/20 of the macrocrack length, and its right tip (for the right half of the plate) coincides with the branch tip projection from the previous problem.

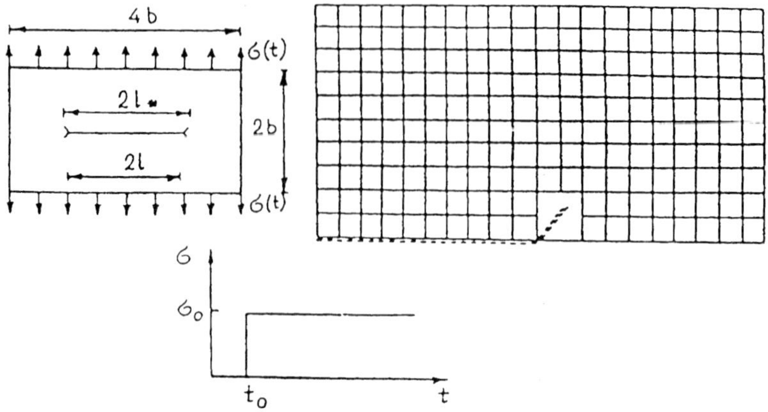

Figure 15. The branched crack in the plate.

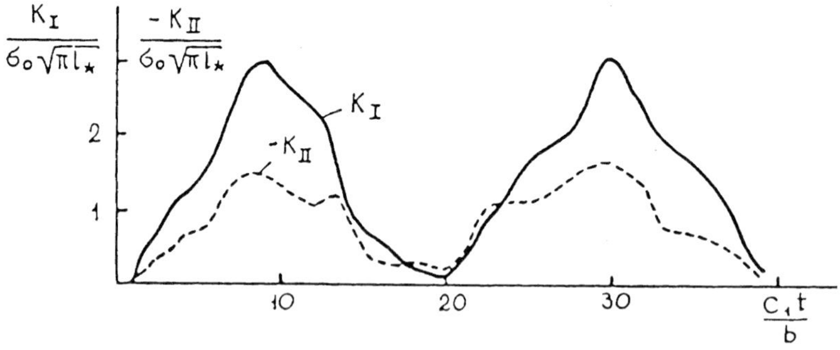

Figure 16. The dynamic SIF dependencies on time.

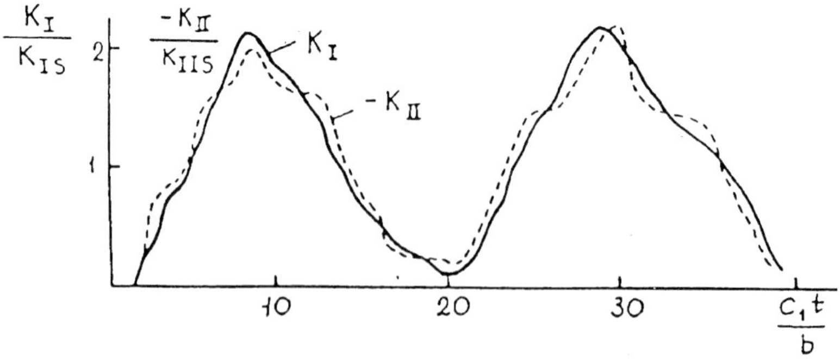

Figure 17. The ratio of the dynamic and static SIF.

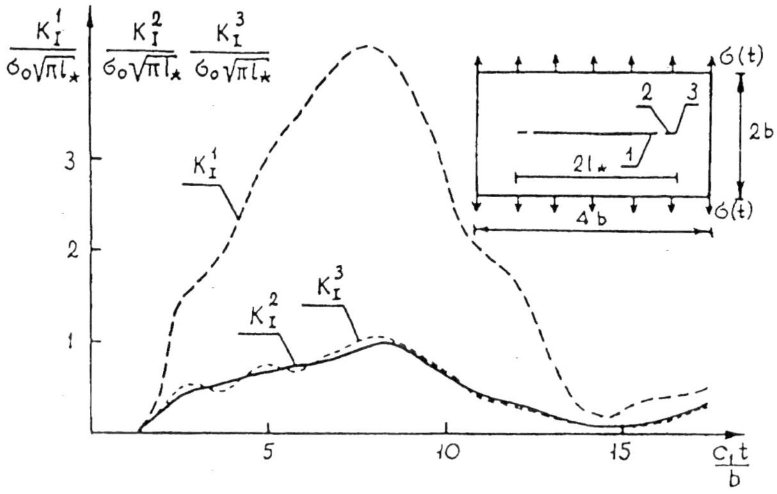

Figure 18. The plate with macrocrack and microcracks.

The distance between the cracks is approximately equal to the 'microcrack' length. The finite element grid contained of 119 elements and 139 nodes. The dependencies of the three SIF are presented in Figure 18. The static values are $K_{I_s}/\sigma_0\sqrt{\pi l_*} = 1.80$; 0.51; 0.52 for the tips 1, 2, 3 respectively.

These dependencies can be used for the detailed analysis of processes in the crack tip with the account of macroscopically observed microbranches and other microdefects. Particularly one can recommend to incorporate such defects behind the crack tip into the calculation models in practical cases and for the evaluation of the experimental data.

For the analysis of the dynamic SIF in the rolls of rolling mills we have considered the configuration shown in Figure 19. The stress state of the rolls is essentially three-dimensional. Usually one can discover the penny-shaped cracks of the radii about 10–50 mm in it. Since the dynamic behavior of the crack is primarily due to the wave propagation in the diametrical cross-sections we can treat the problem as plane one in the first approach. The static analysis shows that the crack displayed in Figure 19 is under the compression. But really the compression stresses are superposed to the initial tensile stresses. The calculated roll was composed of two cylinders, and the ratio of the inner cylinder stiffness to the outer one stiffness was equal to 0.9. The crack length was equal to 0.005 of rolls radii.

The last calculations refer to the SIF in the no-anvil hammers striking the rigid foundation (Figure 20). The ultrasonic examination usually displays that

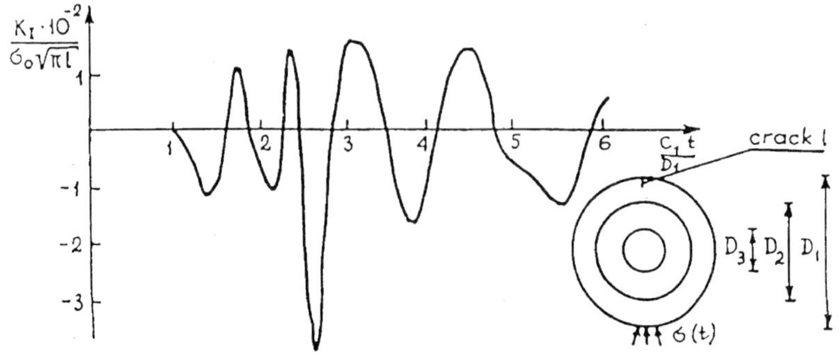

Figure 19. The crack in the roll and the SIF dependency.

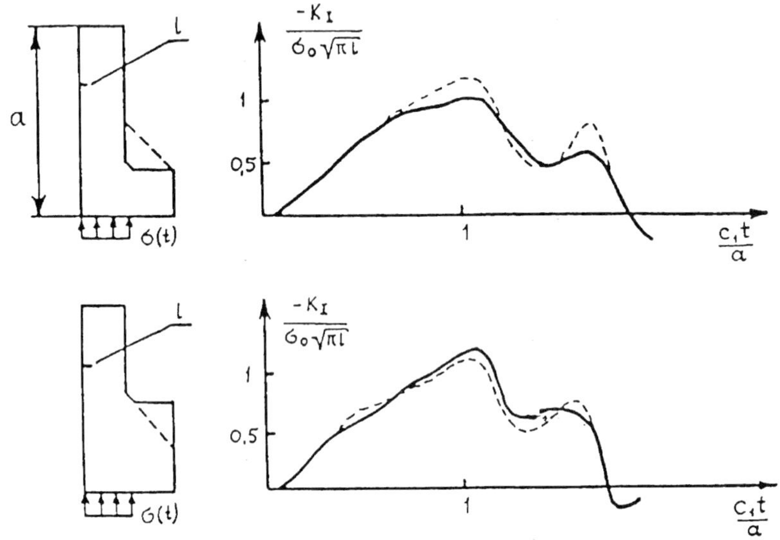

Figure 20. No-anvil hammers and the SIF dependencies.

the cracks are localized near the weld and that they are penny-shaped. The cracks can have quite big size (up to 250 mm in diameter). The experimental results show that the stress state is characterized primarily with the wave propagation along the hammer in vertical direction. This fact allows us to reduce the problem to the plane one.

The detailed analysis should account for the solution of the contact problem, but we have applied to the bottom surface the stresses found from the experiment. Their dependence on time was described with the triangle impulse (duration up to 10^{-2}s), so the SIF had the tendency to decrease.

The presented dependencies refer to the dynamic SIF normalized by means of the static (conditional) value. We can now obtain the approach to the dynamic SIF in three-dimensional case as the product of the displayed function to the static SIF in the three-dimensional case.

2.4. COMPARISON OF THEORETICAL AND EXPERIMENTAL RESULTS IN DYNAMIC FRACTURE MECHANICS

In this section we shall describe the difference between dynamic fracture mechanics and quasi-static mechanics, and also discuss their principal aims and basic assumptions.

The following problems arise in the study of dynamic fracture mechanics. First, we must find the conditions under which a quasi-static or dynamic loading can cause catastrophic growth of a crack of a given size. Secondly, conditions of unloading under which crack growth can be arrested must be determined. Thirdly, the loading parameters and the properties of materials determining crack propagation must be specified. Finally, we must know the conditions under which a propagating crack branches out and the mechanism underlying this phenomenon. These problems are called the problems of starting, arrest, propagation, and branching of a crack, respectively. The task of dynamic fracture mechanics is to find a solution to these problems.

Thus, it is obvious that the range of topics covered by dynamic fracture mechanics is much wider than that of quasi-static fracture mechanics. While quasi-static fracture mechanics deals, as a rule, with formulating just the criterion for unstable crack propagation, in dynamic fracture mechanics a whole range of criteria, viz., the starting criterion, the criterion for crack arrest, bending and branching criteria, etc., have to be established. The phenomenological description of fracture dynamics, through the concept of a sharp arterial crack, results in the appearance of a large number of critical stress intensity factors, like the stress intensity factor for crack start, which depends on the loading rate, the crack arrest stress intensity factor, and the intensity factor depending on the velocity of crack propagation. Some experimental results can be explained in this way, though serious disagreement are observed between theoretical and experimental results in some other cases. However, it must be noted that the experimental results themselves are conflicting. Quite often, one can find an experimental result in literature completely repudiating the result of another experiment.

2.4.1. Idealized Model for Quasi-brittle Fracture and its Inconsistencies

In the prevailing idealized model based on the ideas put forth by Griffith, Irwin, and others, the growth of a rectilinear crack in an elastic plane is usually considered. In this case, infinite stresses appear at the crack tip and the fracture process is assumed to take place at the crack tip itself. Moreover, it is assumed that the energy γ spent in creating a unit new surface area is a constant quantity characteristic of the material. On the basis of this assumption, the elasto-dynamic stress field at the crack tip is calculated and the energy balance equation is formulated. The stress at the crack tip has a $1/\sqrt{r}$-type singularity and the stress intensity factors depend on the crack velocity ϑ. If this dependence is determined by solving the elastodynamic problem with a moving crack and is then substituted into the fracture criterion, the crack velocity can be determined, i.e., the crack behavior can be predicted. Depending on loading conditions, the crack may continue to propagate or it may get arrested. The starting criterion is also derived from the energy balance equation.

Thus, we can formulate the main features of the quasi-brittle dynamic fracture model as follows:

1. Stress fields at the crack tip are described with the help of stress intensity factors.
2. Criteria for start, propagation, and arrest are derived from the constancy of specific fracture energy.

The adequacy of this model can be judged by analyzing experimental data on the stress intensity factor and by comparing the conditions of crack start, propagation, and arrest with theoretical results.

It is found that, in the first place, the 'independence' of the criteria of crack start, propagation, and arrest is established beyond a shadow of doubt. This follows from the existence of a large number of critical stress intensity factors describing the start, propagation, and arrest of cracks, which taken together do not satisfy the energy balance equation. Secondly, the relation between the instantaneous stress intensity factor and instantaneous crack velocity is established from the energy balance equation under the assumption that the surface fracture energy is independent of the loading rate and the past history of the loading process. However, this relation which establishes a one-to-one correspondence between K and ϑ, is not confirmed by experiments. Besides, the idealized model does not provide a satisfactory explanation of the branching phenomenon.

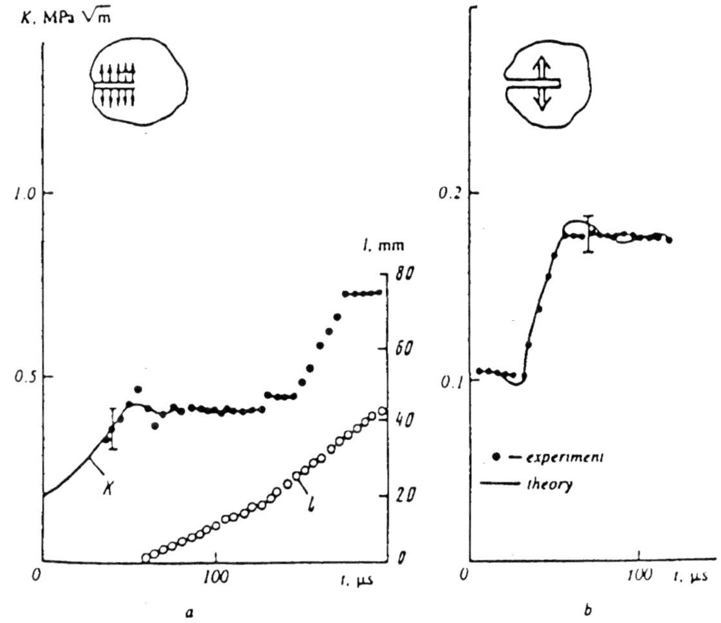

Figure 21. Time dependence of the stress intensity factor upon loading the crack by concentrated (a) and distributed (b) loads comparison of theory and experiment (the vertical mark indicates start of the crack).

A direct comparison of analytical and experimental results of investigations of the K vs. t dependence is carried out in[16] for a stationary crack which is set into motion by an impact load applied to its faces. The results obtained in[16] are discussed in[11].

Experiments were carried out on plates of Homalite-100 (perspex) using the caustics method. A very good agreement between theoretical and experimental values of $K_I(t)$ was obtained both before and after crack initiation (Figure 21).

A similar experiment, carried out on the same material at a much faster loading rate and under higher stresses, showed that such an agreement exists only up to the instant when the crack starts to propagate (Figure 22).

This discrepancy can be explained by taking into account the following circumstance.[13] Under an impact loading of the crack faces, the size of the region around the crack tip where the stress satisfies the theoretical criteria is equal to zero at the initial instant of time and increases with the velocity

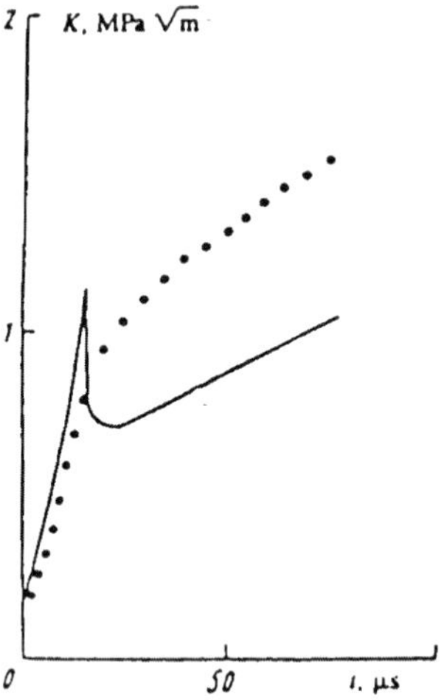

Figure 22. Time dependence of the stress intensity factor for a uniformly distributed load equal to double the load used in Figure 21.

of propagation of elastic waves. Thus, in order to establish a zone size which could provide experimental information about the singular stressed state, a certain amount of time is required. This time is large in comparison with the time scale of the processes taking place during dynamic fracture, and increases with the crack velocity. Since the field is assumed to be steady during a theoretical decomposition of the stresses in the vicinity of the crack tip, the experimental methods based on this decomposition can be used only if the process is stablized in a certain finite interval of time (before the moment of time under consideration). In other words, no stress waves should arrive at the crack tip during this period and there should be no abrupt variation in crack velocity.

Thus, in regions where the theoretical and experimental values of the stress fields coincide, the idealized model of dynamic fracture mechanics can be assumed to be valid except during transient processes. For all other cases, the behavior of real materials under dynamic fracture differs from the idealized approach towards fracture at the crack tip.

2.4.2. Microscopic Structural Aspects of Dynamic Fracture

A number of inconsistencies between the idealized fracture model and the experiment can be explained by assuming that the fracture occurs in a certain region in front of the crack tip, as, for example, in the case of viscous bodies under quasi-static fracture. Usually, it is assumed that the main feature of deviation from idealized behavior is the plastic flow at the crack tip accompanied by the appearance of cavities which grow in the region of strong plastic deformation. This widespread point of view is undoubtedly valid for viscous materials. However, it is worth noting that even a brittle fracture does not lead to an ideal fracture for dynamic crack propagation. The existence of microfracture near the crack tip certainly influences the time dependence of the fracture process, i.e., affects the fracture rate, since a finite amount of time is required for establishing the region where the fracture process takes place.

In the experiments carried out on Homolite-100, which undergoes a brittle fracture upon static and dynamic loadings, a finite fracture region existed, while there was obviously no plastic deformation. At least three different observation lead to this result: the dependence of fracture surface roughness on the instantaneous stress intensity factor, highspeed photography of discrete fracture sources at the crack tip on the real time scale, and the stress waves generated by these discrete microscopic fractures.

Let us consider in detail the dependence of fracture surface roughness on loading level, investigated experimentally. It is convenient to divide the analysis into three stages. Observation of the fracture surface (after fracture) under low magnification reveals the main features of surface quality variation as a result of a variation of the macroscopic parameters of the process. The surface is then studied under an electron microscope, which shows its microscopic characteristics. The basic conclusions drawn after a visual inspection of the surface are confirmed at this stage. Finally, the same behavior is confirmed by photomicrography of discrete fracture sources on real time scale, i.e., during the fracture process.

After the propagation of a crack at a high speed, three regions can be isolated on the fracture surface: 'mirror', 'mist' and 'hackle'. After passing the hackled region, the crack splits into several branches. The mirror zone is characterized by an absolutely smooth surface which completely reflects the light falling on it. In the mist region, the surface becomes rougher, while it becomes hackled in the last region. These results were obtained in four experiments on impact fracture of samples as a result of step-by-step increase in the load.

A constant (time-independent) velocity was observed in all experiments. The stress intensity factor was the only parameter that varied in time. In the first case, the crack propagated along the symmetry line until the waves reflected at the crack faces began interact with the crack tip. Since the waves were reflected from faces that were not fixed in the same way, the crack deviated from its rectilinear propagation path in such a way that the inplane

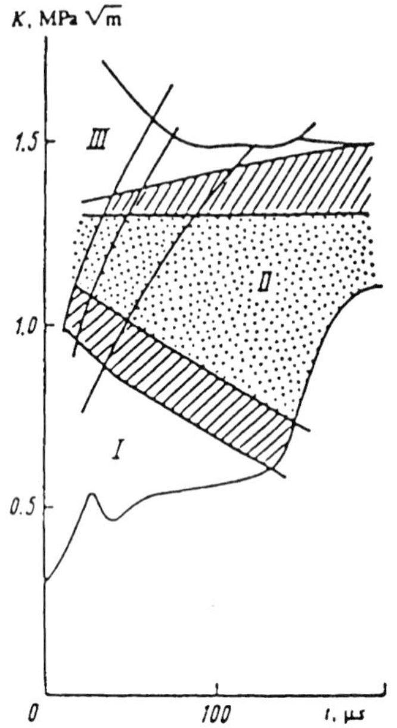

Figure 23. Dependence of K on t for four loading levels. (I) mirror zone; (II) mist zone; (III) hackle zone.

shear stress intensity factor along its path was equal to zero. The fracture surface consisted mainly of the mirror region. Throughout the period of its propagation, the crack had the same velocity, equal to 363 m/s. At the instant of arrival of the scattered wave, i.e., in about 150 μs, the stress intensity factor began to increase sharply (Figure 23) and the fractured surface became misty.

In the second case (for a large value of the load), the crack propagated with a velocity of 410 m/s and the fracture surface consisted of three regions, including the hackled region. When the wave arrived at the crack tip ($\sim$ 150 μs), branching sets in. A similar picture was observed in the last two experiments, but all processes occurred at a higher speed. The size of the mirror region at the crack faces was reduced, but the other two regions became larger in size.

The transition regions that exist between different regions are hatched in Figure 23. It can be stated on the basis of the experimental results that the energy liberated at the crack tip is spent in the formation of a rough surface. It becomes clear that this circumstance may be decisive contradiction in the idealized model of dynamic fracture, since according to this model the

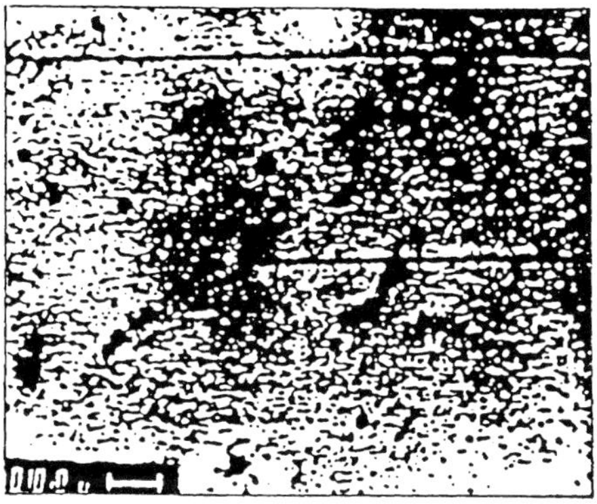

Figure 24. Microphotograph of the mirror zone.

fracture energy is calculated by multiplying the energy density by the area of the fractured surface, which is equal to the product of the crack length and the sample thickness. In actual practice, if we take into account the roughness of the surface, the area of the fractured surface will have a different value. A direct attempt to correct the energy balance makes the fracture energy density a function of the past history of loading, and makes it extremely difficult to solve dynamic fracture mechanics problems.

The mechanism of rough surface formation becomes clear from a microscopic study of the fracture surface (Figures 24–26, the magnification is 7000). In the mirror region (Figure 24), the crack encounters a large number of 10–25 μm cavities on its way. The interaction of the crack with these cavities is responsible for the start of a large number of cracks which, however, do not change the direction of arterial crack propagation. It can be stated that the microscopic cracks originating at the microcavities do not interact in the mirror region. In the mist region, the stress intensity factor increases and the stresses become sufficient for activating isolated cavities and triggering an interaction between them (Figure 25). This is accompanied by the appearance of many parabolic figures, a characteristic feature of the interaction between cavities and cracks propagating at a constant velocity. These parabolas have different sizes and depths, thus indicating their propagation in three dimensions. Hence, even before the arrival of the arterial crack in the foggy region, a large number of microscopic defects appear with their orientations in different planes, endeavoring to change the direction of propagation of the crack. Finally, this process becomes even more intensive in the hackled region and covers an ever increasing region in front of the crack tip. Small 'rivulets' appear and grow in a direction perpendicular to the crack (Figure 26).

Figure 25. Microphotograph of the mist zone.

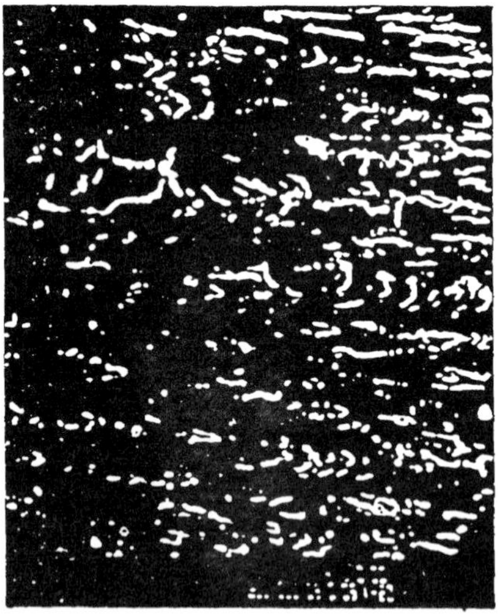

Figure 26. Microphotograph of the hackle zone.

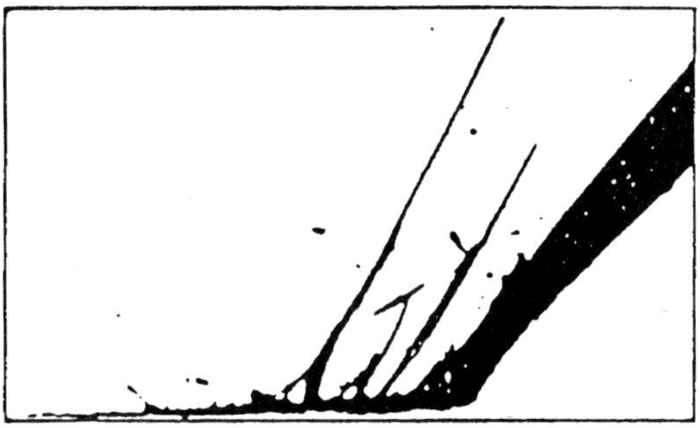

Figure 27. Magnified photograph of the crack branches.

It can be concluded from here that a unit crack propagates in the mirror region in the beginning, and its behavior differs considerably from quasi-static growth. In the mist region, simultaneous uniform propagation of an ensemble of cracks takes place. In the hackled region, crack propagation follows the same physical process, but the size of the micro-fracture region increases. Thus, it can be stated that crack propagation for a high level of stresses is governed by the growth of microscopic cavities and cracks, their confrontation and interaction.

From this point of view it is possible to give an acceptable qualitative description of crack branching as a continuous process of evolution of leading microcracks. Indeed, let us take a look at the photomicrograph of crack branching (Figure 27 for example). It is clearly seen that the branching process is a continuation of the intensive growth and interaction of microscopic defects taking place in the feather region. The onset of final branching is preceded by numerous attempts at branching; the microscopic cracks deviate from the direction of arterial crack propagation and are then arrested. A complex process of wave interaction occurs between these microscopic branches and the arterial crack. At a certain instant of time, the stresses acquire such a value that the crack finally branches out. Without doubt, this process is statistical and three-dimensional, but it also has deterministic features; the spread in the coordinate of the final branching point was just 1 mm in a series of five experiments. It must be emphasized that branching is an evolutionary process and does not involve just the conversion of a single mathematical cut into several cuts, as modelled in the problems of elastodynamics, but involves a gradual qualitative charge in the fracture front according to the mechanism shown in Figure 28.

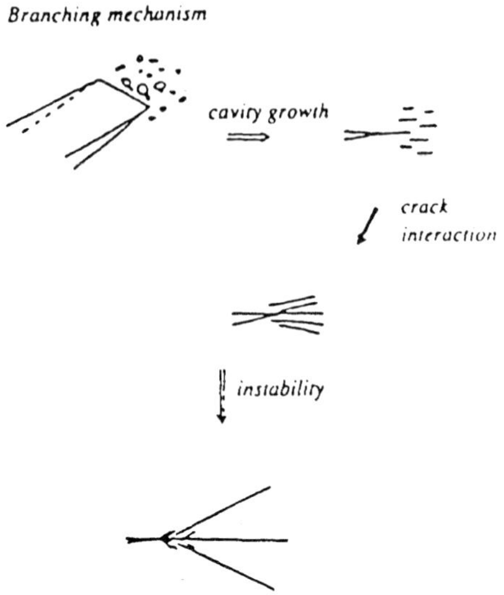

Figure 28. Branching mechanism.

2.5. THERMAL SHOCK RESPONSE OF CRACKED MEDIA

Thermal shock is a problem wherein large magnitude dynamic stresses arise in a structure due to a sudden change in temperature. For example, during high temperature applications of turbine engines and during the re-entry of a spacecraft into the earth's atmosphere, structures like the turbine blade and the nose cone are suddenly exposed to high temperatures that give rise to severe transient stresses. The thermal shock can be described mathematically by jointly solving the equation of heat conduction and the thermoelastic system with the account of the inertia terms.

The problem of thermal shock in an isotropic half space has been well documented. It was first studied by Danilovskaya[17], who treated the problem as a quasi-static case and neglected the thermoelastic coupling term in the heat conduction equation. The effect of the inertia term in the uncoupled thermoelastic theory was subsequently studied by Sternberg and Chakravorty.[18] Exact analytical solutions for the coupled theory including inertial effects[19] was first determined by Boley and Tolins in[20]. This problem was also solved numerically using the finite-element approach in[21,22].

In all of the above-mentioned papers it has been shown that in the case of a single isotropic layer the effect of the inertia term on the magnitude of the thermal stresses is very small as compared to the effect of the coupling term and that in most cases the inertia term can be dropped from the solution. Approximate finite difference solutions for the thermal and elastic responses of a layered half-plane subjected to sudden surface heating have been obtained in[23] on the base of a coupled thermoelastic theory[19] with regard to inertia effects. It has been shown that for very short times the dynamic effect of the thermal stresses are more important than the effect of the thermomechanical coupling, which may be neglected.

The problem of estimating the strength of cracked plates and shells subjected to thermal shock was considered very actively in the last few years by numerical methods. The literature devoted to problems in this field has been cited for example in[24,25].

In this section we shall consider the problem of thermal shock in a plane with a semi-infinite rectilinear cut. While formulating the boundary-value problem, we shall confine ourselves to plane strain.

Let K be a plane with the rejected semi-axis $\Gamma = \{x = (x_1, x_2); x_2 = 0, x_1 \leqslant 0\}$. We assume that its temperature is equal to zero at the initial instant, and then cut Γ instantly acquires a constant temperature T_0. The temperature drop creates a stressed state in K, and we have to determine the stress intensity factors $K_I(t)$ and $K_{II}(t)$.

Mathematically, the problem is formulated as follows. The temperature T is determined from the solution of the boundary value problem

$$\begin{cases} \frac{\partial T}{\partial t} - a^2 \Delta T = 0 & \text{on } K \times (o, \infty) \\ T = T_0 & \text{on } \Gamma \times (0, \infty), \quad T = 0 \text{ for } t = 0 \end{cases}$$

where a is the thermal diffusivity.

The displacement vector u generated by this thermal field is obtained from the solution of the following boundary value problem:

$$-\delta \frac{\partial^2 u}{\partial t^2} + \mu \Delta u + (\lambda + \mu) \text{graddiv}\, u = \gamma \text{grad}\, T \quad \text{on } K \times (o, \infty)$$

$$\begin{aligned} \sigma_{22} = \gamma T, \quad \sigma_{21} = 0 \text{ on } \Gamma \times (o, \infty) \\ u = \frac{\partial u}{\partial t} = 0 \text{ for } t = 0 \end{aligned} \tag{17}$$

where

$$\sigma_{ij} = \lambda \operatorname{div} u + 2\mu(\partial u_i / \partial x_j + \partial u_j / \partial x_i)$$

$\gamma = 2\mu\alpha_T(1+\nu)/(1-2\nu)$, μ is the shear modulus, α_T is the coefficient of linear thermal expansion, and ν is the Poisson's ratio.

For simplicity of calculations, we shall temporarily assume that $\alpha = 1$ and $T_0 = 1$.

Applying the Laplace transform, we obtain the temperature T in the form

$$T(p, x_1, x_2) = \int_{-\infty}^{\infty} T(p, \xi) \exp(ix_1\xi - (p + \xi^2)^{1/2} x_2) d\xi$$

where

$$\tilde{T}(p, \xi) = -\frac{p^{-3/4}}{2\pi i(\xi + i0)} (p^{1/2} + i\xi)^{-1/2}$$

Applying the Laplace transform in variable t to the problem (17), we obtain the boundary value problem

$$\begin{cases} -\delta p^2 u + \mu \Delta u + (\lambda + \mu) \text{graddiv } u = \gamma \text{grad } T \quad \text{on } K \\ \sigma_{22} = \gamma T, \quad \sigma_{21} = 0 \quad \text{on } \Gamma \end{cases}$$

It follows from the results obtained in[26] that the same asymptotic formula is applicable for the displacements u when $\tau \to 0$, as in the case of the plane static problem in the theory of elasticity (i.e., for $p = 0$).

The following representation holds for the stress intensity factors $K_I(p)$ and $K_{II}(p)$ (see Refs 27 and 28):

$$K_j(p) = -\gamma \int_K T \text{ div } \zeta^j dx, \quad j = I, II \tag{18}$$

The vector-functions ζ^j are explicitly constructed in[1]. Substituting them into (18) and releasing the assumption $a = 1$ and $T_0 = 1$, we get the following expression for the tensile stress intensity factor:

$$K_I(t) = T_0 \gamma \frac{x - 1}{x + 1} a^{1/2} t^{1/4} M(b_1 a t^{-1/2}) \tag{19}$$

with the function $M(h)$ explicitly constructed in[1].

Since the function T is even, the stress intensity factor K_{II} is equal to zero.

Let us carry out an asymptotic analysis of the function $K_I(t)$. Using the asymptotic expressions for $M(h)$ from[1] we can write the following asymptotic formulas for the stress intensity factor:

$$K_I(t) \sim -\frac{4}{\pi} \Gamma\left(\frac{3}{4}\right) T_0 \gamma \frac{x - 1}{x + 1} a^{1/2} t^{1/4} \text{ for } b_1 a \ll t^{1/2} \tag{20}$$

$$K_I(t) \sim -\frac{2\sqrt{2}\beta_2^2}{\beta_R \sqrt{\pi}} T_0 \gamma \frac{x - 1}{x + 1} Q(0) \left(\frac{t}{b_1}\right)^{1/2} \quad \text{for } b_1 a \gg t^{1/2} \tag{21}$$

where $x = 3 - 4\nu, b_1 = c_1^{-1}, b_2 = c_1/c_2, \beta_R = c_1/c_R(c_1, c_2$, and c_R are the velocities of dilatational, shear and Rayleigh waves respectively). The quantities b_1, b_2, b_R as well as $Q(0)$ are completely defined by the value of the Poisson's ratio ν.

In both vases, a sudden cooling ($T_0 < 0$) results in tensile stresses, while a sudden heating ($T_0 > 0$) leads to compressive stresses. The first of these asymptotic coincides with the one obtained in[27] for the quasistatic temperature problem, i.e., for $b_1 = 0$.

For $T_0 < 0$, we obtain from the asymptotics $K_I(t)$ the asymptotics t^* of the beginning of crack propagation. Indeed, t^* is the smallest instant of time satisfying the equation $K_I(t*) = K_{I_C}$ where K_{I_C} is the critical value of the tensile stress intensity factor.

Substituting the asymptotics (20) into this equality, we get

$$t^* \sim \left(\frac{\pi(x+1)K_{I_c}}{4\Gamma(3/4)\gamma(x-1)a^{1/2}T_0}\right)^4, \quad |T_0|\gamma a b_1^{1/2} \ll K_{I_c} \tag{22}$$

In turn, (21) leads to the expression

$$t^* \sim \left(\frac{\sqrt{\pi}\beta_R(x+1)K_{I_c}b_2^{1/2}}{2\sqrt{2}\beta_2^2\gamma(x-1)T_0}\right)^2, \quad |T_0|\gamma a b_1^{1/2} \gg K_{I_c} \tag{23}$$

The first of these asymptotics coincides with the quasistatic one[27], while the second depends to a considerable extent on the inertia term in the dynamic equations of the theory of elasticity. It should also be observed that for the large values of the thermal diffusivity a, the fracture time does not depend asymptotically on a.

Finally, let us consider the case of a bounded domain. Let Ω_0 be a plane domain with a smooth boundary Γ_0 (Figure 29). There is a rectilinear cut l in Ω_0, connecting the origin of coordinates $0 \in \Omega_0$ with the point $A \in \Gamma_0$. By Γ we shall mean the contour Γ_0 supplemented by twice-passed segment l, while Ω indicates the domain bounded by Γ.

The temperature $\hat{T}$ is determined from the boundary value problem

$$\frac{\partial \hat{T}}{\partial t} - a^2 \Delta \hat{T} = 0 \ \text{ on } \ \Omega \times (0, \infty)$$
$$\hat{T} = T_0 \ \text{ on } \ \Gamma \times (0, \infty), \quad \hat{T} = 0 \ \text{ for } \ t = 0$$

The displacement vector u generated by this field is obtained from the solution of the boundary value problem similar to (17).

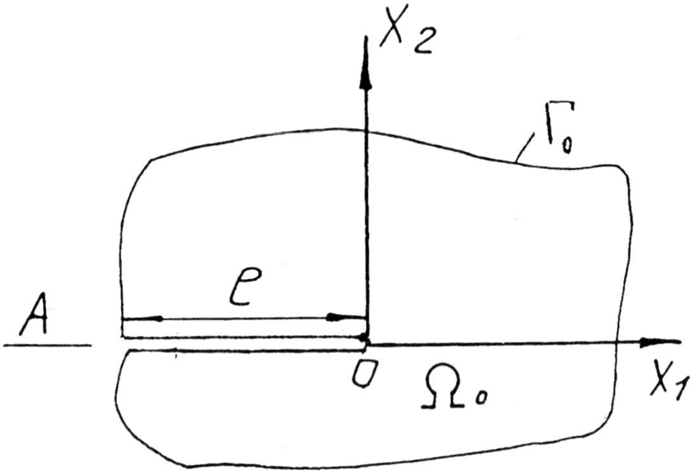

Figure 29. A medium with a crack.

We denote the stress intensity factors at the crack tip l by $\hat{K}j(t)$, $j = I, II$ and, as before, the tensile stress intensity factor for a plane with a semi-infinite cut by $K_I(t)$. Let s be the distance between the crack tip l and the boundary Γ_0. We shall assume that s is the characteristic size of the region Ω.

The basic information about $\hat{K}_I(t)$ and $\hat{K}_{II}(t)$ can be obtained by combining the formula (19) for $K_I(t)$ with the estimates obtained in[27].

$$|\hat{K}_I(t) - K_I(t)| \leqslant c_N T_0 s^{1/2} \gamma \left(\frac{a^2 t}{s^2}\right)^N$$

$$|\hat{K}_{II}(t)| \leqslant c_N T_0 s^{1/2} \gamma \left(\frac{a^2 t}{s^2}\right)^N$$

where $2t \leqslant b_1 s$ and $at^{1/2} \leqslant s$, N is any positive number, and the quantity c_N depends on the Poisson's ration ν, the number N, and the geometry of the boundary Γ_0, but not on T_0, l, a, or t.

Among other things, this means that the asymptotic formula (20) is valid for $\hat{K}_I(t)$ in the region $2a^2 s^{-2} t < b_1^2 a^2 t^{-1} \ll 1$ while the asymptotic formula (21) is valid in the region $a^2 s^{-2} t \ll 1 \ll b_1^2 a^2 t^{-1}$. The asymptotics (22) for the instant of fracture becomes valid for

$$|T_0|\gamma a b_1^{1/2} \ll K_{I_c} \ll |T_0|\gamma a^{1/2}(b_1 s)^{1/4}$$

and the asymptotics (23) for

$$|T_0|\gamma \min\{sa^{-1}b_1^{-1/2}, ab_1^{1/2}\} \gg K_{I_c}.$$

References

1. Parton, V.Z. and V.G. Boriskovsky, 1989, Dynamic Fracture, *Mechanics*, Vol. 1: Stationary cracks, Hemisphere: New York.
2. Sih, G.C. (ed), 1977, *Elastodynamic Crack Problems*, Noordhoff, Leyden.
3. Lancaster, P., 1969, *Theory of Matrices*, Academic Press: New York.
4. Samoilovich, G.S., 1975, *Excitation of Vibrations in Turbine Blades*, Mashinostroyeniye: Moscow, (in Russian).
5. Liebowitz, H. (ed), *Fracture* (Acad. Press, New York, London. Vol. I, 1968; Vol II, 1968; Vol. III, 1971; Vol. IV, 1969; Vol. V, 1969; Vol. VI, 1969; Vol. VII, 1972).
6. Thau, S.A. and T.-H. Lu, 1971, *Int. J. Solids and Struct.*, **7**, 731–750.
7. Cherepanov, G.P., 1979, *Mechanics of Brittle Fracture*, McGraw Hill: New York.
8. Freund, L.B., 1974, *Int. J. Eng. Sci.*, **12**, 179–189.
9. Maue, A.W., 1954, *Z. Angew. Math. und Mech.*, **34**, 1–12.
10. Baker, B.R., 1962, *Trans. ASME.J. Appl. Mech.*, **29**, 449–458.
11. Parton, V.Z. and V.G. Boriskovsky, 1990, *Dynamic Fracture Mechanics, Vol. 2: Propagating Cracks*, Hemisphere: New York.
12. Knauss, W.G. and K. Ravi-Chandar, 1985, *Int. J. Fract.*, **27**, 127–144.
13. Knauss, W.G. and K. Ravi-Chandar, 1986, *Eng. Fract. Mech.*, **23**, 9–20.
14. Laturelle, F.G., 1989, *Computers and Structures*, **32**, 721–735.
15. Bathe, K.J., 1982, *Finite Element Procedures in Engineering Analysis*, Prentice Hall, Englewood Cliffs, NC.
16. Ravi-Chandar, K. and W.G. Knauss, 1982, *Int. J. Fract.*, **20**, 209–222.
17. Danilovskaya, V.I., 1950, *Prikl. Mat. Mekh.*, **14**, 316–318, (in Russian).
18. Sternberg, E. and J.G. Chakravorty, 1959, *J. Appl. Mech.*, **26**, 503–509.
19. Boley, B.A. and J.H. Weiner, 1960, *Theory of Thermal Stresses*, John Wiley and Sons, Inc.: New York.
20. Boley, B.A. and I.S. Tolins, 1962, *J. Appl. Mech.*, **29**, 637–644.
21. Ting, E.C. and H.C. Chen, 1982, *Computers and Structures*, **15**, 165–175.
22. Prevost, J.H. and D. Tao, 1983, *J. Appl. Mech.*, **50**, 817–822.
23. Santhosh. U., 1992, *J. Thermal Stresses*, **15**, 339–353.
24. Erdogan, F., 1992, *Int. J. Fract.*, **53**, 159–185.
25. Noda, N. *et al.*, 1990, *J. Pres. Vessels and Pip.*, **42**, 247–257.
26. Kondrat'ev, V.A., 1967, *Trudy Mos. Matem. Obshch.*, **16**, 209–292.
27. Kozlov, V.A. *et al.*, 1985, *Prikl. Mat. Mekh.*, **49**, 627–636.
28. Maz'ya, V.G. and B.A. Plamenevskii, 1977, *Math. Nachr.*, **76**, 29–60.

3 BOUNDARY INTEGRAL EQUATION METHODS IN DYNAMICS AND FRACTURE

M.H. ALIABADI

Wessex Institute of Technology, Ashurst Lodge, Ashurst, Southampton, UK

In this chapter the fundamental concepts of boundary integral formulations in elastodynamic are described. Also presented are the computational algorithms required to implement these formulations in practical engineering analysis.

The dual boundary element methods are presented as an efficient method for the solution of crack problems in elastodynamic. The method is combined with a time domain, Laplace transform formulation or the dual reciprocity method. The quarter-point elements or the J integral are used to evaluate the dynamic stress intensity factors.

Also presented are indirect boundary element formulations known as the fictitious stress method and the displacement discontinuity method. These formulations are presented in the Laplace transform space.

Several examples of crack problems subjected to dynamic loadings are presented for both two and three-dimensional problems.

3.1. INTRODUCTION

The application of the Boundary Element Method (BEM) to linear elastic fracture mechanics is now well established and widely used in practice. The method offers a clear advantage over other domain type methods. One of the main reasons for this advantage is the possibility of evaluating the Stress Intensity Factors (SIF) accurately. There have been many methods devised for the evaluation of SIF's using BEM. A detailed description of these methods can be found in the text book by Aliabadi and Rooke.[1]

Straightforward applications of the boundary element method to crack problems leads to a mathematical degeneration, if the two crack surfaces are considered co-planar. For symmetrical geometries, it is possible to overcome this difficulty by imposing the symmetry boundary condition and hence modelling only one crack surface. However, for non-symmetrical crack problems, another way must be found. The first widely applicable method for dealing with two co-planar crack surfaces was the subregion method. The subregion method introduces artificial boundaries into the body, which connect the cracks to the boundary, in such a way that each region contains a crack surface. The two regions are then joined together such that equilibrium of tractions and compatibility of displacements are enforced. The main draw back of this technique is that the introduction of the artificial boundaries are not unique, and thus can not be implemented into an automatic procedure. In addition, the method generates a larger system of algebraic equations than is strictly required.

More recently the Dual Boundary Element Method (DBEM) as developed by Portela, Aliabadi and Rooke[2] for two-dimensional problems and Mi and Aliabadi[3] in three-dimensional problems has been shown to be, a general and computationally efficient way of modelling crack problems in BEM. Here the application of the dual boundary element method to dynamic fracture mechanics will be presented.

The boundary element solutions to elastodynamic problems are usually obtained by either the time domain, Laplace or Fourier transform or the dual reciprocity method. Here, the dual boundary element formulations are presented for the three methods. Dynamic stress intensity factors are evaluated using the quarter-point elements and the J-integral. Several examples are presented to demonstrate the efficiency and accuracy of these methods. Also presented are comparisons of the three methods.

The above formulations are known as direct boundary element formulations. In these formulations the displacements and tractions (or stresses) are the primary parameters. An alternative formulation to the direct method is known as the indirect method. Several indirect formulations have been reported in the literature. One of the most efficient of these formulations is known as displacement discontinuity method. Here the application of the displacement discontinuity method to dynamic fracture mechanics is presented using the Laplace transform method. Several examples of cracks in two and three dimensional problems are presented to demonstrate the efficiency of the method.

Reviews of the BEM formulations in dynamics and fracture mechanics can be found in Beskos[11] and Aliabadi and Brebbia[13] respectively. For further information on BEM in dynamics see[4–9] and advances in computational methods for dynamic fracture mechanics are reported in[10].

The boundary element formulations reported here are based on the papers by Fedelinski, Aliabadi and Rooke[14,18,21–23] and Wen, Aliabadi and Rooke[24–27] for direct and indirect methods respectively.

3.2. TIME DOMAIN METHOD (TDM)

The governing equation of dynamic Naviers's equation for elastic body can be written as

$$(\lambda + \mu)u_{i,ij}(\mathbf{x}, t) + \mu u_{j,ii}(\mathbf{x}, t) + \rho\left[b_j(\mathbf{x}, t) - \ddot{u}_j(\mathbf{x}, t)\right] = 0 \qquad (1)$$

where λ and μ are constants denoting the Lame$'$ and shear modulus of elasticity, ρ is the mass density, b_j the body force (per unit mass) and $\ddot{u}_j = \partial^2 u_j/\partial t^2$ is the acceleration. Equation (1) can be rewritten in a form

$$(c_1^2 - c_2^2)u_{i,ij}(\mathbf{x}, t) + c_2^2 u_{j,ii}(\mathbf{x}, t) + b_j(\mathbf{x}, t) - \ddot{u}_j(\mathbf{x}, t) = 0 \qquad (2)$$

where the dilatational and shear wave velocities are given as $c_1^2 = (\lambda + 2\mu)/\rho$ and $c_2^2 = \mu/\rho$ respectively.

Considering a domain Ω with a bounded surface Γ, the boundary condition can be written as

$$u_i(\mathbf{x}, t) = u_i^o \qquad t = 0 \qquad (3)$$

$$\dot{u}_i(\mathbf{x}, t) = \dot{u}_i^o \qquad t = 0 \qquad (4)$$

for the initial condition, and

$$u_i(\mathbf{x}, t) = \hat{u}_i(\mathbf{x}, t) \qquad \text{on } \Gamma_u \qquad (5)$$

$$t_i(\mathbf{x}, t) = \hat{t}_i(\mathbf{x}, t) \qquad \text{on } \Gamma_t \qquad (6)$$

where $t_i = \sigma_{ij}n_j$, with n_j denoting the components of the outward normal to the boundary Γ at $\mathbf{x}$, Γ_u and Γ_t denote parts of the boundary with either displacement or traction boundary conditions. The stress tensors are given by

$$\sigma_{ij}(\mathbf{x}, t) = \rho(c_1^2 - 2c_2^2)\delta_{ij}u_{m,m}(\mathbf{x}, t) + \rho c_2^2(u_{i,j}(\mathbf{x}, t) + u_{j,i}(\mathbf{x}, t)) \qquad (7)$$

The reciprocal theorem in dynamics is due to Graf[12] and is the extension of the well known Bettis reciprocal theorem in elastostatics. Considering two independent elastodynamic states with parameters $(u_i, t_i, u_i^o, \dot{u}_i^o, b_i)$ and $(u_i^*, t_i^*, u_i^{*o}, \dot{u}_i^{*o}, b_i^*)$ defined in a same domain Ω bounded by a surface Γ, the reciprocal theorem for $t \geq 0$ can be written as

$$\int_\Gamma t_i * u_i^* d\Gamma + \int_\Omega \rho\left[b_i * u_i^* + \dot{u}_i^o u_i^* + u_i^o \frac{\partial u_i}{\partial t}^*\right] d\Omega =$$

$$= \int_\Gamma t_i^* * u_i d\Gamma + \int_\Omega \rho\left[b_i^* * u_i + \dot{u}_i^{*o} u_i + u_i^* \frac{\partial u_i}{\partial t}\right] d\Omega \qquad (8)$$

where * denotes Reimann convolution.

The elastic state $(u_i^*, t_i^*, u_i^{*o}, \dot{u}_i^{*o}, b_i^*)$ is chosen to correspond to a unit impulse applied at a time τ at point $\mathbf{x}'$, in a direction e_i, that is

$$\rho b_i^*(\mathbf{x}, t) = \delta(t-\tau)\delta(\mathbf{x}-\mathbf{x}')e_i \tag{9}$$

where δ is the Dirac delta function. The displacement at point $\mathbf{x}$ at time t due to the unit impulse is given as

$$u_i(\mathbf{x}, t) = U_{ij}(\mathbf{x}, t; \mathbf{x}', \tau)e_j \tag{10}$$

and the traction as

$$t_i(\mathbf{x}, t) = T_{ij}(\mathbf{x}, t; \mathbf{x}', \tau)e_j \tag{11}$$

Substituting expressions (9),(10) and (11) into (8) gives

$$\begin{aligned} u_i(\mathbf{x}', t) = &\int_0^t \int_\Gamma \big(U_{ij}(\mathbf{x}, t; \mathbf{x}', \tau)t_i(\mathbf{x}, t) - T_{ij}(\mathbf{x}, t; \mathbf{x}', \tau)u_i(\mathbf{x}, t)\big)\, d\Gamma d\tau + \\ &\rho \int_0^t \int_\Omega U_{ij}(\mathbf{x}, t; \mathbf{x}', \tau) b_i(\mathbf{x}, t) d\Omega d\tau + \\ &\rho \int_0^t \int_\Omega \left(\frac{\partial u_i(\mathbf{x}, 0)}{\partial t} U_{ij}(\mathbf{x}, t; \mathbf{x}', 0) + u_i(\mathbf{x}, 0) \frac{\partial T_{ij}(\mathbf{x}, t; \mathbf{x}', 0)}{\partial t} \right) d\Omega d\tau \end{aligned} \tag{12}$$

Equation (12) relates the interior displacements to the boundary values of displacement and tractions.

For three-dimensional problems the displacement fundamental solution is given as

$$\begin{aligned} U_{ij}(\mathbf{x}, t; \mathbf{x}', \tau) = &\frac{1}{4\pi\rho} \left\{ \left(\frac{3r_ir_j}{r^3} - \frac{\delta_{ij}}{r}\right) \int_{1/c_1}^{1/c_2} \lambda\delta(t-\tau-\lambda r)d\lambda + \right. \\ &\left. + a_{ij}\left[\frac{1}{c_1^2}\delta(t-\tau-\frac{r}{c_1}) - \frac{1}{c_2^2}\delta(t-\tau-\frac{r}{c_2})\right] + \frac{b_{ij}}{c_2^2}\delta(t-\tau-\frac{r}{c_2}) \right\} \end{aligned} \tag{13}$$

where $a_{ij} = \frac{r_ir_j}{r^3}$, $b_{ij} = \frac{\delta_{ij}}{r}$, $r_i = x_i - x_i'$.

The traction fundamental solution in 3D is given as

$$\begin{aligned} T_{ij}(\mathbf{x}, t; \mathbf{x}', \tau) = &\frac{1}{4\pi} \left\{ -6c_2^2(5a_{ij} - b_{ij}) \int_{1/c_1}^{1/c_2} \lambda\delta(t-\tau-\lambda r)d\lambda \right. \\ &+ (12a_{ij} - 2b_{ij}) \left[\delta(t-\tau-\frac{r}{c_2}) - \left(\frac{c_2}{c_1}\right)^2 \delta(t-\tau-\frac{r}{c_1}) \right] \end{aligned}$$

$$+\frac{2ra_{ij}}{c_2}\left[\dot{\delta}(t-\tau-\frac{r}{c_2})-\left(\frac{c_2}{c_1}\right)^3\dot{\delta}(t-\tau-\frac{r}{c_1})\right]$$

$$-c_{ij}\left(1-\frac{2c_2^2}{c_1^2}\right)\left[\delta(t-\tau-\frac{r}{c_1})+\left(\frac{r}{c_1}\right)\dot{\delta}(t-\tau-\frac{r}{c_1})\right]$$

$$-d_{ij}\left[\delta(t-\tau-\frac{r}{c_2})+\left(\frac{r}{c_2}\right)\dot{\delta}(t-\tau-\frac{r}{c_2})\right]\Bigg\} \qquad (14)$$

where $a_{ij}=\frac{r_ir_jr_mn_m}{r^5}$, $c_{ij}=\frac{r_jn_i}{r^3}$, $d_{ij}=\frac{r_in_j+\delta_{ij}r_mn_m}{r^3}$, $b_{ij}=c_{ij}+d_{ij}$.

For two-dimensional problems, the displacement fundamental solution can be obtained from (13) by integration along the third spatial dimension, that is

$$U_{ij}^{2D}=\int_{-\infty}^{\infty}U_{ij}^{3D}dx_3$$

The final expression is

$$U_{ij}(\mathbf{x},t;\mathbf{x}',\tau)=\frac{1}{2\pi\rho}\left\{\frac{1}{c_1}H(c_1t'-r)\left[\frac{2c_1^2t'^2-r^2}{\sqrt{c_1^2t'^2-r^2}}\left(\frac{r_ir_j}{r^4}\right)-\frac{\delta_{ij}}{r^2}\sqrt{c_1^2t'^2-r^2}\right]\right.$$
$$+\frac{1}{c_2}H(c_2t'-r)\left[-\frac{2c_2t'^2-r^2}{\sqrt{c_2^2t'^2-r^2}}\left(\frac{r_ir_j}{r^4}\right)\right.$$
$$\left.\left.+\frac{\delta_{ij}}{r^2}\sqrt{c_2^2t'^2-r^2}+\frac{\delta_{ij}}{\sqrt{c_2^2t'^2-r^2}}\right]\right\} \qquad (15)$$

where $t'=t-\tau$ is the retarded time. In (13) the Heaviside function ensures causality condition (i.e. the term multiplying it will vanish if the wave has not reached the field point).

The traction fundamental solution can be derived from (15) as

$$T_{ij}(\mathbf{x},t;\mathbf{x}',\tau) =$$

$$= \frac{\mu}{2\pi\rho r}\left\{\frac{1}{c_1}H(\frac{c_1t'}{r}-1)\left[\frac{1}{\left(\left(\frac{c_1t'}{r}\right)^2-1\right)^{\frac{3}{2}}}\left(\frac{A_1}{r}\right)\right.\right.$$

$$\left.+\frac{2\left(\frac{c_1t'}{r}\right)-1}{\sqrt{\left(\frac{c_1t'}{r}\right)^2-1}}\left(\frac{2A_2}{r}\right)\right] \tag{16}$$

$$\left.-\frac{1}{c_2}H\left(\frac{c_2t'}{r}-1\right)\left[\frac{1}{\left(\left(\frac{c_1t'}{r}\right)^2-1\right)^{\frac{3}{2}}}\left(\frac{A_3}{r}\right)+\frac{2\left(\frac{c_1t'}{r}\right)^2-1}{\sqrt{\left(\frac{c_1t'}{r}\right)^2-1}}\left(\frac{2A_2}{r}\right)\right]\right\}$$

where $A_1 = (\lambda/\mu)n_ir_{,j} + 2r_{,i}r_{,j}\frac{\partial r}{\partial n}$, $A_2 = n_ir_{,j} + n_jr_{,i} + \frac{\partial r}{\partial n}(\delta_{ij} - 4r_{,i}r_{,j})$ and $A_3 = \frac{\partial r}{\partial n}(2r_{,i}r_{,j} - \delta_{ij}) - n_jr_{,i}$.

The strain field throughout the body may be obtained by differentiating equation (12), which leads to the equation

$$\frac{\partial u_i(\mathbf{x}',t)}{\partial \mathbf{x}'_k} =$$

$$= \int_0^t\int_\Gamma\left(\frac{\partial U_{ij}(\mathbf{x},t;\mathbf{x}',\tau)}{\partial \mathbf{x}'_k}t_i(\mathbf{x},t) - \frac{\partial T_{ij}(\mathbf{x},t;\mathbf{x}',\tau)}{\partial \mathbf{x}'}u_i(\mathbf{x},t)\right)d\Gamma d\tau +$$

$$\rho\int_0^t\int_\Omega \frac{\partial U_{ij}(\mathbf{x},t;\mathbf{x}',\tau)}{\partial \mathbf{x}'_k}b_i(\mathbf{x},t)d\Omega d\tau +$$

$$\rho\int_0^t\int_\Omega\left(\frac{\partial u_i(\mathbf{x},0)}{\partial t}\frac{\partial U_{ij}(\mathbf{x},t;\mathbf{x}',\tau)}{\partial \mathbf{x}'_k} + u_i(\mathbf{x},0)\frac{\partial\left(\frac{\partial T_{ij}(\mathbf{x},t,\mathbf{x}',0)}{\partial t}\right)}{\partial \mathbf{x}'_k}\right)d\Omega d\tau \tag{17}$$

substituting equation (17) into Hooke's law, that is

$$\sigma_{ij} = \lambda\delta_{ij}u_{k,k} + \mu(u_{i,j} + u_{j,i})$$

yields the stresses at an interior point $\mathbf{x}'$:

$$\sigma_{ij}(\mathbf{x}',\tau) = \int_0^t\int_\Gamma \big(U_{kij}(\mathbf{x},t;\mathbf{x}',\tau)t_k(\mathbf{x},t)$$

$$-T_{kij}(\mathbf{x},t;\mathbf{x}',\tau)u_k(\mathbf{x},t)\big)\,d\Gamma d\tau$$

$$+\rho\int_0^t\int_\Omega U_{kij}(\mathbf{x},t;\mathbf{x}',\tau)b_k(\mathbf{x},t)d\Omega d\tau$$

$$+\rho\int_0^t\int_\Omega\left(\frac{\partial u_k(\mathbf{x},0)}{\partial t}U_{kij}(\mathbf{x},t;\mathbf{x}',\tau)+u_k(\mathbf{x},0)\frac{\partial T_{kij}(\mathbf{x},t,\mathbf{x}',0)}{\partial t}\right)d\Omega d\tau \tag{18}$$

where U_{kij} and T_{kij} are given in Appendix A.

The boundary displacement integral equation can be obtained by taking the limiting process as the internal point goes to the boundary, to give

$$C_{ij}(\mathbf{x}')u_j(\mathbf{x}',t)=\int_0^t\int_\Gamma\big(U_{ij}(\mathbf{x},t;\mathbf{x}',\tau)t_k(\mathbf{x},t)$$

$$-T_{ij}(\mathbf{x},t;\mathbf{x}',\tau)u_k(\mathbf{x},t)\big)\,d\Gamma d\tau$$

$$+\rho\int_0^t\int_\Omega U_{ij}(\mathbf{x},t;\mathbf{x}',\tau)b_k(\mathbf{x},t)d\Omega d\tau$$

$$+\rho\int_0^t\int_\Omega\left(\frac{\partial u_k(\mathbf{x},0)}{\partial t}U_{ij}(\mathbf{x},t;\mathbf{x}',0)+u_k(\mathbf{x},0)\frac{\partial T_{ij}(\mathbf{x},t,\mathbf{x}',0)}{\partial t}\right)d\Omega d\tau \tag{19}$$

where C_{ij} is a constant that depends on the position of the source point $\mathbf{x}'$ and can be obtained from the consideration of rigid body condition.

The boundary stress integral equation can be derived in a similar way by considering the limiting form of equation (18) as $\mathbf{x}'$ tends to a smooth boundary to give

$$\frac{1}{2}\sigma_{ij}(\mathbf{x}',\tau)=\int_0^t\int_\Gamma\big(U_{kij}(\mathbf{x},t;\mathbf{x}',\tau)t_k(\mathbf{x},t)$$

$$-T_{kij}(\mathbf{x},t;\mathbf{x}',\tau)u_k(\mathbf{x},t)\big)\,d\Gamma d\tau$$

$$+\rho\int_0^t\int_\Omega U_{kij}(\mathbf{x},t;\mathbf{x}',\tau)b_k(\mathbf{x},t)d\Omega d\tau$$

$$+\rho\int_0^t\int_\Omega\left(\frac{\partial u_k(\mathbf{x},0)}{\partial t}U_{kij}(\mathbf{x},t;\mathbf{x}',\tau)+u_k(\mathbf{x},0)\frac{\partial T_{kij}(\mathbf{x},t,\mathbf{x}',0)}{\partial t}\right)d\Omega d\tau \tag{20}$$

The traction integral equation is obtained by multiplying equation (20) by the components of the outward normal at $\mathbf{x}'$, to give

$$\frac{1}{2}t_j(\mathbf{x}', t) =$$

$$= n_i(\mathbf{x}') \left\{ \int_0^t \int_\Gamma \left(U_{kij}(\mathbf{x}, t; \mathbf{x}', \tau) t_k(\mathbf{x}, \tau) - T_{kij}(\mathbf{x}, t; \mathbf{x}', \tau) u_k(\mathbf{x}, \tau) \right) d\Gamma d\tau \right.$$

$$\left. \rho \int_0^t \int_\Omega U_{kij}(\mathbf{x}, t; \mathbf{x}', \tau) b_k(\mathbf{x}, t) d\Omega d\tau + \right\} \tag{21}$$

$$\rho \int_0^t \int_\Omega \left(\frac{\partial u_k(\mathbf{x}, 0)}{\partial t} U_{kij}(\mathbf{x}, t; \mathbf{x}', \tau) + u_k(\mathbf{x}, 0) \frac{\partial T_{kij}(\mathbf{x}, t, \mathbf{x}', 0)}{\partial t} \right) d\Omega d\tau \right\}$$

where $n_i(\mathbf{x}')$ are components of the outward normal at the collocation point $\mathbf{x}'$; $U_{kij}(\mathbf{x}', t; \mathbf{x}, \tau)$ and $T_{kij}(\mathbf{x}', t; \mathbf{x}, \tau)$ are other fundamental solutions of elastodynamics. The fundamental solution U_{ij} is weakly singular and the corresponding integral can be evaluated accurately using a suitable transformation of variable or Gaussian quadrature formula (see Aliabadi and Rooke[1]). The integral involving the terms T_{ij} and U_{kij} are singular and must be treated as Cauchy principal value integrals. Similarly the integral involving the term T_{kij} is hypersingular and must be treated as a Hadamard principal value integral. For details of appropriate integration method see the cited papers by Aliabadi and his co-workers.

3.2.1. Crack Modelling Strategy

The boundary element formulation presented here uses discontinuous quadratic elements for the crack modelling, as shown in Figure 1. The general modelling strategy, developed can be summarized as follows:

- The displacement integral equation is applied for $\mathbf{x}'$ on one of the crack boundaries and the remaining boundaries.
- The traction integral equation is applied for collocation on the opposite crack boundary.
- The crack boundaries are discretized with discontinuous quadratic elements.
- Continuous quadratic boundary elements are used for discretization along the remaining boundaries of the problem domain, except at the intersection between a crack and an edge, where semi-discontinuous elements are used.

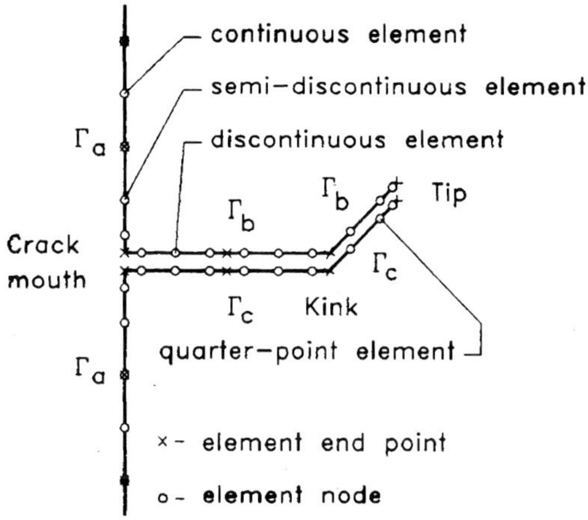

Figure 1. Crack modelling strategy.

This simple strategy is robust and allows the dual boundary element method to effectively model general edge or embedded crack problems; crack tips, crack-edge corners and crack kinks.

3.2.2. Numerical Discretization: Time domain method (TDM)

The numerical solution is obtained after discretizing both space and time variations. The boundary Γ is divided into M boundary elements with P nodes per element. The observation time t is divided into N time steps. The temporal variation of boundary quantities is specified by Q values within the time step. Displacements and tractions are approximated within each element using interpolation functions $N^p(\xi)$, and within each time step using interpolation functions $M^q(\tau)$. The discretized form of the displacement and the traction equation in absence of body forces and assuming the domain is initially at rest, can be written as

$$c_{ij}^b u_j^{bN} = \sum_{m=1}^{M}\sum_{p=1}^{P}\sum_{n=1}^{N}\sum_{q=1}^{Q}\Bigg\{ t_j^{mpnq} \int_{-1}^{1}\Bigg[\int_{\tau^{n-1}}^{\tau^n} U_{ij}^{bN}(\xi,\tau)M^q(\tau)d\tau\Bigg]N^p(\xi)J^m(\xi)d\xi$$

$$-u_j^{mpnq}\int_{-1}^{1}\Bigg[\int_{\tau^{n-1}}^{\tau^n} T_{ij}^{bN}(\xi,\tau)M^q(\tau)d\tau\Bigg]N^p(\xi)J^m(\xi)d\xi\Bigg\}, \quad b=1,2,\ldots,B_1 \tag{22}$$

and

$$\frac{1}{2}t_j^{bN} = n_i^b \sum_{m=1}^{M}\sum_{p=1}^{P}\sum_{n=1}^{N}\sum_{q=1}^{Q}\Big\{t_k^{mpnq}\int_{-1}^{1}\Big[\int_{\tau^{n-1}}^{\tau^n} U_{kij}^{bN}(\xi,\tau)M^q(\tau)d\tau\Big]N^p(\xi)J^m(\xi)d\xi$$

$$-u_k^{mpnq}\int_{-1}^{1}\Big[\int_{\tau^{n-1}}^{\tau^n} T_{kij}^{bN}(\xi,\tau)M^q(\tau)d\tau\Big]N^p(\xi)J^m(\xi)d\xi\Big\}, \quad b = 1, 2, ..., B_2 \tag{23}$$

where B_1 and B_2 are respectively the numbers of collocation points for which the displacement and the traction equations are applied; $B_1+B_2 = B$, the total number of nodes; J^m is the Jacobian of transformation from the global system to the local system and ξ is the local coordinate $(-1 \leq \xi \leq 1)$. A distinct set of boundary integral equations is obtained by applying the displacement Equation (23) for collocation points along the external boundary and along one of the crack faces, and the traction Equation (24) for the opposite surface of the crack.

The expressions defining the spatial shape functions are given in the Table 1 below, where ξ_p' is the local coordinate of the node p.

The displacements are approximated within each time step by using linear interpolating functions and the tractions are piecewise constant. This mixed variation give a better solution when the structure is subjected to impact loads.

The constant temporal shape function is

$$M^1(\tau) = 1 \tag{24}$$

and the linear temporal shape functions are:

$$M^1(\tau) = \frac{\tau - \tau^{n-1}}{\Delta\tau}, \text{ and } M^2(\tau) = \frac{\tau^n - \tau}{\Delta\tau}, \tag{25}$$

where $\tau^{n-1} \leq \tau \leq \tau^n$, $\Delta\tau$ is the time step ($\tau^n = n\Delta\tau$), and the superscripts 1, 2 denote the forward and the backward local time node, respectively.

Table 1.

Element	Coordinates			Spatial shape functions		
	ξ_1'	ξ_2'	ξ_3'	N^1	N^2	N^3
continuous	-1	0	1	$\frac{1}{2}\xi(\xi-1)$	$(1+\xi)(1-\xi)$	$\frac{1}{2}\xi(\xi+1)$
semi-	$-\frac{2}{3}$	0	1	$\frac{9}{10}\xi(\xi-1)$	$\frac{3}{2}(\frac{2}{3}+\xi)(1-\xi)$	$\frac{3}{5}\xi(\xi+\frac{2}{3})$
discontin.	-1	0	$\frac{2}{3}$	$\frac{3}{5}\xi(\xi-\frac{2}{3})$	$\frac{3}{2}(1+\xi)(\frac{2}{3}-\xi)$	$\frac{9}{10}\xi(\xi+1)$
discontin.	$-\frac{2}{3}$	0	$\frac{2}{3}$	$\frac{9}{8}\xi(\xi-\frac{2}{3})$	$\frac{9}{4}(\frac{2}{3}+\xi)(\frac{2}{3}-\xi)$	$\frac{9}{8}\xi(\xi+\frac{2}{3})$

The time integrals in Equations (23) and (24) with simple temporal shape functions in Equations (25) and (26) can be calculated analytically. The following simplifying notation is used:

$$\varphi_\alpha = \frac{r}{c_\alpha \Delta\tau (N-n+1)}\,, \tag{26}$$

$$\chi_\alpha = \frac{r}{c_\alpha \Delta\tau (N-n)}\,, \tag{27}$$

$$\psi_\alpha = \frac{r}{c_\alpha \Delta\tau (N-n-1)}\,; \tag{28}$$

For the piecewise constant interpolation of tractions the convoluted fundamental solution $\tilde{U}_{ij}^{Nn}$ has the form:

$$\tilde{U}_{ij}^{Nn} = \int_{\tau^{n-1}}^{\tau^n} U_{ij}^N d\tau =$$

$$\sum_{\alpha=1}^{2} \frac{1}{4\pi\rho c_\alpha^2} \left[\delta_{ij} \left(\ln \frac{1+\sqrt{1-\varphi_\alpha^2}}{\varphi_\alpha} - \ln \frac{1+\sqrt{1-\chi_\alpha^2}}{\chi_\alpha} \right) \right.$$

$$\left. +(-1)^\alpha (\delta_{ij} - 2r_{,i}r_{,j}) \left(\frac{\sqrt{1-\varphi_\alpha^2}}{\varphi_\alpha^2} - \frac{\sqrt{1-\chi_\alpha^2}}{\chi_\alpha^2} \right) \right], \tag{29}$$

For the piecewise linear interpolation of displacements the convoluted fundamental solution $\tilde{T}_{ij}^{Nn}$ has the form:

$$\tilde{T}_{ij}^{Nn} = \int_{\tau^{n-1}}^{\tau^n} T_{ij}^N M^1 d\tau + \int_{\tau^n}^{\tau^{n+1}} T_{ij}^N M^2 d\tau =$$

$$\sum_{\alpha=1}^{2} \frac{\mu}{2\pi\rho c_\alpha^2} \frac{1}{c_\alpha \Delta\tau} \left\{ \frac{2A_\alpha}{3} \left[\frac{(1-\varphi_\alpha^2)^{\frac{3}{2}}}{\varphi_\alpha^3} - 2\frac{(1-\chi_\alpha^2)^{\frac{3}{2}}}{\chi_\alpha^3} + \frac{(1-\psi_\alpha^2)^{\frac{3}{2}}}{\psi_\alpha^3} \right] \right.$$

$$\left. -B_\alpha \left[\frac{\sqrt{1-\varphi_\alpha^2}}{\varphi_\alpha} - 2\frac{\sqrt{1-\chi_\alpha^2}}{\chi_\alpha} + \frac{\sqrt{1-\psi_\alpha^2}}{\psi_\alpha} \right] \right\} \tag{30}$$

The coefficients of the convoluted fundamental solution are:

$$A_\alpha = -(-1)^\alpha \left[n_j r_{,i} + n_i r_{,j} + \frac{\partial r}{\partial n} (\delta_{ij} - 4r_{,i}r_{,j}) \right],$$

$$B_1 = \frac{\lambda}{\mu} n_j r_{,i}, +2\frac{\partial r}{\partial n} r_{,i} r_{,j},$$

$$B_2 = \frac{\partial r}{\partial n}(\delta_{ij} - 2r_{,i}r_{,j}) + n_i r_{,j},$$

where λ is the Lamé constant. For the linear interpolation functions the contribution of the integration over the time interval before and after the time node is taken into account.

For the piecewise constant interpolation of tractions the convoluted fundamental solution $\tilde{U}_{kij}^{Nn}(\mathbf{x}', \mathbf{x})$ has the form:

$$\tilde{U}_{kij}^{Nn}(\mathbf{x}', \mathbf{x}) = \int_{\tau^{n-1}}^{\tau^n} U_{kij}^N(\mathbf{x}', \mathbf{x}) d\tau =$$

$$\sum_{\alpha=1}^{2} \frac{(-1)^\alpha \mu}{2\pi\rho c_\alpha^2} \frac{1}{r} \Big\{ -C_\alpha \Big[\frac{1}{\sqrt{1-\varphi_\alpha^2}} - \frac{1}{\sqrt{1-\chi_\alpha^2}} \Big]$$

$$+2D \Big[\frac{\sqrt{1-\varphi_\alpha^2}}{\varphi_\alpha^2} - \frac{\sqrt{1-\chi_\alpha^2}}{\chi_\alpha^2} \Big] \Big\}.$$

The coefficients of the convoluted fundamental solution are:

$$C_1 = \frac{\lambda}{\mu} \delta_{ij} r_{,k} + 2r_{,i} r_{,j} r_{,k},$$

$$C_2 = 2r_{,i} r_{,j} r_{,k} - r_{,i}\delta_{jk} - r_{,j}\delta_{ik},$$

$$D = r_{,i}\delta_{jk} + r_{,j}\delta_{ik} + r_{,k}\delta_{ij} - 4r_{,i} r_{,j} r_{,k}.$$

For the piecewise linear interpolation of displacements the convoluted fundamental solution $\tilde{T}_{kij}^{Nn}$ has the form:

$$\tilde{T}_{kij}^{Nn} = \int_{\tau^{n-1}}^{\tau^n} T_{kij}^N M^1 d\tau + \int_{\tau^n}^{\tau^{n+1}} T_{kij}^N M^2 d\tau =$$

$$\sum_{\alpha=1}^{2} \frac{(-1)^\alpha \mu^2}{2\pi\rho c_\alpha^2} \frac{1}{r c_\alpha \Delta\tau} \Big\{ 2(E_\alpha + F_\alpha) \Big[\frac{\sqrt{1-\varphi_\alpha^2}}{\varphi_\alpha} - 2\frac{\sqrt{1-\chi_\alpha^2}}{\chi_\alpha} + \frac{\sqrt{1-\psi_\alpha^2}}{\psi_\alpha} \Big]$$

$$-\frac{4}{3} G \Big[\frac{(1-\varphi_\alpha^2)^{\frac{3}{2}}}{\varphi_\alpha^3} - 2\frac{(1-\chi_\alpha^2)^{\frac{3}{2}}}{\chi_\alpha^3} + \frac{(1-\psi_\alpha^2)^{\frac{3}{2}}}{\psi_\alpha^3} \Big]$$

$$+E_\alpha\Big[\frac{\varphi_\alpha}{\sqrt{1-\varphi_\alpha^2}}-2\frac{\chi_\alpha}{\sqrt{1-\chi_\alpha^2}}+\frac{\psi_\alpha}{\sqrt{1-\psi_\alpha^2}}\Big]\Big\}.$$

The coefficients of the convoluted fundamental solution are:

$$E_1=(\frac{\lambda}{\mu}\delta_{ij}+2r_{,i}r_{,j})(\frac{\lambda}{\mu}n_k+2\frac{\partial r}{\partial n}r_{,k}),$$

$$E_2=\frac{\partial r}{\partial n}(4r_{,i}r_{,j}r_{,k}-r_{,i}\delta_{jk}-r_{,j}\delta_{ik})-n_ir_{,j}r_{,k}-n_jr_{,i}r_{,k}$$

$$F_1=-n_k\Big[\frac{\lambda}{\mu}\delta_{ij}(2+\frac{\lambda}{\mu})+2r_{,i}r_{,j}\Big]-2n_ir_{,j}r_{,k}-2n_jr_{,i}r_{,k}$$
$$-2\frac{\partial r}{\partial n}(r_{,i}\delta_{jk}+r_{,j}\delta_{ik}+r_{,k}\delta_{ij}-6r_{,i}r_{,j}r_{,k})$$

$$F_2=n_i(\delta_{jk}-2r_{,j}r_{,k})+n_j(\delta_{ik}-2r_{,i}r_{,k})-2n_kr_{,i}r_{,j}$$
$$-2\frac{\partial r}{\partial n}(r_{,i}\delta_{jk}+r_{,j}\delta_{ik}+r_{,k}\delta_{ij}-6r_{,i}r_{,j}r_{,k})$$

$$G=n_i(-\delta_{jk}+4r_{,j}r_{,k})+n_j(-\delta_{ik}+4r_{,i}r_{,k})+n_k(-\delta_{ij}+4r_{,i}r_{,j})$$
$$+4\frac{\partial r}{\partial n}(r_{,i}\delta_{jk}+r_{,j}\delta_{ik}+r_{,k}\delta_{ij}-6r_{,i}r_{,j}r_{,k}).$$

In evaluating the above variables φ_α, χ_α and ψ_α the causality condition must be satisfied. That is, if r is greater than the distance travelled by the wave at the given time then the value of the variable φ_α, χ_α and ψ_α is greater than 1. In this case the terms in $\tilde{U}_{ij}^{Nn}$, $\tilde{T}_{ij}^{Nn}$, $\tilde{U}_{kij}^{Nn}$ and $\tilde{T}_{kij}^{Nn}$ which contain that variable must be put equal to zero.

The set of discretized boundary equations can be written in matrix form at the time step N as

$$\tilde{\mathbf{H}}^{NN}\mathbf{u}^N=\tilde{\mathbf{G}}^{NN}\mathbf{t}^N+\sum_{n=1}^{N-1}(\tilde{\mathbf{G}}^{Nn}\mathbf{t}^n-\tilde{\mathbf{H}}^{Nn}\mathbf{u}^n);\tag{31}$$

where $\mathbf{u}^n$, $\mathbf{t}^n$ contain nodal values of displacements and tractions at the time step n; $\tilde{\mathbf{H}}^{Nn}$ and $\tilde{\mathbf{G}}^{Nn}$ depend on the integrals of the fundamental solutions and interpolating functions. The superscripts Nn emphasize that the matrix depends on the difference between the time step N and n. The columns of matrices $\tilde{\mathbf{H}}^{NN}$, $\tilde{\mathbf{G}}^{NN}$ are reordered according to the boundary conditions, giving new matrices $\tilde{\mathbf{A}}^{NN}$ and $\tilde{\mathbf{B}}^{NN}$. The matrix $\tilde{\mathbf{A}}^{NN}$ is multiplied by the vector $\mathbf{x}^N$ of unknown displacements and tractions and the matrix $\tilde{\mathbf{B}}^{NN}$ by the vector $\mathbf{y}^N$ of known boundary conditions, as follows

$$\tilde{\mathbf{A}}^{NN}\mathbf{x}^N=\tilde{\mathbf{B}}^{NN}\mathbf{y}^N+\sum_{n=1}^{N-1}(\tilde{\mathbf{G}}^{Nn}\mathbf{t}^n-\tilde{\mathbf{H}}^{Nn}\mathbf{u}^n).\tag{32}$$

In each time step only the matrices, which correspond to the maximum difference $N - n$ are computed. The rest of the matrices are known from the previous steps. The matrices $\tilde{\mathbf{A}}^{NN}$ and $\tilde{\mathbf{B}}^{NN}$ are calculated in the first step only since they are the same at each time step; $\tilde{\mathbf{A}}^{NN} = \tilde{\mathbf{A}}$ and $\tilde{\mathbf{B}}^{NN} = \tilde{\mathbf{B}}$. The matrix Equation (32) can be written in a simpler form as

$$\tilde{\mathbf{A}}\mathbf{x}^N = \mathbf{f}^N, \tag{33}$$

where

$$\mathbf{f}^N = \tilde{\mathbf{B}}\mathbf{y}^N + \sum_{n=1}^{N-1} (\tilde{\mathbf{G}}^{Nn}\mathbf{t}^n - \tilde{\mathbf{H}}^{Nn}\mathbf{u}^n), \tag{34}$$

is a known vector. The matrix equation is solved step-by-step giving the unknown displacements and tractions at each time step. During the initial steps the fundamental solutions are non-zero only in the neighbourhood of the collocation point; they are therefore integrated only over that part of the boundary. The solution process becomes progressively slower at later times because the vector $\mathbf{f}^N$ depends on all the matrices from the previous steps.

3.2.3. Dynamic Stress Intensity Factors

One of the most important parameters in dynamic fracture mechanics is the dynamic stress intensity factor (DSIF), since it characterizes the stress field in the vicinity of the crack, and controls crack growth. Here, two different methods are presented based on the quarter point elements and the J−integral method.

3.2.3.1. Quarter-point elements

The DSIF can be calculated directly from the displacements of the standard discontinuous elements on the crack faces. However more accurate displacements and DSIF are obtained if the position of nodes of the straight discontinuous element adjacent to the crack tip is changed, as shown in Figure 2.

The new distances of the pairs of nodes: B-C, D-E and F-G from the crack tip A are: $\frac{1}{36}l$, $\frac{9}{36}l$ and $\frac{25}{36}l$, respectively, where l is the length of the element. The local coordinates of the nodes are the same as other discontinuous elements i.e. ($\xi' = -\frac{2}{3}, 0, \frac{2}{3}$). For the distorted elements the local coordinate ξ is the square-root function of the distance R from the crack tip:

$$\xi = \pm 1 \mp 2\sqrt{\frac{R}{l}}, \quad \text{for cracks tip at } \xi = \pm 1, \tag{35}$$

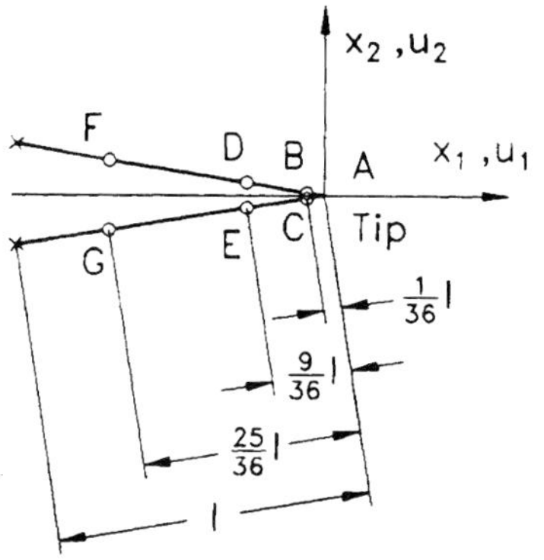

Figure 2. Quarter point elements.

After the above modification, the displacements of the crack tip elements are approximated by a function

$$u = a_1 + a_2\sqrt{R} + a_3 R, \tag{36}$$

where $a_1,\ a_2,\ a_3$ are coefficients, which depend on the displacements of nodes of the element. This type of function better represents the displacement field near the crack tip.

The DSIF are calculated by minimizing the sum of squared differences between the analytical and numerical values of crack opening displacements for two pairs of nodes B-C and D-E. The analytical expression Δv for the crack opening is

$$\Delta v = K\frac{\kappa+1}{2\mu}\sqrt{\frac{2}{\pi}}\sqrt{R}, \tag{37}$$

where K is the DSIF; $\kappa = 3 - 4\nu$ for plane strain and $\kappa = (3-\nu)/(1+\nu)$ for plane stress; μ is the shear modulus; and ν is Poisson's ratio. The sum of squared differences is

$$\varepsilon = (\Delta v^{BC} - \Delta u^{BC})^2 + (\Delta v^{DE} - \Delta u^{DE})^2, \tag{38}$$

where Δu denotes the numerical value of the crack opening and the superscripts the pairs of nodes.

The parameter ε can be expressed in terms of the DSIF from Equation (37). The sum of squared differences between the numerical and analytical opening displacements, defined by ε, is a minimum, if the DSIF, for mode I and mode II have the following values:

$$K_I = \frac{6\mu}{5(\kappa+1)}\sqrt{\frac{\pi}{2l}}(\Delta u_2^{BC} + 3\Delta u_2^{DE}), \tag{39}$$

$$K_{II} = \frac{6\mu}{5(\kappa+1)}\sqrt{\frac{\pi}{2l}}(\Delta u_1^{BC} + 3\Delta u_1^{DE}), \tag{40}$$

where Δu_1 and Δu_2 are the crack opening displacements along and perpendicular to the crack, respectively.

3.2.3.2. *Path independent integrals*

For structures subjected to dynamic loads, the dynamic stress intensity factors can be calculated from the $\hat{J}$-integral which represents an energy release rate for a stationary crack and has the physical meaning of a crack driving force. The $\hat{J}$ -integral is defined as:

$$\hat{J} = \int_{S+S_c} (Wn_1 - t_i u_{i,1})dS + \int_A \rho\ddot{u}_i u_{i,1} dA; \tag{41}$$

where S is an arbitrary curve surrounding the crack tip; S_c are the crack surfaces; A is the area enclosed by S and S_c; W the strain energy density; and n_1 a component of the unit outward normal to the boundary of A. The variables in Equation (41) are expressed in the local crack reference system, shown in Figure 3. In order to obtain stress intensity factors for a mixed mode case the $\hat{J}$ -integral is decoupled as follows:

$$\hat{J} = \hat{J}^I + \hat{J}^{II}, \tag{42}$$

where

$$\hat{J}^\beta = \int_{S+S_c} (W^\beta n_1 - t_i^\beta u_{i,1}^\beta)dS + \int_A \rho\ddot{u}_i^\beta u_{i,1}^\beta dA, \quad \beta = I,\ II; \tag{43}$$

the superscripts β denote the deformation modes I or II.

Displacements are decomposed into components for the symmetric mode I and antisymmetric mode II as:

$$\begin{bmatrix} u_1^I \\ u_2^I \end{bmatrix} = \frac{1}{2}\begin{bmatrix} u_1 + u_1' \\ u_2 - u_2' \end{bmatrix}, \quad \begin{bmatrix} u_1^{II} \\ u_2^{II} \end{bmatrix} = \frac{1}{2}\begin{bmatrix} u_1 - u_1' \\ u_2 + u_2' \end{bmatrix}, \tag{44}$$

and stresses as

$$\begin{bmatrix} \sigma_{11}^I \\ \sigma_{22}^I \\ \sigma_{12}^I \end{bmatrix} = \frac{1}{2}\begin{bmatrix} \sigma_{11} + \sigma_{11}' \\ \sigma_{22} + \sigma_{22}' \\ \sigma_{12} - \sigma_{12}' \end{bmatrix}, \quad \begin{bmatrix} \sigma_{11}^{II} \\ \sigma_{22}^{II} \\ \sigma_{12}^{II} \end{bmatrix} = \frac{1}{2}\begin{bmatrix} \sigma_{11} - \sigma_{11}' \\ \sigma_{22} - \sigma_{22}' \\ \sigma_{12} + \sigma_{12}' \end{bmatrix}, \tag{45}$$

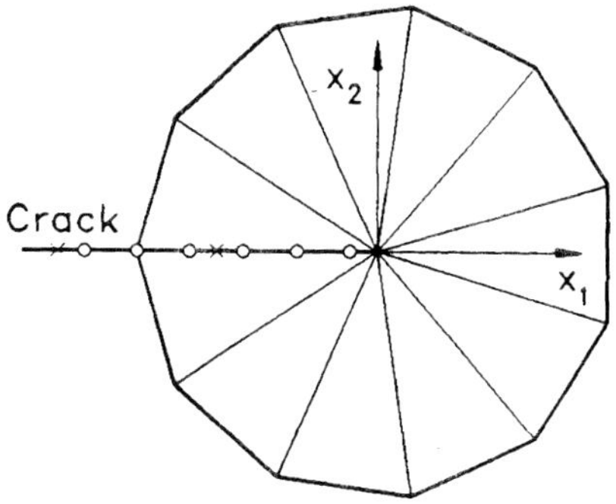

Figure 3. Integration path for the J-integral.

where u_i, σ_{ij}, u'_i and σ'_{ij} are displacements and stresses at the points $P(x_1, x_2)$ and $P'(x_1, -x_2)$ symmetric with respect to the crack.

Strains and derivatives of displacements with respect to coordinates are decomposed in the same way as stresses, while tractions and accelerations are decomposed as displacements. Knowing the $\hat{J}^\beta$ -integral we can calculate stress intensity factors, as follows:

$$K_I = \sqrt{\frac{8\mu}{\kappa+1}\hat{J}^I} \quad \text{and} \quad K_{II} = \sqrt{\frac{8\mu}{\kappa+1}\hat{J}^{II}}. \tag{46}$$

The sign of K_I and K_{II} is determined by the relative displacements of the crack surfaces. A regular polygonal path with the centre at the crack tip is assumed for the contour. The first and the last point of the path are nodes on the crack faces. The domain enclosed by the path is divided into triangles, as shown in Figure 3. The corners of the triangles are located symmetrically relative to the crack. The field variables required for the $\hat{J}$-integral, namely derivatives of displacements, strains, stresses and tractions are calculated using appropriate boundary integral equations. The accelerations can be calculated from the displacements at different times in a finite difference formula; or more accurately, from derivatives of stresses and the equations of motion. The displacements in the first method and the derivatives of stresses in the second method are computed using boundary integral equations. The boundary term in Equation (71) is computed using the trapezoidal rule and the domain term by Gaussian quadrature.

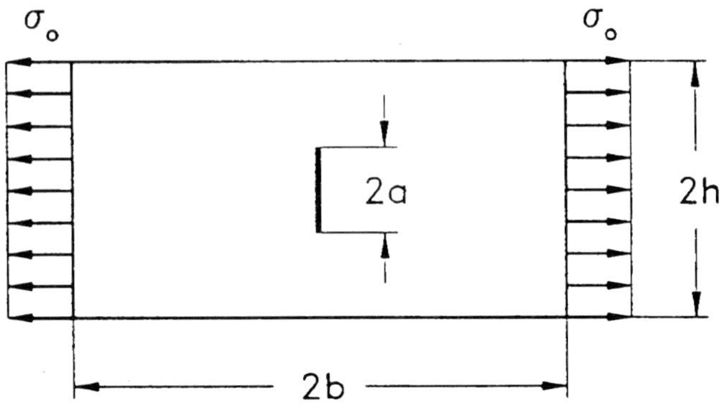

Figure 4. Central crack in a rectangluar sheet.

3.2.4. Numerical Examples

The time domain formulation presented above was applied to several problems by Fedelinski, Aliabadi and Rooke.[14] Here, the problems of a central crack in a rectangular sheet in[14] is reported. The plate is of length 2b = 40 mm and height 2h = 20 mm with a crack of length 2a = 4.8 mm as shown in Figure 4. The material properties are: the shear modulus $\mu = 76.92 \times 10^9$ Pa; Poisson ratio $\upsilon = 0.3$; the density $\rho = 5000$ Km^{-3} and the plate is under a state of plane strain. The stress σ_o with Heaviside-function time dependence is applied at the ends of the plate. Convergence solutions were obtained with 32 elements and the time step $\Delta\tau = 0.3\mu s$.

The normalized DSIF K_I/K_o ($K_o = \sigma_o\sqrt{\pi a}$) is plotted in Figure 5 and compared with that of Chen[15], who used finite difference method, and Aberson *et al.*[16], who used a finite element method.

3.3. LAPLACE TRANSFORM METHOD (LTM)

Application of the Laplace transform to the equation of motion (2) will give

$$(c_1^2 - c_2^2)\bar{u}_{i,jj}(\mathbf{x}, s) + c_2^2\bar{u}_{j,ii}(\mathbf{x}, s) + \bar{b}_j(\mathbf{x}, s) - s^2\bar{u}_j(\mathbf{x}, s) = 0 \quad (47)$$

where $\bar{u}(\mathbf{x}, s)$ is a component of the transformed displacement of a point $\mathbf{x}$ and s is a Laplace parameter. The Laplace transform of function $f(\mathbf{x}, t)$ is defined as

$$\mathfrak{L}[f(\mathbf{x}, t)] = \bar{f}(\mathbf{x}, s) - \int_0^\infty f(\mathbf{x}, t)e^{-st}dt \quad (48)$$

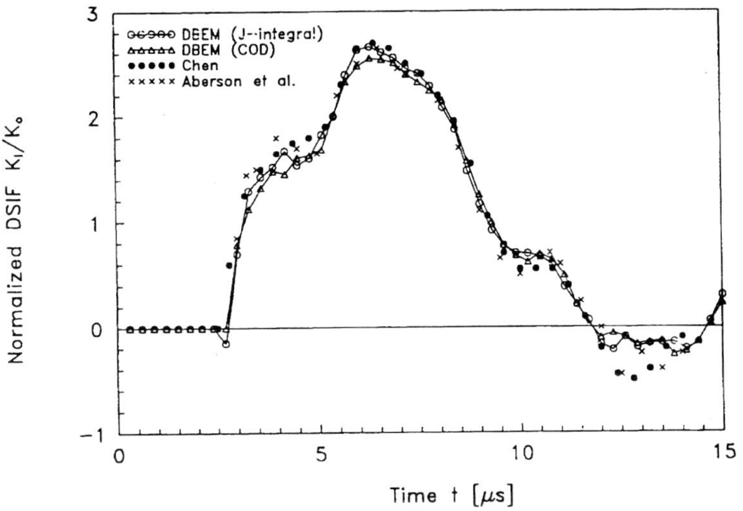

Figure 5. Stress intensity factors for a central crack in a rectangular plate.

The transformed displacements and tractions satisfy the following boundary conditions:

$$\bar{u}_i(\mathbf{x}, t) = \hat{u}_i(\mathbf{x}, t) \quad \text{on} \quad \Gamma_u$$
$$\bar{t}_i(\mathbf{x}, t) = \hat{t}_i(\mathbf{x}, t) \quad \text{on} \quad \Gamma_t \tag{49}$$

The Laplace transform of the stress tensors are given by

$$\bar{\sigma}_{ij}(\mathbf{x}, s) = \rho(c_1^2 - 2c_2^2)\delta_{ij}\bar{u}_{m,m}(\mathbf{x}, s) + \rho c_2^2(\bar{u}_{i,j}(\mathbf{x}, t) + \bar{u}_{j,i}(\mathbf{x}, s)) \tag{50}$$

Assuming zero initial displacements and velocities, the reciprocal identity for two solutions $\bar{u}_i(\mathbf{x}, s)$ and $\bar{u}_i^*(\mathbf{x}, s)$ can be written as

$$\int_\Gamma \bar{t}_i(\mathbf{x}, s)\bar{u}_i^*(\mathbf{x}, s)d\Gamma + \int_\Omega \bar{b}_i(\mathbf{x}, s)\bar{u}_i^*(\mathbf{x}, s)d\Omega =$$
$$= \int_\Gamma \bar{t}_i^*(\mathbf{x}, s)\bar{u}_i(\mathbf{x}, s)d\Gamma + \int_\Omega \bar{b}_i^*(\mathbf{x}, s)\bar{u}_i(\mathbf{x}, s)d\Omega \tag{51}$$

The elastic state $(\bar{u}_i^*, \bar{t}_i^*, \bar{b}_i^*)$ is chosen to correspond to a unit harmonic body force. The Somigliana's identity for displacement at an internal point $\mathbf{x}'$ can be written as

$$\bar{u}_i(\mathbf{x}', s) = \int_\Gamma \left(\bar{U}_{ij}(\mathbf{x}, , \mathbf{x}', s)\bar{t}_i(\mathbf{x}, s) - \bar{T}_{ij}(\mathbf{x}, \mathbf{x}', s)\bar{u}_i(\mathbf{x}, s)\right) d\Gamma +$$

$$\int_{\Omega} \bar{U}_{ij}(\mathbf{x}, \mathbf{x}', s)\bar{b}_i(\mathbf{x}, s)d\Omega \tag{52}$$

where the fundamental solutions $\bar{U}_{ij}$ and $\bar{T}_{ij}$ for three-dimensional problems are given as

$$\bar{U}_{ij}(\mathbf{x}, \mathbf{x}', s) = \frac{1}{4\pi\mu}(\Lambda\delta_{ij} - \Theta r_{,i}r_{,j}) \tag{53}$$

and

$$\bar{T}_{ij}(\mathbf{x}, \mathbf{x}', s) = \frac{1}{4\pi}\left[\left(\frac{\partial\Lambda}{\partial r} - \frac{\Theta}{r}\right)\left(\delta_{ij}\frac{\partial r}{\partial n} + r_{,i}n_i\right) - \right.$$
$$\left. -\frac{2\Theta}{r}\left(n_i r_{,j} - 2r_{,i}r_{,j}\frac{\partial r}{\partial n}\right) - 2\frac{\partial\Theta}{\partial r}r_{,i}r_{,j}\frac{\partial r}{\partial n} + \Xi r_{,j}n_i\right] \tag{54}$$

where

$$\Lambda = \left(\frac{c_2^2}{s^2r^2} + \frac{c_2}{sr} + 1\right)\left(\frac{e^{-\frac{sr}{c_2}}}{r}\right) - \frac{c_2^2}{c_1^2}\left(\frac{3c_1^2}{s^2r^2} + \frac{c_1}{sr}\right)\frac{e^{-\frac{sr}{c_1}}}{r}$$

$$\Theta = \left(\frac{3c_2^2}{s^2r^2} + \frac{3c_2}{sr} + 1\right)\frac{e^{-\frac{sr}{c_2}}}{r} - \frac{c_2^2}{c_1^2}\left(\frac{3c_1^2}{s^2r^2} + \frac{3c_1}{sr} + 1\right)\frac{e^{-\frac{sr}{c_1}}}{r}$$

$$\Xi = \left[\left(\frac{c_1}{c_2}\right)^2 - 2\right]\left(\frac{\partial\Lambda}{\partial r} - \frac{\partial\Theta}{\partial r} - \frac{\Theta}{r}\right)$$

For two-dimensional problems, they are given as

$$\bar{U}_{ij}(\mathbf{x}, \mathbf{x}', s) = \frac{1}{2\pi\mu}(\psi\delta_{ij} - \chi r_{,i}r_{,j}) \tag{55}$$

and

$$\bar{T}_{ij}(\mathbf{x}, \mathbf{x}', s) = \frac{1}{2\pi}\left[\left(\frac{\partial\psi}{\partial r} - \frac{\chi}{r}\right)\left(\delta_{ij}\frac{\partial r}{\partial n} + r_{,i}n_i\right)\right.$$
$$\left. -\frac{2\chi}{r}\left(r_{,i}n_j - 2r_{,i}r_{,j}\frac{\partial r}{\partial n}\right) - 2\frac{\partial\psi}{\partial r}r_{,i}r_{,j}\frac{\partial r}{\partial n} + \Xi r_{,i}n_j\right] \tag{56}$$

where

$$\psi = K_o(z_2) + \left(\frac{1}{z_2}\right)\left[K_1(z_2) - \left(\frac{c_2}{c_1}\right)K_1(z_1)\right]$$

$$\chi = K_2(z_2) - \left(\frac{c_2}{c_1}\right)^2 K_2(z_1)$$

$$\Xi = \left[\left(\frac{c_1}{c_2}\right)^2 - 2\right]\left(\frac{\partial\psi}{\partial r} - \frac{\partial\chi}{\partial r} - \frac{\chi}{r}\right)$$

The stresses at an interior point $\mathbf{x}'$ can be obtained by differentiating the displacement integral equation and the application of the Hooke's law to give

$$\bar{\sigma}_{ij}(\mathbf{x}',s) = \int_\Gamma \big(\bar{U}_{kij}(\mathbf{x},\mathbf{x}',s)\bar{t}_i(\mathbf{x},s) - \bar{T}_{kij}(\mathbf{x},\mathbf{x}',s)\bar{u}_i(\mathbf{x},s)\big)\,d\Gamma +$$

$$+ \int_\Omega \bar{U}_{kij}(\mathbf{x},\mathbf{x}',s)\bar{b}_i(\mathbf{x},s)d\Omega \tag{57}$$

where $\bar{U}_{kij}$ and $\bar{T}_{kij}$ fundamental solutions of for two-dimensional problems are:

$$\bar{U}_{kij}(\mathbf{x}',\mathbf{x},s) = \frac{1}{2\pi}\Big\{\Big[2\frac{\chi}{r} - \frac{\lambda}{\mu}(\psi_{,r} - \chi_{,r} - \frac{\chi}{r})\Big]\delta_{ij}r_{,k}$$

$$-(\psi_{,r} - \frac{\chi}{r})(\delta_{ik}r_{,j} + \delta_{jk}r_{,i}) + 2(\chi_{,r} - 2\frac{\chi}{r})r_{,i}r_{,j}r_{,k}\Big\}$$

$$\bar{T}_{kij}(\mathbf{x}',\mathbf{x},s) = \frac{\mu}{2\pi}\Big\{\frac{\partial r}{\partial n}\Big[4(\chi_{,rr} - 5\frac{\chi_{,r}}{r} + 8\frac{\chi}{r^2})r_{,i}r_{,j}r_{,k}$$

$$-(\psi_{,rr} - \frac{\psi_{,r}}{r} - 3\frac{\chi_{,r}}{r} + 6\frac{\chi}{r^2})(\delta_{ik}r_{,j} + \delta_{jk}r_{,i})$$

$$+2[2\frac{\chi_{,r}}{r} - 4\frac{\chi}{r^2} + \frac{\lambda}{\mu}(\chi_{,rr} - 2\frac{\chi}{r^2} - \psi_{,rr} + \frac{\psi_{,r}}{r})]\delta_{ij}r_{,k}\Big]$$

$$+2[2\frac{\chi_{,r}}{r} - 4\frac{\chi}{r^2} + \frac{\lambda}{\mu}(\chi_{,rr} - 2\frac{\chi}{r^2} - \psi_{,rr} + \frac{\psi_{,r}}{r})]r_{,i}r_{,j}n_k$$

$$-(\psi_{,rr} - \frac{\psi_{,r}}{r} - 3\frac{\chi_{,r}}{r} + 6\frac{\chi}{r^2})(r_{,j}n_i + r_{,i}n_j)r_{,k}$$

$$+\Big[4\frac{\chi}{r^2} + 4\frac{\lambda}{\mu}(\frac{\chi_{,r}}{r} + \frac{\chi}{r^2} - \frac{\psi_{,r}}{r}) + (\frac{\lambda}{\mu})^2(\chi_{,rr} + 2\frac{\chi_{,r}}{r} - \psi_{,rr} - \frac{\psi_{,r}}{r})\Big]\delta_{ij}n_k$$

$$-2(\frac{\psi_{,r}}{r} - \frac{\chi}{r^2})(\delta_{kj}n_i + \delta_{ki}n_j)\Big\}.$$

The function ψ and its derivatives are:

$$\psi = \Big[-\frac{c_2}{c_1}K_1(z_1) + K_1(z_2)\Big]\frac{1}{z_2} + K_0(z_2)$$

$$\frac{d\psi}{dr} = \Big\{\frac{c_2}{c_1}\Big[K_0(z_1)z_1 + 2K_1(z_1)\Big]$$

$$-\Big[K_0(z_2)z_2 + 2K_1(z_2)\Big] - K_1(z_2)z_2^2\Big\}\frac{1}{rz_2}$$

$$\frac{d^2\psi}{dr^2} = \Big\{ - \frac{c_2}{c_1}\Big[3K_0(z_1)z_1 + K_1(z_1)(z_1^2+6)\Big]$$

$$+\Big[3K_0(z_2)z_2 + K_1(z_2)(z_2^2+6)\Big] + \Big[K_0(z_2)z_2 + K_1(z_2)\Big]z_2^2\Big\}\frac{1}{z_2 r^2}.$$

The function χ and its derivatives are:

$$\chi = -(\frac{c_2}{c_1})^2\Big[K_0(z_1) + \frac{2}{z_1}K_1(z_1)\Big] + K_0(z_2) + \frac{2}{z_2}K_1(z_2)$$

$$\frac{d\chi}{dr} = \Big\{(\frac{c_2}{c_1})^2\Big[K_1(z_1)z_1 + 2[K_0(z_1) + \frac{2}{z_1}K_1(z_1)]\Big]$$

$$-\Big[K_1(z_2)z_2 + 2[K_0(z_2) + \frac{2}{z_2}K_1(z_2)]\Big]\Big\}\frac{1}{r}$$

$$\frac{d^2\chi}{dr^2} = \Big\{ - (\frac{c_2}{c_1})^2\Big[K_0(z_1)z_1^2 + 3K_1(z_1)z_1 + 6[K_0(z_1) + \frac{2}{z_1}K_1(z_1)]\Big]$$

$$+\Big[K_0(z_2)z_2^2 + 3K_1(z_2)z_2 + 6[K_0(z_2) + \frac{2}{z_2}K_1(z_2)]\Big]\Big\}\frac{1}{r^2}$$

where K_0 and K_1 are the modified Bessel functions of the second kind of orders zero and one respectively; $z_1 = sr/c_1$ and $z_2 = sr/c_2$. The three-dimensional fundamental solutions for the stress integral equations can be obtained in a similar way by differentiating the fundamental solutions in the displacement fundamental solutions with respect to the source point and the application of Hookes' law.

The displacement boundary integral equation can be obtained by considering the limiting form of the Equation (52) as $\mathbf{x}'$ tends to the boundary, to give

$$c_{ij}(\mathbf{x}')\bar{u}_j(\mathbf{x}', s) = \int_\Gamma \bar{U}_{ij}(\mathbf{x}', \mathbf{x}, s)\bar{t}_j(\mathbf{x}, s)d\Gamma(\mathbf{x})$$

$$\int_\Gamma \bar{T}_{ij}(\mathbf{x}', \mathbf{x}, s)\bar{u}_j(\mathbf{x}, s)d\Gamma(\mathbf{x}), \quad i, j = 1, 2, \tag{58}$$

Similarly the traction integral equation can be obtained from (57), to give

$$\frac{1}{2}\bar{t}_j(\mathbf{x}', s) = n_i(\mathbf{x}')\int_\Gamma \bar{U}_{kij}(\mathbf{x}', \mathbf{x}, s)\bar{t}_k(\mathbf{x}, s)d\Gamma(\mathbf{x})$$

$$\int_{\Gamma} \bar{T}_{kij}(\mathbf{x}', \mathbf{x}, s)\bar{u}_k(\mathbf{x}, s)d\Gamma(\mathbf{x})\Big], \quad i, j, k = 1, 2, \tag{59}$$

It is also possible to arrive at the solution of $\bar{u}_i$ via the Fourier transform. In this case the Laplace parameter s should be replaced with $-i\omega$, where ω is the angular frequency. The above formulation results in the reduction of the transient problem to a steady state one. To obtain the transient solution it is necessary to invert back to the real space by some efficient numerical technique.

3.3.1. Numerical Discretization

The boundary is discretized in the same way as for the time domain method. As a result of the approximation of spatial variations of transformed boundary displacements and tractions the displacement and traction equation are

$$c_{ij}^{b}\bar{u}_j^{b}(s) = \sum_{m=1}^{M}\sum_{p=1}^{P}\Big[\bar{t}_j^{\,mp}(s)\int_{-1}^{1}\bar{U}_{ij}^{b}(\xi, s)N^{p}(\xi)J^{m}(\xi)d\xi$$

$$-\bar{u}_j^{\,mp}(s)\int_{-1}^{1}\bar{T}_{ij}^{b}(\xi, s)N^{p}(\xi)J^{m}(\xi)d\xi\Big], \quad b = 1, 2, \ldots, B_1 \tag{60}$$

and

$$\frac{1}{2}\bar{t}_j^{b}(s) = n_i^{b}\sum_{m=1}^{M}\sum_{p=1}^{P}\Big[\bar{t}_k^{\,mp}(s)\int_{-1}^{1}\bar{U}_{kij}^{b}(\xi, s)N^{p}(\xi)J^{m}(\xi)d\xi$$

$$-\bar{u}_k^{\,mp}(s)\int_{-1}^{1}\bar{T}_{kij}^{b}(\xi, s)N^{p}(\xi)J^{m}(\xi)d\xi\Big], \quad b = 1, 2, \ldots, B_2, \tag{61}$$

where the same notation is used as for the time domain method.

The boundary Equations (60) and (61) are applied for the boundary nodes. The set of discretized boundary integral equations can be written in matrix form as

$$\bar{\mathbf{H}}\bar{\mathbf{u}} = \bar{\mathbf{G}}\bar{\mathbf{t}}, \tag{62}$$

where $\bar{\mathbf{u}}$ and $\bar{\mathbf{t}}$ contain nodal values of the transformed displacements and tractions respectively, and $\bar{\mathbf{H}}$ and $\bar{\mathbf{G}}$ depend on integrals of the transformed fundamental solutions and the interpolating functions. The matrices $\bar{\mathbf{H}}$ and $\bar{\mathbf{G}}$ are reordered according to the boundary conditions, in the same way as in elastostatics, to give new matrices $\bar{\mathbf{A}}$ and $\bar{\mathbf{B}}$. The matrix $\bar{\mathbf{A}}$ is multiplied by the vector $\bar{\mathbf{x}}$ of unknown transformed displacements and tractions and $\bar{\mathbf{B}}$ by the vector $\bar{\mathbf{y}}$ of the known transformed boundary conditions, as follows

$$\bar{\mathbf{A}}\bar{\mathbf{x}} = \bar{\mathbf{B}}\bar{\mathbf{y}}, \tag{63}$$

or

$$\bar{\mathbf{A}}\bar{\mathbf{x}} = \bar{\mathbf{f}}, \tag{64}$$

where $\bar{\mathbf{f}} = \bar{\mathbf{B}}\bar{\mathbf{y}}$ is a known vector.

The matrix Equation (64) is solved giving the unknown transformed displacements and tractions for a particular Laplace parameter. For a simple temporal variation of the prescribed boundary conditions their Laplace transforms can be calculated analytically. In order to obtain the unknown displacements and tractions as functions of time, the unknown transformed variables must be computed for a series of Laplace parameters. The final time-dependent solution can be obtained from a numerical inversion, for instance the Durbin method.[17]

3.3.2. Numerical Examples

The Laplace transform dual boundary element method described above was applied to several problems by Fedelinski, Aliabadi and Rooke.[18] They used the quarter-point elements and the crack opening displacements similar to that described for the time domain formulation to obtain the DSIFs.

Here, the problems of an inclined edge crack in rectangular sheet as reported in[??] is presented. The crack is slanted at an angle $\alpha = 45^o$ as shown in Figure 6. The sheet is of width $b = 44$ mm, and height $h = 32$ mm, and the distance of the crack root from one end of the plate is $c = 6$ mm. The crack length of $a = 22.63$ mm is chosen and the sheet is instantaneously loaded by a uniform tensile stress σ_o at time $\tau = 0$. This problem was solved using two types of discretization: 29 boundary elements and 25 Laplace parameters, or 58 elements and 50 Laplace parameters. The solutions were compared (see Figure 7) with the results obtained by Nishioka and Atluri[19], who used the finite element method.

3.4. DUAL RECIPROCITY METHOD (DRM)

The differential equation of motion can be expressed as follows:

$$\sigma_{jk,k} = \rho \ddot{u}_j \tag{65}$$

The differential Equation (65) is solved subjected to the boundary conditions given in (3),(4) and (5),(6).

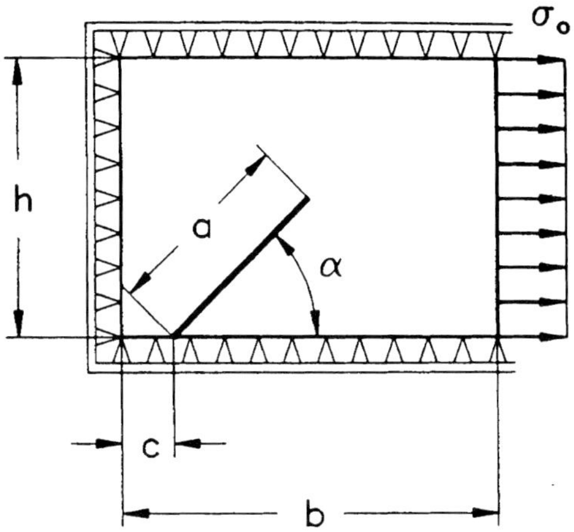

Figure 6. Inclined edge crack in a rectangular sheet.

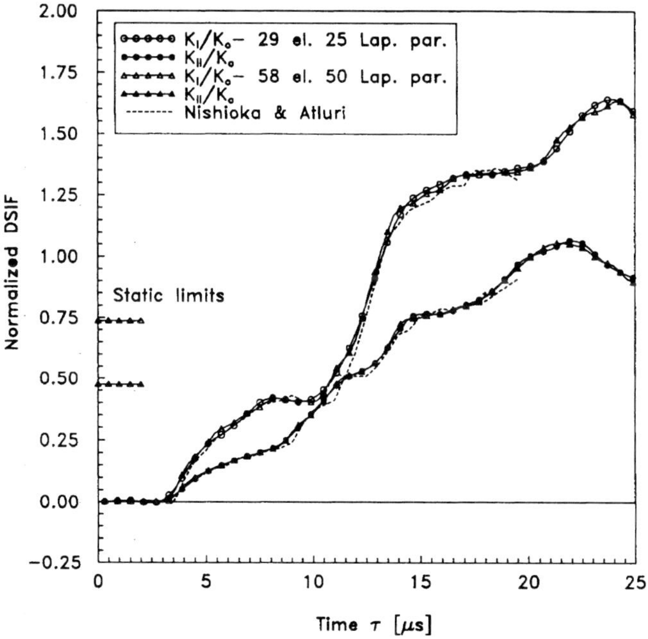

Figure 7. Stress intensity factors for the inclined edge crack problem.

The boundary element formulation for elastodynamic problems can be derived from the Maxwell-Betti reciprocal theorem for two independent states (u, t) and (u^*, t^*) and expressed as

$$\int_{\Omega} u_j^*(\mathbf{x}, \mathbf{x}')\sigma_{jk,k}(\mathbf{x}, \mathbf{x}')d\Omega - \int_{\Omega} u_j(\mathbf{x}, \mathbf{x}')\sigma_{jk,k}^*(\mathbf{x}, \mathbf{x}')d\Omega =$$

$$= \int_{\Gamma} u_j^*(\mathbf{x}, \mathbf{x}')t_j(\mathbf{x}, \mathbf{x}')d\Gamma - \int_{\Gamma} u_j(\mathbf{x}, \mathbf{x}')t_j^*(\mathbf{x}, \mathbf{x}')d\Gamma \tag{66}$$

If the state (u^*, t^*) is chosen to correspond to the solution of static problem due to a unit force, Equation (66) can be rewritten as

$$u_i(\mathbf{x}') = \int_{\Gamma} \left[U_{ij}(\mathbf{x}, \mathbf{x}')t_j(\mathbf{x}) - T_{ij}(\mathbf{x}, \mathbf{x}')u_j(\mathbf{x})\right] d\Gamma - \int_{\Omega} U_{ij}(\mathbf{x}, \mathbf{x}')\sigma_{jk,k}(\mathbf{x})d\Omega \tag{67}$$

where

$$U_{ij}(\mathbf{x}', \mathbf{x}) = \frac{1}{16\pi\mu(1-\nu)r}\left[(3-4\nu)\delta_{ij} + r_{,i}r_{,j}\right]$$

$$T_{ij}(\mathbf{x}', \mathbf{x}) = \frac{-1}{8\pi(1-\nu)r^2}\left\{\left[(1-2\nu)\delta_{ij}\frac{\partial r}{\partial n} + 3r_{,i}r_{,j}\right]\right.$$
$$\left. -(1-2\nu)(n_j r_{,i} - n_i r_{,j})\right\}$$

for three-dimensional problems and

$$U_{ij}(\mathbf{x}', \mathbf{x}) = \frac{1}{8\pi\mu(1-\nu)}\left[(4\nu-3)\ln(r)\delta_{ij} + r_{,i}r_{,j}\right]$$

$$T_{ij}(\mathbf{x}', \mathbf{x}) = \frac{-1}{4\pi(1-\nu)r}\left\{\left[(1-2\nu)\delta_{ij} + 2r_{,i}r_{,j}\right]\frac{\partial r}{\partial n} - \right.$$
$$\left. -(1-2\nu)(r_{,i}n_j - r_{,j}n_i)\right\}$$

Substituting $\sigma_{jk,k}$ from Equation (65), results in

$$u_i(\mathbf{x}') = \int_{\Gamma} \left[U_{ij}(\mathbf{x}, \mathbf{x}')t_j(\mathbf{x}) - T_{ij}(\mathbf{x}, \mathbf{x}')u_j(\mathbf{x})\right] d\Gamma - \rho\int_{\Omega} U_{ij}(\mathbf{x}, \mathbf{x}')\ddot{u}_j(\mathbf{x}, \tau)d\Omega \tag{68}$$

Similarly stresses at an internal point can be written as

$$\begin{aligned}\sigma_{ij}(\mathbf{x}') = & \int_{\Gamma} \left[U_{kij}(\mathbf{x}, \mathbf{x}')t_k(\mathbf{x}) - T_{kij}(\mathbf{x}, \mathbf{x}')u_k(\mathbf{x})\right] d\Gamma \\ & - \rho\int_{\Omega} U_{kij}(\mathbf{x}, \mathbf{x}')\ddot{u}_k(\mathbf{x}, \tau)d\Omega\end{aligned} \tag{69}$$

where the fundamental solutions are given as:

$$U_{kij}(\mathbf{x}',\mathbf{x}) = \frac{1}{8\pi\,(1-\nu)\,r^2}\Big[(1-2\nu)(\delta_{ki}r_{,j}+\delta_{kj}r_{,i}-\delta_{ij}r_{,k})+3r_{,i}r_{,j}r_{,k}\Big]$$

$$T_{kij}(\mathbf{x}',\mathbf{x}) = \frac{\mu}{4\pi(1-\nu)r^3}\Big\{3\frac{\partial r}{\partial n}\Big[(1-2\nu)\delta_{ij}r_{,k}+\nu(\delta_{ik}r_{,j}+\delta_{jk}r_{,i})-5r_{,i}r_{,j}r_{,k}\Big]$$
$$+3\nu(n_i r_{,j}r_{,k}+n_j r_{,k}r_{,i})$$
$$+(1-2\nu)(2n_k r_{,i}r_{,j}+n_j\delta_{ik}+n_i\delta_{jk})-(1-4\nu)n_k\delta_{ij}\Big\}.$$

for three-dimensional problems, and

$$U_{kij}(\mathbf{x}',\mathbf{x}) = \frac{1}{4\pi(1-\nu)r}\Big[(1-2\nu)(\delta_{ki}r_{,j}+\delta_{kj}r_{,i}-\delta_{ij}r_{,k})+2r_{,i}r_{,j}r_{,k}\Big]$$

$$T_{kij}(\mathbf{x}',\mathbf{x}) = \frac{\mu}{2\pi(1-\nu)r^2}\Big\{2\frac{\partial r}{\partial n}\Big[(1-2\nu)\delta_{ij}r_{,k}+\nu(\delta_{ik}r_{,j}+\delta_{jk}r_{,i})-4r_{,i}r_{,j}r_{,k}\Big]$$
$$+2\nu(n_i r_{,j}r_{,k}+n_j r_{,k}r_{,i})$$
$$+(1-2\nu)(2n_k r_{,i}r_{,j}+n_j\delta_{ik}+n_i\delta_{jk})-(1-4\nu)n_k\delta_{ij}\Big\}.$$

for two-dimensional problems assuming plane strain conditions.

The displacement and traction boundary integral equations are obtained through the usual limiting process and can be written as

$$C_{ij}(\mathbf{x}')u_j(\mathbf{x}')+\int_\Gamma T_{ij}(\mathbf{x},\mathbf{x}')u_j(\mathbf{x})d\Gamma = \int_\Gamma U_{ij}(\mathbf{x},\mathbf{x}')t_j(\mathbf{x})+$$
$$+\rho\int_\Omega U_{ij}(\mathbf{x},\mathbf{x}')\ddot{u}_j(\mathbf{x},\tau)d\Omega \tag{70}$$

and

$$\frac{1}{2}t_j(\mathbf{x}')+\int_\Gamma T_{kij}(\mathbf{x},\mathbf{x}')u_k(\mathbf{x})d\Gamma = \int_\Gamma U_{kij}(\mathbf{x},\mathbf{x}')t_k(\mathbf{x})+$$
$$+\rho\int_\Omega U_{kij}(\mathbf{x},\mathbf{x}')\ddot{u}_{kj}(\mathbf{x},\tau)d\Omega \tag{71}$$

To transform the domain integral in (70) and (71) into boundary ones the DRM will be used. The DRM was originally developed by Brebbia and Nardini[20] for the displacement integral equation. The method is essentially a generalized way of constructing particular solutions that can be used to solve time-dependent problems. In this method the equations of motion are expressed in a boundary integral form using the fundamental solutions of elastostatics. This can be achieved by approximating the acceleration of a point $\mathbf{x}$ of the body by a sum of N coordinate functions $f^n(\mathbf{x}'',\mathbf{x})$ multiplied by unknown time-dependent coefficients $\ddot{\alpha}_l^n(\tau)$:

$$\ddot{u}_l(\mathbf{x},\tau) = \sum_{n=1}^{N}\ddot{\alpha}_l^n(\tau)f^n(\mathbf{x}'',\mathbf{x}), \tag{72}$$

where the dot above the variable denotes the derivative with respect to time. The approximation function $f^n(\mathbf{x}'', \mathbf{x}) = 1 + r(\mathbf{x}'', \mathbf{x})$ is chosen, where c is a constant and $r(\mathbf{x}'', \mathbf{x})$ is the distance between a defining point $\mathbf{x}''$ and the point $\mathbf{x}$. The defining point can be a boundary or a domain point.

Using this assumption the displacement boundary equation of motion, for a homogeneous and isotropic linear elastic body can be written as

$$c_{ij}(\mathbf{x}')u_j(\mathbf{x}', \tau) - \int_\Gamma U_{ij}(\mathbf{x}, \mathbf{x}')t_j(\mathbf{x}, \tau) + \int_\Gamma T_{ij}(\mathbf{x}, \mathbf{x}')u_j(\mathbf{x}, \tau)d\Gamma =$$

$$\sum_{n=1}^{N} \rho\ddot{\alpha}_l^n(\tau)\Big[c_{ij}(\mathbf{x}')\hat{u}_{lj}^n(\mathbf{x}', \mathbf{x}'') - \int_\Gamma U_{ij}(\mathbf{x}, \mathbf{x}')\hat{t}_{lj}^n(\mathbf{x}, \mathbf{x}'')d\Gamma$$

$$+ \int_\Gamma T_{ij}(\mathbf{x}, \mathbf{x}')\hat{u}_{lj}^n(\mathbf{x}, \mathbf{x}'')d\Gamma)\Big] \tag{73}$$

The traction integral equation, for a point, which belongs to a smooth boundary, has the form

$$\frac{1}{2}t_j(\mathbf{x}', \tau) - n_i(\mathbf{x}')\Big[\int_\Gamma U_{kij}(\mathbf{x}, \mathbf{x}')t_k(\mathbf{x}, \tau)d\Gamma$$

$$- \int_\Gamma T_{kij}(\mathbf{x}, \mathbf{x}')u_k(\mathbf{x}, \tau)d\Gamma\Big] =$$

$$\sum_{n=1}^{N} \rho\ddot{\alpha}_l^n(\tau)\left\{\frac{1}{2}\hat{t}_{lj}^n(\mathbf{x}', x'') - n_i(\mathbf{x}')\Big[\int_\Gamma U_{kij}(\mathbf{x}, \mathbf{x}')\hat{t}_{lk}(\mathbf{x}, \mathbf{x}')d\Gamma(\mathbf{x})\right.$$

$$\left.\int_\Gamma T_{kij}(\mathbf{x}, \mathbf{x}')\hat{u}_{lk}^n(\mathbf{x}, \mathbf{x}'')d\Gamma\Big]\right\} \tag{74}$$

where ρ is the mass density; $U_{ij}(\mathbf{x}', \mathbf{x})$, $T_{ij}(\mathbf{x}', \mathbf{x})$, $U_{kij}(\mathbf{x}', \mathbf{x})$ and $T_{kij}(\mathbf{x}', \mathbf{x})$ are fundamental solutions of elastostatics; $\hat{u}_{lj}^n(\mathbf{x}'', \mathbf{x})$ and $\hat{t}_{lj}^n(\mathbf{x}'', \mathbf{x})$ are particular displacements and tractions, which correspond to the function $f^n(\mathbf{x}'', \mathbf{x})$; they are given as

$$\hat{u}_{lj}(\mathbf{x}, \mathbf{x}') = \frac{1-2\nu}{(6-4\nu)\mu}r_{,l}r_{,j}r^2 + \frac{1}{48(1-\nu)\mu}\left[\frac{11-12\nu}{3}\delta_{ij} - r_{,l}r_{,j}\right]r^3$$

$$\hat{t}_{lj}(\mathbf{x}, \mathbf{x}') = \frac{1-2\nu}{3-2\nu}\left[\frac{1+2\nu}{1-2\nu}r_{,l}n_j + \frac{1}{2}\left(r_{,j}n_l + \delta_{lj}\frac{\partial r}{\partial n}\right)\right]r +$$

$$\frac{1}{24(1-\nu)}\left[(5-6\nu)r_{,j}n_l - (1-6\nu)r_{,l}n_j + \left[(5-6\nu)\delta_{lj} - r_{,l}r_{,j}\right]\frac{\partial r}{\partial n}\right]r^2$$

for three-dimensional problems, and

$$\hat{u}_{lj}(\mathbf{x},\mathbf{x}') = \frac{1-2\nu}{(5-4\nu)\mu} r_{,l} r_{,j} r^2 + \frac{1}{30(1-\nu)\mu}\left[\frac{9-10\nu}{3}\delta_{lj} - r_{,l} r_{,j}\right] r^3$$

$$\hat{t}_{lj}(\mathbf{x},\mathbf{x}') = \frac{2(1-2\nu)}{5-4\nu}\left[\frac{1+\nu}{1-2\nu} r_{,l} n_j + \frac{1}{2}\left(r_{,j} n_l + \delta_{lj}\frac{\partial r}{\partial n}\right)\right] r +$$

$$\frac{1}{15(1-\nu)}\left[(4-5\nu) r_{,j} n_l - (1-5\nu) r_{,l} n_j + \left[(5-4\nu)\delta_{lj} + r_{,l} r_{,j}\right]\frac{\partial r}{\partial n}\right] r^2$$

3.4.1. Numerical Discretization

The boundary of the body is discretized as in the previous approaches. The displacements and the tractions, $u_j(\mathbf{x},\tau)$, $t_j(\mathbf{x},\tau)$ and $\hat{u}^n_{lj}(\mathbf{x}'',\mathbf{x})$, $\hat{t}^n_{lj}(\mathbf{x}'',\mathbf{x})$ within each element are approximated using the same interpolation functions. As a result of the approximation the following displacement and traction equations are obtained:

$$c^b_{ij} u^b_j(\tau) - \sum_{m=1}^{M}\sum_{p=1}^{P}\left[t^{mp}_j(\tau)\int_{-1}^{1} U^b_{ij}(\xi) N^p(\xi) J^m(\xi) d\xi\right.$$

$$\left. - u^{mp}_j(\tau)\int_{-1}^{1} T^b_{ij}(\xi) N^p(\xi) J^m(\xi) d\xi\right] =$$

$$\sum_{n=1}^{N}\rho\ddot{\alpha}^n_l(\tau)\left\{c^b_{ij}\hat{u}^{bn}_{lj} - \sum_{m=1}^{M}\sum_{p=1}^{P}\left[\hat{t}^{nmp}_{lj}\int_{-1}^{1} U^b_{ij}(\xi) N^p(\xi) J^m(\xi) d\xi\right.\right.$$

$$\left.\left. - \hat{u}^{nmp}_{lj}\int_{-1}^{1} T^b_{ij}(\xi) N^p(\xi) J^m(\xi) d\xi\right]\right\}, \quad b = 1, 2, ..., B_1, \tag{75}$$

and

$$\frac{1}{2} t^b_j(\tau) - n^b_i \sum_{m=1}^{M}\sum_{p=1}^{P}\left[t^{mp}_k(\tau)\int_{-1}^{1} U^b_{kij}(\xi) N^p(\xi) J^m(\xi) d\xi\right.$$

$$\left. - u^{mp}_k(\tau)\int_{-1}^{1} T^b_{kij}(\xi) N^p(\xi) J^m(\xi) d\xi\right] =$$

$$\sum_{n=1}^{N}\rho\ddot{\alpha}^n_l(\tau)\left\{\frac{1}{2}\hat{t}^{bn}_{lj} - n^b_i\sum_{m=1}^{M}\sum_{p=1}^{P}\left[\hat{t}^{nmp}_{lk}\int_{-1}^{1} U^b_{kij}(\xi) N^p(\xi) J^m(\xi) d\xi\right.\right.$$

$$\left.\left. - \hat{u}^{nmp}_{lk}\int_{-1}^{1} T^b_{kij}(\xi) N^p(\xi) J^m(\xi) d\xi\right]\right\}, \quad b = 1, 2, ..., B_2. \tag{76}$$

The boundary equations are applied at the boundary nodes as in the other approaches. The displacement equations are applied at the domain points, when they are used to improve the approximation of accelerations. The set of equations can be written in matrix form as

$$\mathbf{Hu} - \mathbf{Gt} - \rho(\mathbf{H\hat{u}} - \mathbf{G\hat{t}})\ddot{\boldsymbol{\alpha}} = 0, \tag{77}$$

where $\mathbf{H}$ and $\mathbf{G}$ depend on integrals of fundamental solutions and interpolating functions; they are the same as in elastostatics. The vectors $\mathbf{u}$, $\mathbf{t}$, $\hat{\mathbf{u}}$ and $\hat{\mathbf{t}}$ contain nodal values of real and particular displacements and tractions. The relationship between $\ddot{\mathbf{u}}$ and $\ddot{\alpha}$ is established by applying Equation (72) to every boundary and domain node. The resulting set of equations can be written in matrix form:

$$\ddot{\mathbf{u}} = \mathbf{F}\ddot{\boldsymbol{\alpha}}, \tag{78}$$

where the elements of the matrix $\mathbf{F}$ are the values of the function $f^n(\mathbf{x}'', \mathbf{x})$ at all N nodes. The unknown coefficients $\ddot{\alpha}$ can be expressed in terms of the accelerations $\ddot{\mathbf{u}}$ as follows:

$$\ddot{\boldsymbol{\alpha}} = \mathbf{E}\ddot{\mathbf{u}}. \tag{79}$$

If the coincident nodes of the crack are used in the approximation of the acceleration field, the system of Equation (78) is singular; hence the coefficients $\ddot{\alpha}$ cannot be calculated by the direct solution of this system. Substitution of Equation (79) into (77) gives

$$\mathbf{Hu} - \mathbf{Gt} - \rho(\mathbf{H\hat{u}} - \mathbf{G\hat{t}})\mathbf{E}\ddot{\mathbf{u}} = 0, \tag{80}$$

or

$$\mathbf{Hu} - \mathbf{Gt} + \mathbf{M}\ddot{\mathbf{u}} = 0, \tag{81}$$

where $\mathbf{M} = -\rho(\mathbf{H\hat{u}} - \mathbf{G\hat{t}})\mathbf{E}$ and is the mass matrix of the structure. The system of equations of motion (81) is modified, according to the boundary conditions, and can be solved using a direct integration method. The Houbolt method, which possesses strong artificial damping and can damp out undesired high-frequency responses, has been used by Fedelinski *et al.*[21] for structures subjected to impact loads.

3.4.2. Numerical Example

Fedelinski, Aliabadi and Rooke[21] considered a circular disc of radius $r_1 = 100$ mm, shown in Figure 8, with four interior holes of radius $r_3 = 20$ mm, contains a single radial crack of length $a = 20$ mm at the edge of one of the holes. The distance of the centres of the holes from the centre of

Figure 8. A circular disc with four radial holes and a radial crack.

the disc is $r_2 = 60$ mm. The disc has the following material properties: Young modulus $E = 0.2 \times 10^{12}$ Pa; Poisson's ratio $\upsilon = 0.3$; and the mass density $\rho = 8000$ kgm^{-3}. The state of plane strain is assumed. Two of the holes, diametrically opposite in the disc are loaded by a suddenly applied normal pressure, as shown in Figure 8; the pressure distribution is represented by a sine function and has a maximum value of σ_o.

The boundary of the disc was divided into 64 quadratic elements and 28 domain points were used. The time step $\Delta\tau = 0.1\mu s$ was chosen. The $\hat{J}$ integral was used to evaluate the stress intensity factor and the normalised $K_I/\sigma\sqrt{\pi a}$ as a function of time is shown in Figure 9.

3.4. COMPARISON OF METHODS

In order to compare the different approaches described in the previous sections Fedelinski, Aliabadi and Rooke[23] considered a mixed-mode problem. The DSIF were obtained from COD calculations for the time domain and the Laplace transform methods; and from the $\hat{J}$-integral for the dual reciprocity method. The structure is instantaneously loaded by a stress σ_o at time $\tau = 0$. The DSIF are normalized with respect to

$$K_o = \sigma_o\sqrt{\pi a}, \tag{82}$$

where a defines the length of the crack. The static limits are calculated by assuming a limiting small density of the material.

3.5.1. Rectangular Plate with an Inclined Crack

A rectangular plate of length $2b = 60$ mm and width $2h = 30$ mm contains a central inclined crack of length $2a = 14.14$ mm slanted at an angle $\alpha = 45^o$, as shown in Figure 10.

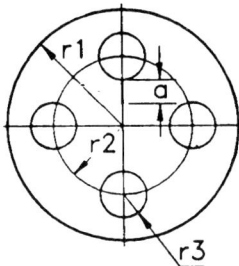

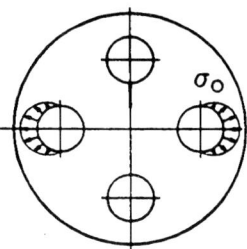

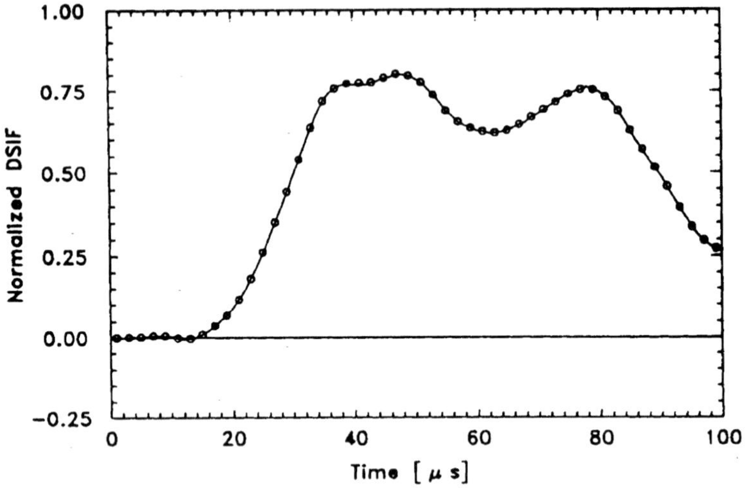

Figure 9. Stress intensity factors for the cracked circular disc.

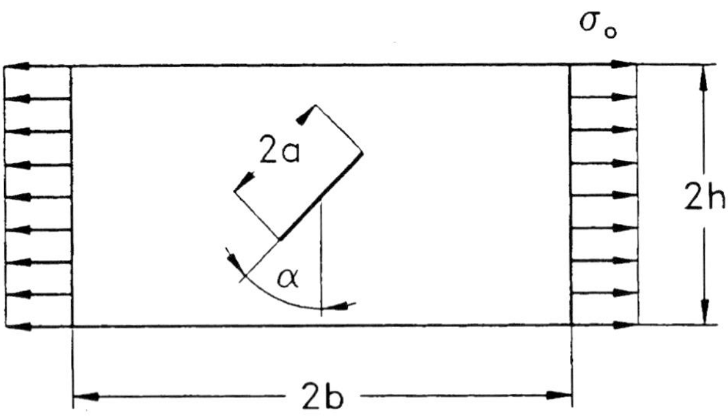

Figure 10. Rectangular plate with an internal inclined crack.

The material properties are $\mu = 76.92 \times 10^{-9}$ Pa, $\nu = 0.3$, and $\rho = 5000$ kgm^{-1}. The opposite ends of the plate are loaded by the stress σ_o at $\tau = 0$.

The boundary is divided into 64 elements for the TDM; 32 elements are used for the LTM; and 40 boundary elements and 40 additional domain points are used for the DRM. The time step $\Delta\tau = 0.4$ μs is used for the TDM and $\Delta\tau = 0.2$ μs for the DRM; 25 Laplace parameters are used for the LTM. For this discretization in space and time similar accuracy of the results was obtained.

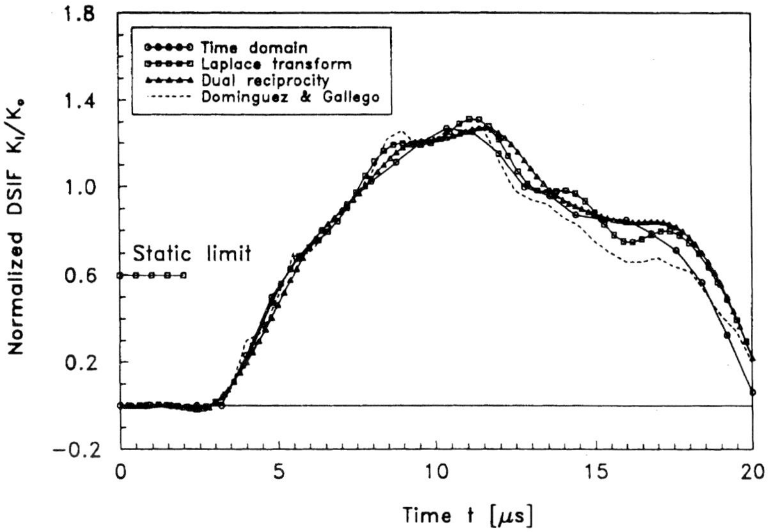

Figure 11. Mode I stress intensity factors for the plate with an inclined crack.

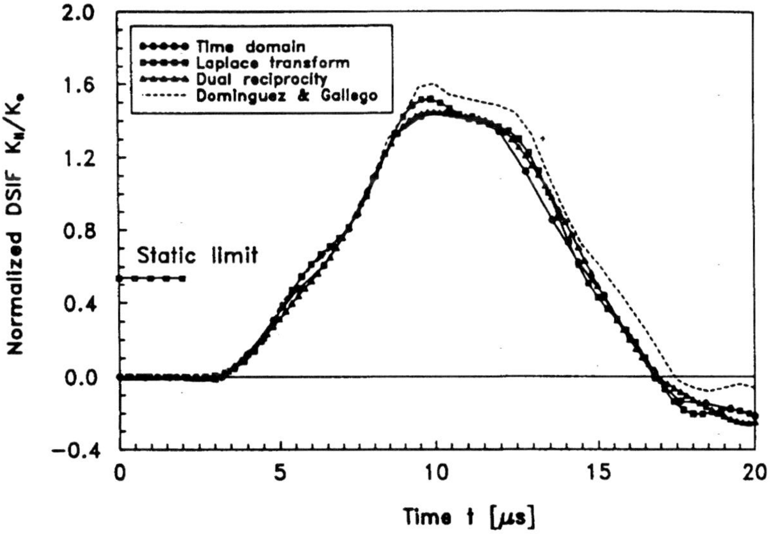

Figure 12. Mode II stress intensity factors for the plate with an inclined crack.

The normalized DSIF K_I/K_o and K_{II}/K_o are plotted in Figures 11 and 12 respectively, and compared with those of Dominguez and Gallego[32], who used the time domain formulation and a subregion technique in the BEM. The solutions obtained by the three methods are similar. The normalized K_I/K_o are bigger and K_{II}/K_o are smaller at later times, than those obtained by Dominguez and Gallego.

Fedelinski, Aliabadi and Rooke[23] compared the computational time of the three methods on a SUN Sparc IPC workstation. The total computational time and the computational time per time step, or per Laplace parameter, is plotted against the number of steps, for the TDM and DRM and verses the number of Laplace parameter for the LTM in Figures 13 and 14.

The TDM is fast during the initial steps because the time-dependent fundamental solutions are integrated along part of the boundary only. For this problem, the TDM requires integration of kernels along the whole boundary after 11 steps. From then on, the solution needs about 2 min per time step, as shown in Figure 14; this time slowly increases, approximately linearly, with the number of steps because the solution depends on all the previous time steps.

The time required for the solution of each Laplace parameter is approximately constant and equals about 5 min, as we can see from Figure 13. Small variations are caused by the different number of terms used to calculate the Bessel functions, which depends on the value of the Laplace parameter.

The DRM requires about 11 min to formulate the system of equations and to decompose the matrix coefficients, but the solution in each time step takes about 2 sec only.

The system of equations of motion is solved faster by the TDM than by the DRM during the first 10 steps. The total time for the TDM is similar function of number of steps as the time for the LTM is function of number of Laplace parameters.

The solutions of the present example were obtained[23] using the same number of boundary elements in order to compare memory and time requirements. However similar accuracy was obtained using different number of boundary elements for each method, namely: 32 for LTM, 40 for DRM and 64 for TDM. The computational times required for the new discretizations are shown in Figure 15.

3.6. INDIRECT BOUNDARY ELEMENT FORMULATION

Two indirect formulations known as the Fictitious Stress and the Displacement Discontinuity methods were developed by Crouch and Starfield[30] to study fracture of rocks. The extension of these formulations to elastodynamic

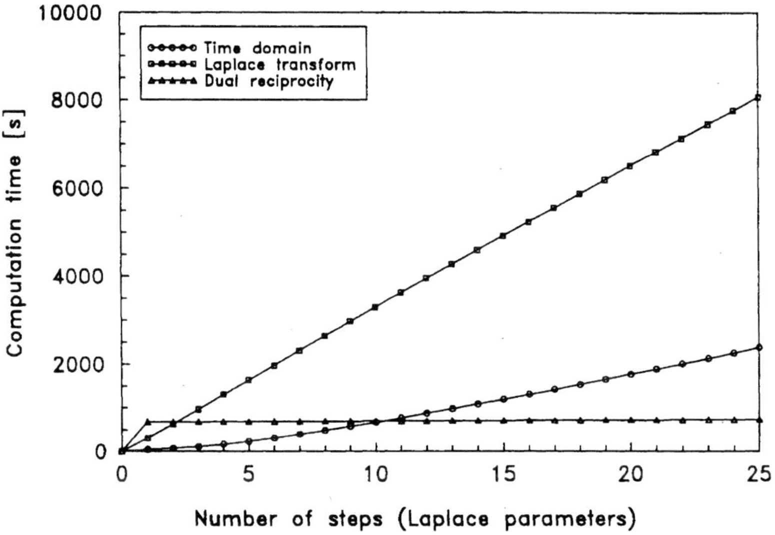

Figure 13. Computational time vs number of steps (number of Laplace parameters) for the plate with an inclined central crack.

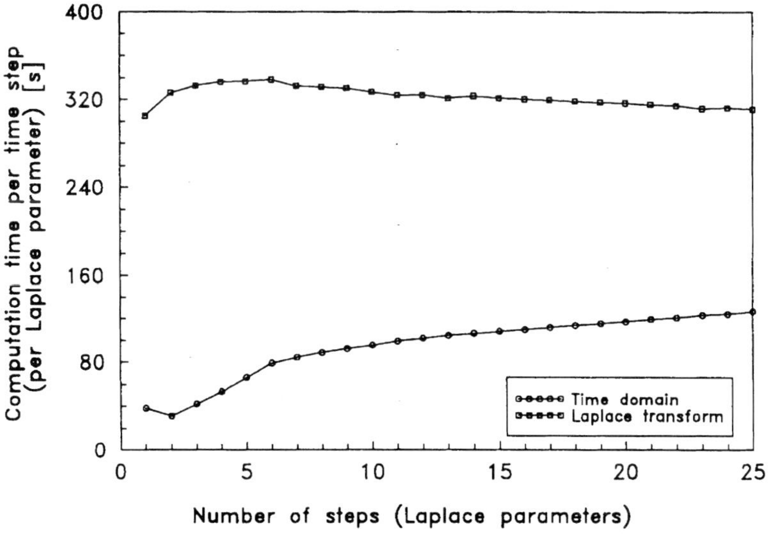

Figure 14. Computational time per step vs number of steps (number of Laplace parameters) for the plate with an inclined central crack.

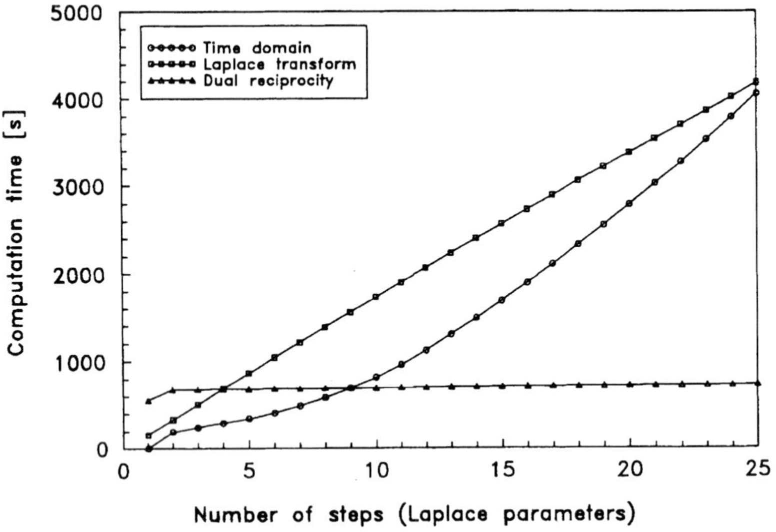

Figure 15. Computational time vs number of steps (number of Laplace parameters) for the plate with an inclined central crack.

crack problems had been developed by Wen, Aliabadi and Rook[24,25] in Laplace transform domain for two- and three-dimensional problems and Siebrits and Crouch[31] in time domain for two-dimensional problems. In this chapter the formulation in Laplace transform domain will be presented. The integral equation for the fictitious stresses are used on the uncracked boundary and the displacement discontinuity integral equation is used on the crack boundary.

3.6.1. Fictitious Stress Method

The displacements and tractions at a general point $\mathbf{x}'$ in the domain can be written in terms of a fictitious sources densities ϕ_j as

$$u_i(\mathbf{x}') = \int_\Gamma U_{ij}(\mathbf{x}', \mathbf{x}, s)\phi_j(\mathbf{x})d\Gamma + \int_{\Gamma_c} U_{ij}(\mathbf{x}', \mathbf{x}, s)\phi_j(\mathbf{x})d\Gamma_c \tag{83}$$

and

$$t_i(\mathbf{x}') = \int_\Gamma T_{ij}(\mathbf{x}', \mathbf{x}, s)\phi_j(\mathbf{x})d\Gamma + \int_{\Gamma_c} T_{ij}(\mathbf{x}', \mathbf{x}, s)\phi_j(\mathbf{x})d\Gamma_c \tag{84}$$

where U_{ij} and T_{ij} are the Laplace transform fundamental solutions as in the direct formulation. Taking $\mathbf{x}'$ to the boundary, the above equations can be written as

$$u_i(\mathbf{x}') = \int_\Gamma U_{ij}(\mathbf{x}', \mathbf{x}, s)\phi_j(\mathbf{x})d\Gamma + \int_{\Gamma_c} U_{ij}(\mathbf{x}', \mathbf{x}, s)\phi_j(\mathbf{x})d\Gamma_c \tag{85}$$

and

$$t_i(\mathbf{x}') = C_{ij}(\mathbf{x}')\phi_i(\mathbf{x}') + \int_\Gamma T_{ij}(\mathbf{x}', \mathbf{x}, s)\phi_j(\mathbf{x})d\Gamma + \int_{\Gamma_c} T_{ij}(\mathbf{x}', \mathbf{x}, s)\phi_j(\mathbf{x})d\Gamma_c \tag{86}$$

where $C_{ij}(\mathbf{x}')$ as in the direct formulation is a constant, and for smooth boundaries is given as $C_{ij} = \frac{\delta_{ij}}{2}$.

3.6.2. Displacement Discontinuity Method (DDM)

If displacement discontinuities ψ_j acts on the curve of the outer boundary Γ of a finite body in an infinite body (transformed), then the displacements and tractions in the domain at $\mathbf{x}'$ can be written as

$$u_i(\mathbf{x}') = \int_\Gamma D_{ij}(\mathbf{x}', \mathbf{x}, s)\psi_j(\mathbf{x})d\Gamma + \int_{\Gamma_c} D_{ij}(\mathbf{x}', \mathbf{x}, s)\psi_j(\mathbf{x})d\Gamma_c \tag{87}$$

and

$$t_i(\mathbf{x}') = \int_\Gamma S_{ij}(\mathbf{x}', \mathbf{x}, s)\psi_j(\mathbf{x})d\Gamma + \int_{\Gamma_c} S_{ij}(\mathbf{x}', \mathbf{x}, s)\psi_j(\mathbf{x})d\Gamma_c \tag{88}$$

where D_{ij} and S_{ij} are respectively the displacement and traction fundamental solutions for a unit displacement discontinuity (per unit length). Taking $\mathbf{x}'$ to the boundary, the two integral equations become

$$u_i(\mathbf{x}') = \frac{\psi_i}{2} + \int_\Gamma D_{ij}(\mathbf{x}', \mathbf{x}, s)\psi_j(\mathbf{x})d\Gamma + \int_{\Gamma_c} D_{ij}(\mathbf{x}', \mathbf{x}, s)\psi_j(\mathbf{x})d\Gamma \tag{89}$$

and

$$t_i(\mathbf{x}') = \int_\Gamma S_{ij}(\mathbf{x}', \mathbf{x}, s)\psi_j(\mathbf{x})d\Gamma + \int_{\Gamma_c} S_{ij}(\mathbf{x}', \mathbf{x}, s)\psi_j(\mathbf{x})d\Gamma \tag{90}$$

3.6.3. Numerical Implementation

The uncracked boundary is divided into N constant elements and the crack line into M segments (generally the element length is the same). The fictitious load ($\phi_i^n, n = 1, 2, \ldots, N; i = 1, 2$) act on the outer boundary elements and the displacement discontinuities ($\psi_i^m, m = 1, 2, \ldots, M; i = 1, 2$) on the crack elements. The tractions on the outer boundary element β and on the crack elements are the summation of all contributions from the fictitious loads on the outer elements and from the discontinuity displacements on the crack elements, that is

$$\sum_{n=1}^{N} A_{ij}^{\beta n} \phi_j^n + \sum_{m=1}^{M} B_{ij}^{\beta m} \psi_j^m = \bar{t}_i^{\beta}\ , \beta = 1, 2, ..N \tag{91}$$

and

$$\sum_{n=1}^{N} A_{ij}^{\prime\beta n} \phi_j^n + \sum_{m=1}^{M} B_{ij}^{\prime\beta m} \psi_j^m = \bar{t}_i^{\beta}\ , \beta = 1, 2, ..M \tag{92}$$

where

$$A_{ij}^{\beta n} = \int_{\Delta\Gamma_n} T_{ij}(\mathbf{x}^{\beta}, \mathbf{x}, s) d\Gamma\ ,$$

$$B_{ij}^{\beta m} = \int_{\Delta\Gamma_m} T_{ij}(\mathbf{x}^{\beta}, \mathbf{x}, s) d\Gamma_c$$

$$A_{ij}^{\prime\beta n} = \int_{\Delta\Gamma_n} S_{ij}(\mathbf{x}^{\beta}, \mathbf{x}, s) d\Gamma$$

and

$$A_{ij}^{\prime\beta m} = \int_{\Delta\Gamma_m} S_{ij}(\mathbf{x}^{\beta}, \mathbf{x}, s) d\Gamma_c$$

3.6.4. Numerical Examples

Wen, Aliabadi and Rooke[24–27] have applied the displacement discontinuity method to several problems in two- and three-dimensional problems. Here, some examples reported by them are included to demonstrate the efficiency of their technique.

3.5.4.1. Two-dimensional problems

A circular hole and two radial edge cracks — A rectangular plate of length $2b = 60$ mm and width $2h = 30$ mm contains a central hole of radius

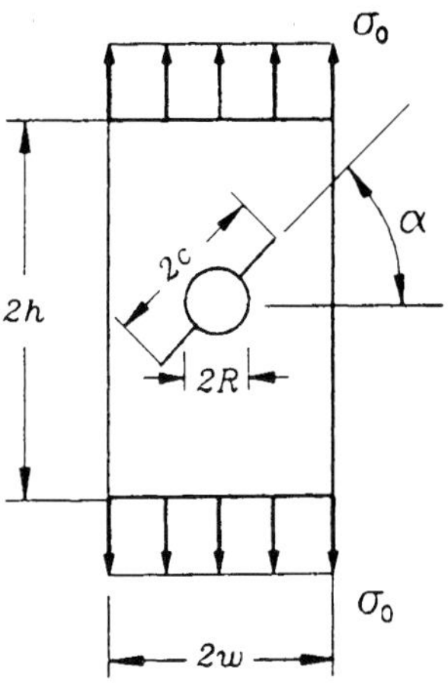

Figure 16. A rectangular plate with a circular hole and two radial edge cracks.

$r = 3.75$ mm, with two equal length cracks, diametrically opposite and the crack is inclined at an angle α as shown in Figure 16. The plate is subjected to a uniform uniaxial tensile stress σ_o with step-function time dependence. The plate has the following material properties: the shear modulus $\mu = 76.92 \times 10^9$ Pa; density $\rho = 5000$ kgm^{-3}; and Poisson's ratio $\upsilon = 0.3$. A state of plane strain was assumed. The plate with a crack length: $a/h = 0.5$ for $\alpha = 0^\circ, 30^\circ, 45^\circ, 60^\circ$ was studied. The fictitious load method was used on the boundary of the plate and the hole with 80 constant elements. The crack boundaries were modelled with the displacement discontinuity method and 20 constant elements. The total number of Laplace parameter was 25. The normalized stress intensity factors $K_I/\sigma\sqrt{\pi c}$ and $K_{II}/\sigma\sqrt{\pi c}$ obtained by Wen, Aliabadi and Rooke[24] are shown in Figure 17.

The normalized stress intensity factors K_I/K_o and K_{II}/K_o evaluated by Fedelinski, Aliabadi and Rooke[23] are also presented in Figure 17. In[23] the plate was discretized using 64 boundary elements and 38 domain points; a time step of $\Delta\tau = 0.2\mu s$ was used.

It can be seen from the figures that the increase of the angle α decreases the maximum value of K_I/K_o, and that K_{II}/K_o are shown to have similar time dependence. In all examples, the stress intensity factors remain zero until the dilatational wave from the loaded portion of the boundary arrives at the crack tip.

Interaction between two cracks — Two collinear cracks AB and CD are shown in Figure 18 with a uniform opening load $\sigma_o H(\tau)$ acting on the crack

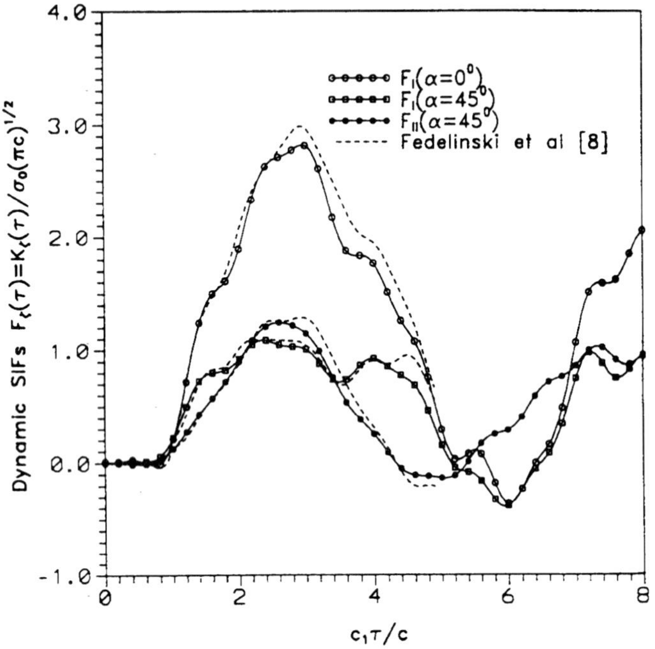

Figure 17. Normalized stress intensity factors for a plate with circular hole and two radial edge cracks.

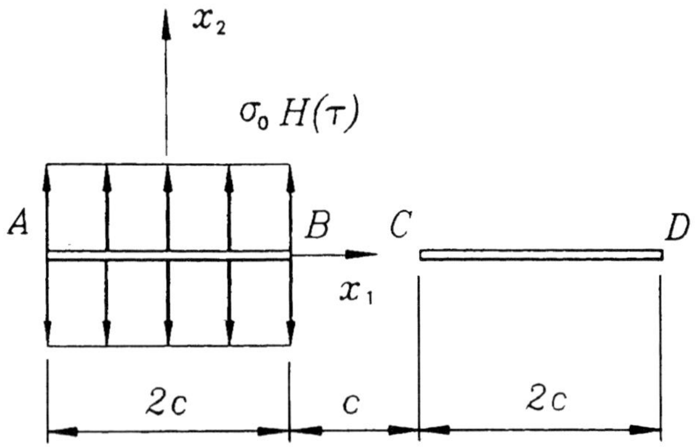

Figure 18. Two collinear cracks in an infinite sheet.

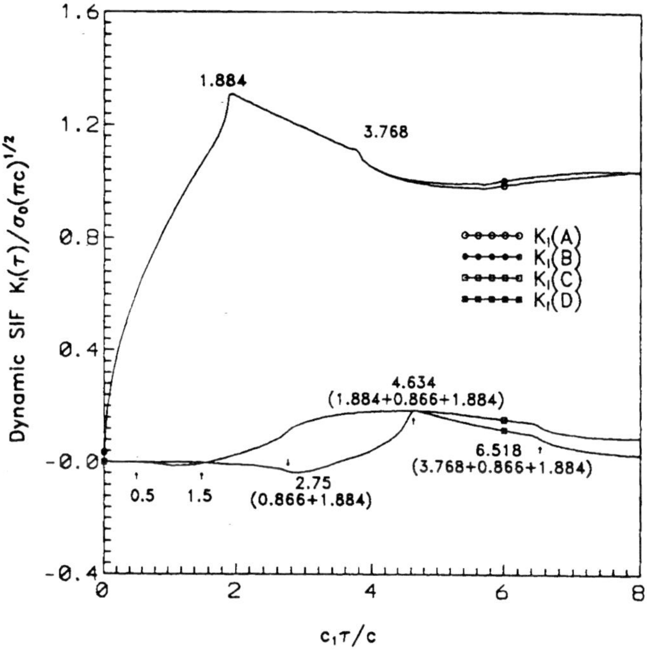

Figure 19. Stress intensity factors for two cracks in an infinte sheet.

faces of AB. The length of each crack is $2c$, and the distance between crack tips B and C is c. The stress intensity factors at these four crack tips are shown in Figure 19. When $c_1\tau/(2c) < 3.768$ there is no difference between K_I^A and K_I^B. The other two K_I^C and K_D^I are zero before dilatation wave starting from crack tip B arrives at points C and D. After this wave arrives, the calculated stress intensity factors K_I^C and K_I^D are less than zero, which means that crack closure has occurred.

3.6.4.2. Three-dimensional problems

Square crack in an infinite body under shear load — Consider a square crack of length $2a$ subjected to a dynamic uniform shear load $s_o H(\tau)$ along the x_1-direction as shown in Figure 20. Wen, Aliabadi and Rooke[27] solved this problem with the grid of 15×15 constant elements modelling the crack. In this case only the displacement discontinuity method is required to model the crack. The number of Laplace parameters was chosen as 70 and the Poisson's

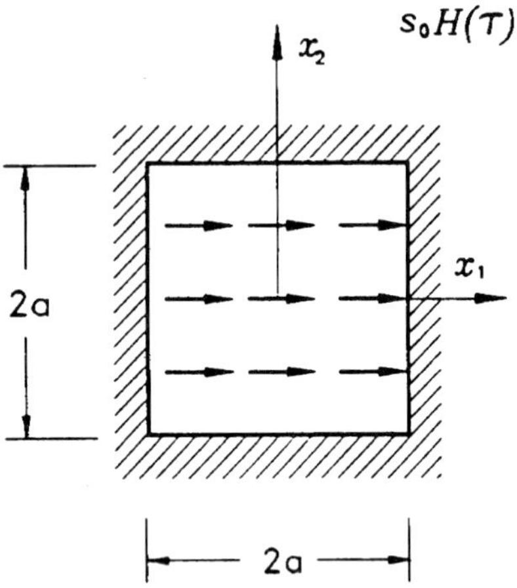

Figure 20. Square crack under shear load.

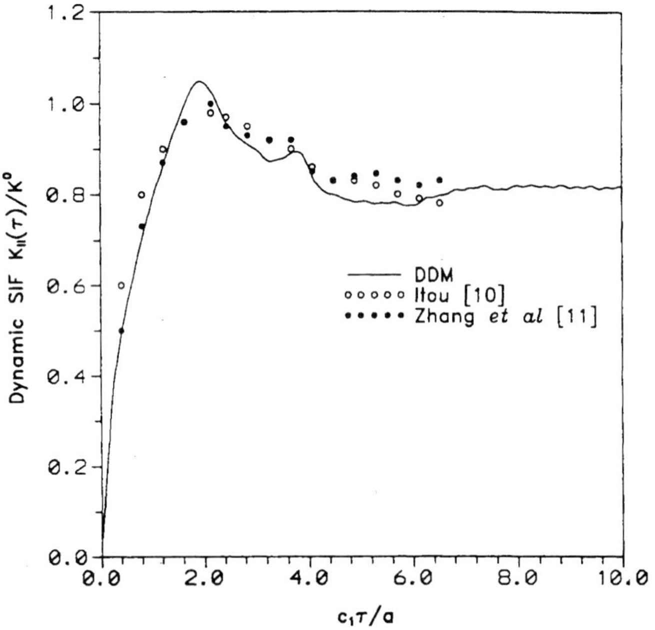

Figure 21. Mode II stress intensity factors for a square crack under shear load.

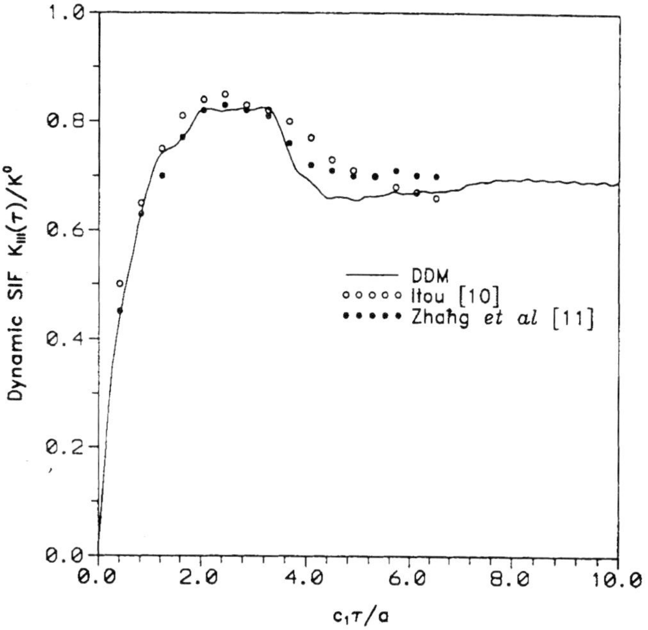

Figure 22. Mode III stress intensity factors for a square crack under shear load.

ratio $\upsilon = 0.2$. The normalised stress intensity factors $K_{II}/s_o\sqrt{\pi a}$, and $K_{III}/s_o\sqrt{\pi a}$ are shown in Figures 21 and 22. The numerical results given by Itou[28] and Zhang *et al.*[29] are also plotted for comparison.

Elliptical crack in a square bar — A square bar ($2w \times 2w \times 4w$) containing a central circular or elliptical crack is loaded on the ends in a direction perpendicular to the crack plane by a Heaviside load as shown in Figure 23. The radius of the circular crack is a and the length of the two principle axes of the elliptical crack are a and b. The radius $a/w = 0.5$ for the circular crack and $b/a = 0.8$ with $a/w = 0.5$ for the elliptical crack; Poisson's ratio $\upsilon = 0.2$. The value of the normalized dynamic stress intensity factors ($K^o = 2\sigma_o\sqrt{\frac{a}{\pi}}$)at point A in Figure 23 were evaluated by Wen, Aliabadi and Rooke[25] are plotted in Figure 24. The number of elements was 69 and 83 for the circular and the elliptical crack respectively.

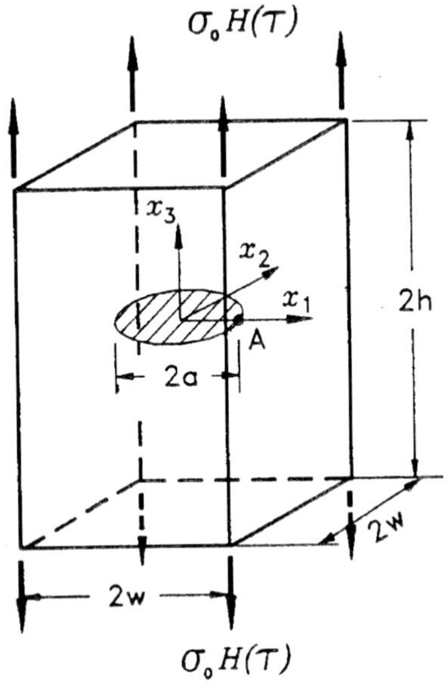

Figure 23. A rectangular bar with an elliptical crack.

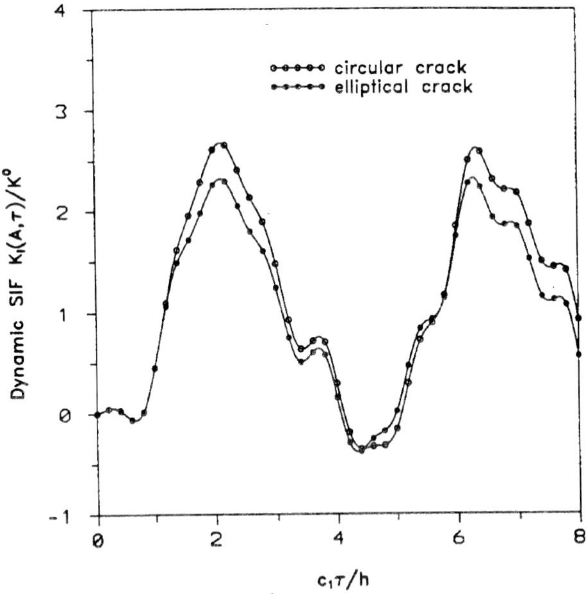

Figure 24. Stress intensity factors for an elliptical crack.

3.7. APPENDIX A

The derivative fundamental solutions for 3D are given as

$$U_{kij}(\mathbf{x},t;\mathbf{x}',\tau)=\frac{1}{4\pi}\left[6c_2^2(a_{kij}-b_{kij})\int_{1/c_1}^{1/c_2}\lambda\delta(t-\tau-\lambda r)d\lambda\right.$$

$$-(12a_{kij}-2b_{kij})\left[\delta(t-\tau-\frac{r}{c_2})-\left(\frac{c_2}{c_1}\right)^2\delta(t-\tau-\frac{r}{c_1})\right]-$$

$$-\frac{2ra_{kij}}{c_2}\left[\dot{\delta}(t-\tau-\frac{r}{c_2})-\left(\frac{c_2}{c_1}\right)^3\dot{\delta}(t-\tau-\frac{r}{c_1})\right]+$$

$$c_{kij}(1-\frac{2c_2^2}{c_1^2})\left[\delta(t-\tau-\frac{r}{c1})+\left(\frac{r}{c_1}\right)\dot{\delta}(t-\tau-\frac{r}{c_1})\right]$$

$$+d_{kij}\left[\delta(t-\tau-\frac{r}{c_2})+\left(\frac{r}{c_2}\right)\dot{\delta}(t-\tau-\frac{r}{c_2})\right]$$

where $a_{kij}=r_kr_ir_j/r^5$, $c_{kij}=\delta_{ij}r_k/r^3$, $d_{kij}=(\delta_{ik}y_j+\delta_{kj}r_i)/r^3$ and $b_{kij}=c_{kij}+d_{kij}$.

$$T_{kij}(\mathbf{x},t;\mathbf{x}',\tau)=-\frac{\rho}{4\pi}\left[12c_2^4(35a_{kij}-5b_{kij}+c_{kij})\int_{1/c_1}^{1/c_2}\lambda\delta(t-\tau-\lambda r)d\lambda\right.$$

$$-4c_2^2(45a_{kij}-6b_{kij}+c_{kij})\left[\delta(t-\tau-\frac{r}{c_2})-\left(\frac{c_2}{c_1}\right)^2\delta(t-\tau-\frac{r}{c_1})\right]-$$

$$-4c_2r(10a_{kij}-b_{kij})\left[\dot{\delta}(t-\tau-\frac{r}{c_2})-\left(\frac{c_2}{c_1}\right)^3\dot{\delta}(t-\tau-\frac{r}{c_1})\right]+$$

$$+2c_2^2(1-2\frac{c_2^2}{c_1^2})(3d_{kij}-2e_{kij})\left[\delta(t-\tau-\frac{r}{c1})+\left(\frac{r}{c_1}\right)\dot{\delta}(t-\tau-\frac{r}{c_1})\right]+$$

$$+c_2^2(3f_{kij}-2g_{kij})\left[\delta(t-\tau-\frac{r}{c_2})+\left(\frac{r}{c_2}\right)\dot{\delta}(t-\tau-\frac{r}{c_2})\right]-$$

$$-4a_{kij}r^2\left[\ddot{\delta}t-\tau-\frac{r}{c_2})-\left(\frac{c_2}{c_1}\right)^4\ddot{\delta}(t-\tau-\frac{r}{c_1})\right]+$$

$$+r^2(1-\frac{2c_2^2}{c_1^2})\left(2c_2^2\frac{d_{kij}}{c_1^2}+(1-\frac{2c_2^2}{c_1^2})e_{kij}\right)\Big[\ddot{\delta}(t-\tau-\frac{r}{c_1})$$

$$+r^2 f_{kij}\ddot{\delta}(t-\tau-\frac{r}{c_2})\Big]$$

where $a_{kij} = r_k r_i r_j r_m n_m / r^7$, $d_{kij} = (r_j r_i n_k + r_k r_m n_m \delta_{ij})/r^5$, $e_{kij} = \delta_{ik} n_k / r^3$, $f_{kij} = \{r_k r_j n_i + r_k r_i n_j + r_m n_m(\delta_{ik} r_j + \delta_{kj} r_i)\}/r^5$, $g_{kij} = (\delta_{kj} n_i + \delta_{ik} n_j)/r^3$, $b_{kij} = d_{kij} + f_{kij}$ and $c_{kij} = e_{kij} + g_{kij}$.

References

1. Aliabadi, M.H. and Rooke, D.P., 1991, *Numerical Fracture Mechanics*, Computational Mechanics Publications, Southampton, Kluwer Academic Publishers, Dordrecht.
2. Portela, A., Aliabadi, M.H. and Rooke, D.P., 1992, The dual boundary element method: effective implementation for crack problems, *Int. J. Numer. Methods in Eng.*, **33**(6), 1269–1287.
3. Mi, Y. and Aliabadi, M.H., 1992, Dual boundary element method for three-dimensional fracture mechanics analysis, *Engng Anal. Bound. Elem.*, **10**, 161–171.
4. Manolis, G.D. and Beskos, D.E., 1988, *Boundary element methods in elastodynamics*, Unwin Hyman, London (1988).
5. Mansur, W.J. and Brebbia, C.A., 1984, Further developments on the solution of the transient scalar wave equation. Chapter 4, in *Topics in boundary element research*, Springer Verlag, Berlin.
6. Dominguez, J., 1993, *Boundary elements in dynamics*, Computational Mechanics Publications, Southampton.
7. Brebbia, C.A., Telles, J. and Wrobel, L., 1984, *Boundary element techniques*, Springer Verlag, Berlin.
8. Israil, A.S.M. and Banerjee, P.K., 1990, Interior stress calculations in 2-D time-domain transient BEM analysis, *Int. J. Solids BEM*, *Int. J. Solids Struc.*, **26**, 851–864.
9. Ahmad, S. and Banerjee, P.K., 1988, Time domain transient analysis of 3-D solids by BEM, *Int.J.Solids and Struc.*, **26**, 1709–1728.
10. Aliabadi, M.H. (Editor), 1984, *Dynamic Fracture Mechanics*, Computational Mechanics Publication, Southampton.
11. Beskos, D., Boundary element methods in dynamic analysis part II (1986–1996), to appear in Applied Mech. Review.
12. Graf, K.F., 1975, *Wave motion in elastic solids*, Clarendon Press, Oxford.
13. Aliabadi, M.H. and Brebbia, C.A., 1996, Boundary element formulation in fracture mechanics: a review, *Localised Damage IV*, edited by Nisitani, H. *et al.*, 1–21.
14. Fedelinski, P., Aliabadi, M.H. and Rooke, D.P., 1995, A single-region time-domain BEM for dynamic crack problems, *Int. J. Solids and Structures*, **32**, 3555–3571.
15. Chen, Y.M., 1975, Numerical computation of dynamic stress intensity factors by a Lagrangian finite-difference method (the HEMP code), *Engng. Fracture Mech.*, **7**, 653–660.
16. Aberson, J.A., Anderson, J.M. and King, W.W., 1977, Dynamic analysis of cracked structures using singularity finite elements. In *Mechanics of Fracture*, Vol. 4: Elastodynamic crack problems (edited by G.C. Sih), pp. 249–294, Noordhoff International Publishing, Leyden.
17. Durbin, F., 1944, Numerical inversion of Laplace transforms: an efficient improvement to Dubner and Abate's method, *Comp. J.*, **17**, 371–376.
18. Fedelinski, P., Aliabadi, M.H. and Rooke, D.P., 1996, The Laplace transform DBEM method for mixed-mode dynamic crack analysis, *Computers and Structures*, **59**, 1021–1031.

19. Nishioka, T. and Atluri, S.N., 1980, Numerical modelling of dynamic propagation in finite bodies, by moving singular elements. Part 2: results *J. Appl. Mech., ASME*, **47**, 577–582.
20. Brebbia, C.A. and Nardini, D., 1983, Dynamic analysis in solid mechanics by an alternative boundary element procedure, *Soil Dyn. Earth. Engng.*, **2**, 228–233.
21. Fedelinski, P., Aliabadi, M.H. and Rooke, D.P., 1993, Dual boundary element method: inertial stress intensity factors, in Proc. *Boundary Element Technology VIII*, Eds Pina, H. and Brebbia, C.A., Computational Mechanics Publications, Southampton, pp. 267–276.
22. Fedelinski, P., Aliabadi, M.H. and Rooke, D.P., 1993, The dual boundary element method in dynamic fracture mechanics, *Engng Anal. Bound. Elem.*, **12**, 203–210.
23. Fedelinski, P., Aliabadi, M.H. and Rooke, D.P., 1996, Boundary element formulations for the dynamic analysis of cracked structures, *Engineering Analysis with Bound. Elem.*, **17**, 45–56.
24. Wen, P.H., Aliabadi, M.H. and Rooke, D.P., 1995, An indirect boundary element method for three-dimensional dynamic problems, *Engineering Analysis with Bound. Elem.*, **16**, 351–362.
25. Wen, P.H., Aliabadi, M.H. and Rooke, D.P., 1996, The influence of waves on dynamic stress intensity factors (two-dimensional problems), *Archives of Applied Mechanics*, **66**, 326–355.
26. Wen, P.H., Aliabadi, M.H. and Rooke, D.P., 1996, The influence of waves on dynamic stress intensity factors (three-dimensional problems), *Archives of Applied Mechanics*, **66**, 385–394.
27. Wen, P.H., Aliabadi, M.H. and Rooke, D.P., 1994, A fictitious stress and displacement discontinuity method for dynamic crack problems. *Proc. of 16th International Conference on Boundary Elements*, edited by C.A. Brebbia, Computational Mechanics Publications, Southampton, pp. 469–476.
28. Itou, S., 1991, Transient dynamic stresses around a rectangular crack under an impact shear load, *Engineering Fracture Mechanics*, **39**, 487–492.
29. Zhang, C.H. and Gross, D., 1993, A non-hypersingular time-domain BIEM for 3-D transient elastodynamic crack analysis. *Int. J. Numer. Methods in Eng.*, **36**, 2997–3017.
30. Crouch, S.L. and Starfield, A.M., 1993, *Boundary element method in solid mechanics*, George Allen and Unwin.
31. Siebrits, E. and Crouch, S.L., 1994, Two-dimensional elastodynamic displacement discontinuity method. *Int. J. Numer. Methods in Eng.*, **37**, 3229–3250.
32. Dominguez, J. and Gallego, R., 1992, Time domain boundary element method for dynamic stress intensity factor computation. *Int. J. Numer. Methods in Eng.*, **33**, 635–647.

4 SEISMIC RETROFITTING OF CONCRETE COLUMNS WITH FIBER COMPOSITE WRAP: AN ANALYTICAL AND EXPERIMENTAL STUDY

HAMID SAADATMANESH

Department of Civil Engineering and Engineering Mechanics,
University of Arizona, Tucson, Arizona 85718, USA
E-mail: hamid@ccit.arizona.edu

4.1. INTRODUCTION

The 1971 San Fernando earthquake, the 1987 Whittier earthquake, and the 1989 Loma Prieta earthquake inflicted substantial damage to a number of older bridge structures in California. One of the major causes of the failure of those bridges was the substandard detailing of the structural components designed before the current seismic design provisions had been adopted. The inadequate detailing of these structures has resulted in many bridges having columns with low flexural strength, low shear strength, and low flexural ductility. The inadequate starter bar lap lengths and insufficient lateral ties in these columns are the major contributors to their insufficiency to resisting earthquake forces.

The work of many researchers has indicated that increasing the confinement in the potential plastic hinge regions of the column will increase the compressive strength, ultimate compression strain and ductility of the core concrete. Therefore, strengthening techniques typically involve methods for increasing the confining forces either in the potential plastic hinge regions or over the entire column.[1]

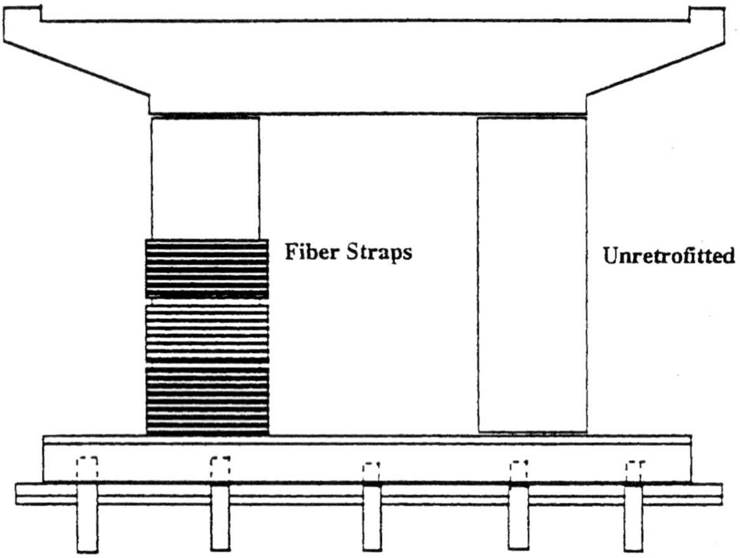

Figure 1. Column wrapped with composite straps.

In this chapter, a new technique for seismic strengthening of concrete columns is presented. Columns in existing structures are externally reinforced by means of high-strength fiber composite straps. The reinforcement is performed by wrapping straps of desired width and thickness around the columns. The straps can be wrapped in a continuous spiral and/or in discontinuous rings. The straps are constructed from high-strength fibers woven to form a flexible fabric-like material. The fabrics can be made very thin, resulting in sufficient flexibility for them to be wrapped around circular as well as rectangular columns. For improved structural performance as well as protection against environmental factors, the straps can be impregnated with resin either before or after the wrapping operation. The ends of the straps can be mechanically coupled or they can be epoxy-bonded to the column. Figure 1 shows typical concrete columns externally confined with fiber composite straps. This method is a spinoff of another study performed by the author and his colleagues, where concrete girders were strengthened by means of composite laminates bonded to the tension face of the girders.[2–4] In some of the girders tested, the ultimate strength was increased by a factor of four. The salient benefits of strengthening concrete columns with fiber composite straps are summarized in the following[5]:

Increased ductility

As a result of the confinement provided by the straps, the concrete will fail at a larger strain than if unconfined. Depending on the degree of confinement, significant increases in ductility can be achieved.

Increased strength

The lateral pressure exerted by the straps will increase the compressive strength of the concrete in both the core and shell regions, resulting in higher load-carrying capacity. The lateral confinement provided by the straps will also provide additional support against buckling of the longitudinal bars.

Circular and square sections

The flexibility of the straps allows wrapping around circular as well as rectangular columns.

Low maintenance

Because of their resistance to electrochemical deterioration, fiber composites do not corrode and they are not affected by salt spray and other aggressive environmental factors. Ultraviolet light, however, can adversely affect some fiber composites. This problem can be eliminated by providing a protective coating for the straps during or after the manufacturing process.

Low weight

The low density of composites (typically one-fifth that of steel) simplifies the construction procedure and reduces cost.

Temporary versus permanent

The proposed method will cause no disturbance to the integrity of the existing structure; i.e., no anchor bolts, dowels, etc., will be required.

Esthetics

The straps are very thin, i.e., less than 100 mils thick; therefore, they will not alter the appearance of the structure.

In this chapter, analytical models are presented for the analysis of concrete columns under monotonic loading and externally confined with fiber composite straps. A parametric study is also conducted to examine the effectiveness of this technique for increasing the strength and ductility of concrete columns. Furthermore, the results of an experimental study on circular and rectangular columns subjected to simulated seismic loading are also presented.

4.2. PREVIOUS WORK

External confinement with steel (steel jacketing) has been used to improve the shear strength and ductility of columns.

To investigate the performance of the columns retrofitted with steel jacketing, six large-scale column models were tested at the University of California at San Diego.[1] The columns were 0.4-scale models of a prototype 1524 mm (60 in.) diameter bridge column. They were 610 mm (24 in.) in diameter and 3657 mm (12 ft) in height. The test columns were constructed with a footing to allow foundation influence or interaction to be monitored. The tests included models with the pre-1971 reinforcing details without retrofitting, columns retrofitted with steel jackets, and a post-damage retrofitted column to determine whether a damaged column can be salvaged after an earthquake. The longitudinal steel reinforcement ratio was 2.53 percent. Transverse reinforcement consisted of circular hoops (No. 2 Grade 40 plain bars) placed at 127 mm (5 in.) on centers uniformly along the height of the column. The confining steel reinforcement ratio was 0.18 percent. The hoops were spliced with a lap length of 305 mm (12 in.) in the cover concrete. Steel jackets for the columns were fabricated from 4.76 mm (3/16 in.) thick A36 hot-rolled steel. A 6.3 mm (1/4 in.) gap was provided between the column and jacket. The gap was pressure-injected with water/cement grout.

From the test results, it was concluded that a lap length of 20 times the longitudinal bar diameter was insufficient to develop yield stress of the longitudinal bars; columns without retrofit degraded rapidly due to bond failure. The confinement provided by fully grouted steel jacket could completely contain the cover concrete to eliminate bond failure. Also, because there was only a 10 to 20 percent increase in lateral stiffness due to additional confinement from the steel jacket, a ductile mode of flexural failure with good energy dissipation could be achieved. The steel jacket enabled a displacement ductility factor of greater than 6 to be achieved.

Katsumata *et al.*[6], tested ten one-quarter-scale column specimens with square cross-sections of 200 $\times$ 200 mm (7.87 $\times$ 7.87 in.). The columns were strengthened with carbon fiber wraps before testing. The columns were tested under cyclic lateral loads and a constant axial load. It was concluded that winding of carbon fiber has ample effect on increasing seismic capacity. In particular, the following factors were determined: 1) ultimate displacement and energy dissipation was increased approximately linearly in accordance with carbon fiber quantity; 2) The earthquake-resistant capacity of a carbon fiber-strengthened column can be correlated roughly with an ordinary reinforced concrete column having only hoop reinforcement; and 3) carbon fiber quantity and steel hoop reinforcement quantity could be mutually convertible by an effective strength ratio.

Recently, a test conducted at the University of California, San Diego involved a new type of fiberglass-epoxy composite material[7], made of high modular and regular glass fibers, and wrapped like a blanket in layers around a 3657 mm (12 ft) high test column. The column size and reinforcement were representative of many of the columns built on the California State Highway system prior to 1971.

Two tests of retrofitted columns were conducted. For the first test, the column was wrapped with eight layers of the material [2.44 mm (0.096 in.) nominal thickness] and pressure-grouted to 1.76 MPa (250 psi) by pumping epoxy between the concrete and the wrap. The second involved wrapping a test column with four layers of the material pressurized to 1.06 MPa (150 psi). A vertical load of 1816 kN (400 kips) was applied to the column in each test.

Application of lateral load and displacement consisted of a series of cycles of imposed inelastic displacements from 25.4 to 254 mm (1 to 10 in.). The columns were pushed and pulled far beyond their yield point, which probably exceeds the displacement expected in maximum credible earthquakes. Even though the concrete crushed and some steel reinforcing bars ruptured in tension during the test, the fiber wrap remained intact.

4.3. FIBER COMPOSITES

The use of composites for a variety of industrial applications has been rapidly increasing in recent years. The main reasons for using these types of materials are their superior strength-to-weight ratio, and durability in corrosive environments, as compared with conventional materials. In addition to the superior strength properties, many composites have shown much better fatigue performance than structural metals.

Fiber composites have been used extensively in the aircraft and aerospace industries. The first recorded application of glass-fiber composites in the aircraft industry dates as far back as 1944.[8] Since then, a variety of composites have been used in other industries such as ship building, chemical processing, medical, automotive, etc.

Composites are made up of short fibers or filaments of glass, carbon, etc., bonded together with a resin matrix. The fibers provide the composites with their unique structural properties. The matrix serves as the bonding and protective agent.

Two types of resin-impregnated unidirectional composite straps will be used in the analytical part of this study; namely E-glass and carbon composites. In the experimental part of this study, only E-glass composite straps will be used. The following is a brief description of the mechanical properties of the two types of straps: E-glass and carbon fiber straps.

4.3.1. E-glass Strap

Glass fiber reinforced composites are among the oldest and least expensive of all composites.[5] Fiberglass is widely used in automotive, marine, sporting goods, and aerospace applications. E-glass is the most common type of glass fiber used in resin matrix composite structures. The individual fibers of E-glass have tensile strength in excess of 3.45 GPa (500 ksi).[9] These fibers range from 3 to 5 microns in diameter. The principal advantages of E-glass are low cost, high tensile and impact strengths, and relatively good chemical resistance. The disadvantages of E-glass, compared to other structural fibers, are lower modulus, lower fatigue resistance, and higher fiber self-abrasion characteristics. The modulus of elasticity of E-glass in the fiber form is 72.4 GPa (10.5×10^3 ksi). The density and coefficient of thermal expansion of E-glass are 2.5 g/cm^3 (0.092 lb/in.^{3}) and 5×10^{-6} cm/cm/C (2.8×10^{-6} in./in./F), respectively.

In general, fiber composites, in the fiber direction, behave linearly elastic to failure. The tensile strength and modulus of elasticity of composites, i.e., resin plus fiber, based on gross cross-sectional area, are smaller than the strength and modulus of the constituent fiber itself. The manufacturer provided the following information on unidirectional E-glass composite (resin plus fiber) tapes that could be used to wrap concrete columns[10]: tensile strength = 1103 MPa (160 ksi) and modulus of elasticity = 48.2 GPa (7×10^3 ksi). Figure 2 shows the stress-strain behavior of E-glass tapes used in the parametric study.

4.3.2. Carbon Fiber Strap

Carbon fiber is another suitable material for retrofitting of concrete columns. It is convenient to classify carbon fiber into four different performance groups, based on tensile modulus, strength, or precursor type.[5] They are standard modulus, intermediate modulus, high modulus, and pitch fibers.[11]

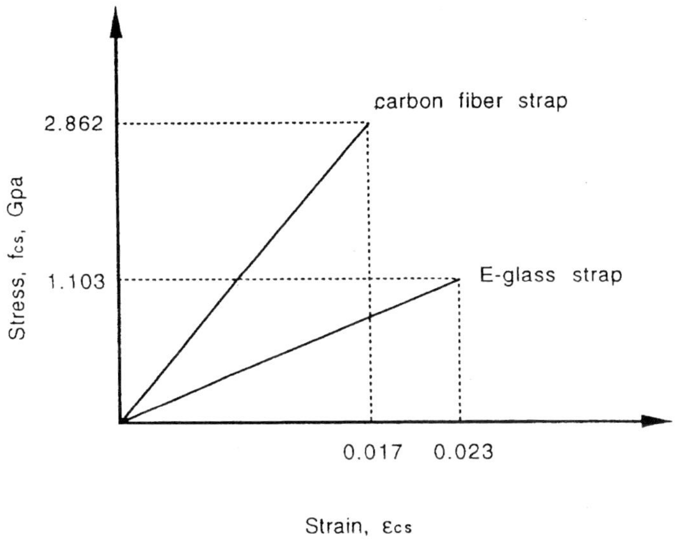

Figure 2. Stress-strain curves of composite straps.

The following is a brief description of the properties of those four groups.

Standard modulus

Standard modulus fibers have tensile strengths in the 2.4 to 3.1 GPa (350 to 450 ksi) range. Advances in fiber technology have brought about high-strength fibers, those with tensile strength greater that 3.445 GPa (500 ksi). Ultimate strain-to-failure ranges from 1.4 to 1.8 percent.

Intermediate modulus

Intermediate modulus fiber has typical modulus values ranging from 274 to 315 GPa (40 to 46 $\times$ 10^3 ksi). Ultimate strain-to-failure ranges from 1.5 to 2.0 percent.

High modulus

High modulus fiber's typical modulus values extend over 345 GPa (50 $\times$ 10^3 ksi). Ultimate strain-to-failure is low, ranging from 0.3–0.5 percent.

Pitch fiber

Pitch fibers generally fall into categories of high modulus, 345 to 485 GPa (50 to 70 × 10^3 ksi), or ultra high modulus, 485 to 825 GPa (70 to 120 × 10^3 ksi). Ultimate strain-to-failure ranges from 0.3 to 0.5 percent.

In order to obtain higher confinement pressure and ductility in concrete columns, unidirectional tapes of standard modulus carbon fiber will also be used in this study. The following information is obtained from the data provided by the manufacturer for intermediate carbon tapes[11]: tensile strength = 2.862 GPa (415 ksi); modulus of elasticity = 172 GPa (25 × 10^3 ksi). Figure 2 shows the stress-strain curve of the carbon tape used in the study.

4.4. ANALYTICAL MODELS

Stress-strain models for confined concrete, developed by Mander *et al.*[12], and based on an equation proposed by Popovics[13], are used in the analysis of circular and rectangular columns confined with composite straps. The models were included in a computer program developed to predict the ultimate moment and curvature at failure of the columns from pure compression to pure bending.[14] The following assumptions were made in the analysis: linear strain distribution through full depth of the cross-section; small deformations; no creep and shrinkage deformations; no shear deformation; no tensile strength for concrete; complete composite action between confining composite materials and concrete column, i.e., no slip; no confinement contribution from the original stirrups; and uniform confinement at corners and along sides of rectangular sections.

4.4.1. Stress-Strain Relationships of Materials

4.4.1.1. Concrete for circular columns

The stress-strain models of confined and unconfined concrete in circular sections in compression proposed by Mander *et al.*[12,15], and based on the work of several other researchers[13,16–18] will be summarized here and adopted for the analysis of concrete columns externally reinforced with fiber composite straps.[5] The stress-strain model shown in Figure 3 is based on an equation proposed by Popovics.[13] For a slow strain rate and monotonic loading, the longitudinal compressive concrete stress f_c is defined by the following equation:

$$f_c = \frac{f'_{cc} x r}{r - 1 + x^r} \quad (1)$$

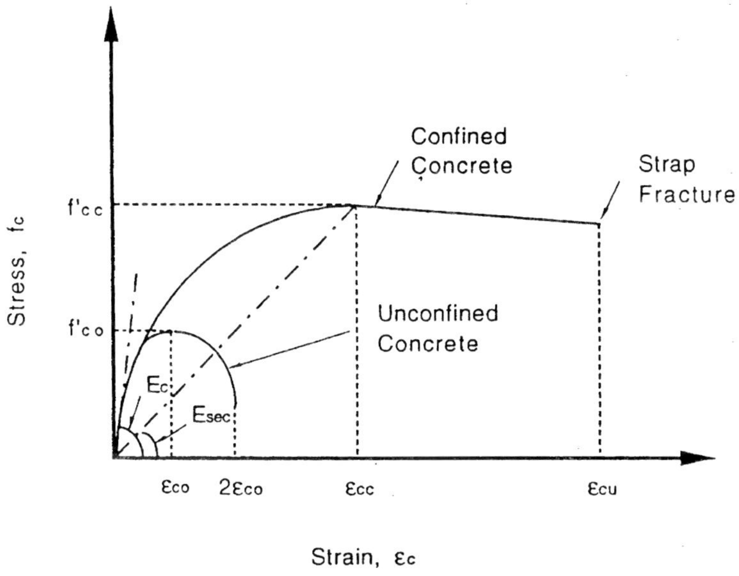

Figure 3. Stress-strain model proposed for unconfined and confined concrete.

where

$$x = \frac{\epsilon_c}{\epsilon_{cc}} \tag{2}$$

$$\epsilon_{cc} = \epsilon_{co}\left[1 + 5\left(\frac{f'_{cc}}{f'_{co}} - 1\right)\right] \tag{3}$$

$$r = \frac{E_c}{E_c - E_{\text{sec}}} \tag{4}$$

$$E_{\text{sec}} = \frac{f_{cc}}{\epsilon_{cc}} \tag{5}$$

$$f'_{cc} = f'_{co}\left(-1.254 + 2.254\sqrt{1 + \frac{7.94 f'_\ell}{f'_{co}}} - 2\frac{f'_\ell}{f'_{co}}\right) \tag{6}$$

where f_c' = compressive strength of confined concrete; f_o' = unconfined concrete strength; ϵ_c = longitudinal compressive strain of concrete; ϵ_{cc} = strain at maximum concrete stress, f_{cc}', of confined concrete; $\epsilon_{co} = 0.002$, strain at maximum concrete stress f_{co}' of unconfined concrete; E_c = tangent modulus of elasticity of concrete; E_{sec} = secant modulus of confined concrete at peak stress; and f_l' = effective lateral confining pressure from transverse reinforcement, assumed to be uniformly distributed over the surface of the concrete core.

Mander *et al.*[12] proposed an effective lateral confining pressure by transverse reinforcements on the circular concrete section. This effective pressure is defined as

$$f_\ell' = f_\ell k_e \tag{7}$$

where

$$k_e = \frac{A_e}{A_{cc}} \tag{8}$$

where f_l = lateral pressure from transverse reinforcement; k_e = confinement effectiveness coefficient; A_e = area of effectively confined concrete core; and A_{cc} = effective area of concrete enclosed by composite strap given by

$$A_{cc} = A_c(1 - \rho_{cc}) \tag{9}$$

where ρ_{cc} = ratio of area of longitudinal reinforcement to gross area of concrete; and A_c = area of concrete enclosed by composite strap.

A technique proposed by Sheikh and Uzumeri[19] is used to determine the area of effectively confined concrete between the composite straps. As shown in Figure 4, it is assumed that an arching action occurs between straps in the form of a second-degree parabola with an initial tangent slope of 45 deg. The concrete within this parabola is assumed to be ineffective. The smallest area of confined concrete occurs at midway between the straps and is calculated from:

$$A_e = \frac{\pi}{4}\left(d_s - \frac{s'}{2}\right)^2 = \frac{\pi}{4}d_s^2\left(1 - \frac{s'}{2d_s}\right)^2 \tag{10}$$

where s' = clear vertical spacing between straps; and d_s = diameter of column.

From Equation (8) through (10), the confinement effectiveness coefficient for circular sections can be calculated as

$$k_e = \frac{\left(1 - \frac{s'}{2d_s}\right)^2}{1 - \rho_{cc}} \tag{11}$$

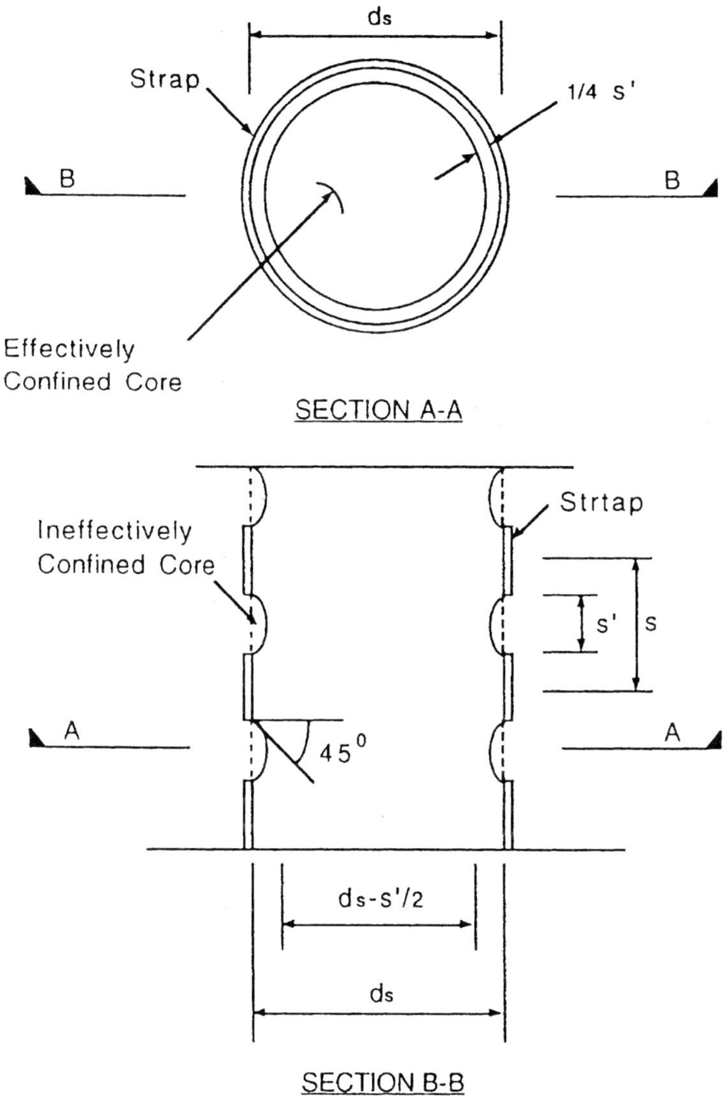

Figure 4. Confinement of circular column.

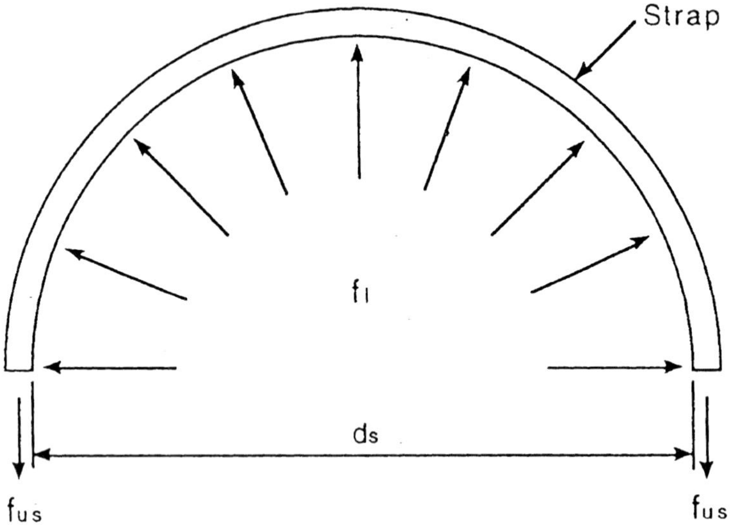

Figure 5. Confining action of composite strap.

The confining pressure induced on the concrete core by the composite strap is calculated by considering the free body of the circular cross-section confined by strap, as shown in Figure 5. The outward expansion of the core concrete is prevented by the action of the strap placed in circumferential tension. From equilibrium of forces, the confining stress f_ℓ can be calculated as

$$f_\ell = \frac{2 f_{us} A_{st}}{d_s s} \tag{12}$$

where f_{us} = ultimate strength of composite strap; A_{st} = cross-sectional area of strap; and s = width of strap. The volumetric ratio of confining strap to concrete core, ρ_s, is given by

$$\rho_s = \frac{A_{st} \pi d_s}{\frac{\pi}{4} d_s^2 s} = \frac{4 A_{st}}{d_s s} \tag{13}$$

Solving Equation (13) for the ratio, $A_{st}/d_s s$ and substituting into Equation (12) results in

$$f_\ell = \frac{1}{2} \rho_s f_{us} \tag{14}$$

The lateral pressure, f_ℓ, calculated from Equation (14) can be substituted into Equation (7) to determine the effective lateral confining pressure, f'_ℓ.

To calculate the longitudinal compressive strain of confined concrete at failure, ϵ_{cu}, the approach proposed by Mander *et al.*[12] and Scott *et al.*[18], based on an energy balance concept, is used. In this approach, the additional ductility available when concrete is confined is considered to be due to energy stored in the confining composite straps.

In Figure 3, for unconfined and confined concrete, the area under each stress-strain curve represents the total strain energy per unit volume of concrete at failure. The difference between these two areas is provided for by the confining effect of the composite strap, as given by the following equation

$$U_{st} = U_{cc} + U_{s\ell} - U_{co} \tag{15}$$

where U_{co} = ultimate strain energy per unit volume of unconfined concrete given by

$$U_{co} = A_c \int_0^{2\epsilon_{co}} f_{uc} d\epsilon_c \tag{16}$$

U_{sl} = energy required to maintain yield in longitudinal steel in compression given by

$$U_{s\ell} = \rho_{cc} A_c \int_0^{\epsilon_{cu}} f_{s\ell} d\epsilon_{s\ell} \tag{17}$$

U_{cc} = ultimate strain energy per unit volume of confined concrete

$$U_{cc} = A_c \int_0^{\epsilon_{cu}} f_c d\epsilon_c \tag{18}$$

U_{st} = ultimate strain energy per unit volume of composite strap given by

$$U_{st} = \rho_s A_c \int_0^{\epsilon_{us}} f_{st} d\epsilon_{st} \tag{19}$$

where ϵ_{us} = ultimate strain of composite strap; ϵ_{cu} = strain in concrete at the point when composite strap ruptures; f_{st} and ϵ_{st} = stress and strain in composite strap; and f_{sl} = stress in longitudinal reinforcement.

Substituting Equation (16) through (19) into Equation (15) and solving for ϵ_{cu}, the ultimate compression strain of concrete at the point of fracture of the confining composite strap can be calculated, resulting in complete determination of the stress-strain curve of the confined concrete throughout the entire range of loading, up to the fracture of composite strap and consequent failure of the column.

4.4.1.2. *Concrete for rectangular sections*

To extend the stress-strain relationship of concrete described in the previous section to that for rectangular cross-sections, it will be necessary to modify the effective lateral confining pressure, f'_ℓ. Assuming again that arching action occurs in the form of a second-degree parabola, as was shown in Figure 4 for circular columns, the area of effectively confined concrete core at midway between the levels of straps can be calculated from

$$A_e = \left(h - \frac{s'}{2}\right)\left(b - \frac{s'}{2}\right) = hb\left(1 - \frac{s'}{2h}\right)\left(1 - \frac{s'}{2b}\right) \qquad (20)$$

where b and h = cross-sectional dimensions.

Substituting Equation (9) and (20) into Equation (8) results in the confinement effectiveness coefficient for rectangular sections given by the following

$$k_e = \frac{\left(1 - \frac{s'}{2h}\right)\left(1 - \frac{s'}{2b}\right)}{1 - \rho_{cc}} \qquad (21)$$

The rest of the procedure for determining the effective lateral confining pressure, f'_ℓ, is similar to that for the circular columns and will not be repeated here.

The stress-strain behavior of steel is idealized as elastic-perfectly plastic.

4.4.1.3. *Composite straps*

Composite straps behave linearly elastic to failure. Figure 2 shows the stress-strain relationships for E-glass and carbon fiber composite straps used in this study.

The strain compatibility method was used to calculate the strains across the depth of the cross-section. Given the value of concrete strain in the extreme compression fiber, the steel strain in reinforcing bars can be described in terms of the concrete strain and depth of the neutral axis, c. The depth of the neutral axis, c, is then obtained by considering the equilibrium of forces across the cross-section. Knowing the location of the neutral axis, the strains in the concrete and steel reinforcing bars can be calculated from similar triangles in the strain diagram. The stresses in concrete and steel reinforcing bars may then be found from the stress-strain curves of concrete and steel. The forces in the concrete and steel reinforcing bars are calculated by multiplying their stresses by their corresponding areas. Knowing the forces in the concrete and steel reinforcing bars, the axial load is obtained from equilibrium of forces. The moment is calculated by multiplying the forces in the concrete and steel reinforcing bars by distances between their corresponding locations and the plastic centroid of the column. The curvature is obtained by dividing the strain in the extreme compression fiber of concrete by the depth of the neutral axis. A computer program was developed to carry out the numerical calculations.[14] Additional detailes on analytical procedures can be obtained from[5].

4.5. PARAMETRIC STUDY

A parametric study was conducted on the behavior of circular and rectangular columns strengthened with composite straps.[14]

The following are the major variable parameters used in the study.

Three values of unconfined concrete compressive strengths were used f'_{cc} = 20.67, 27.56, and 34.45 MPa (3000, 4000, and 5,000 psi).

Three strap thicknesses were used in this study: t = 5, 10, and 15 mm (0.197, 0.394, and 0.591 in.).

Three values of clear spacings were used: s' = 0.0, 152.4, 305 mm (0.0, 6.0, and 12.0 in.).

Two types of straps were used in the study: E-glass and carbon fiber. The stress-strain relationship to failure for these straps are shown in Figure 2.

Axial load-moment-curvature diagrams as well as plots of ductility factor ϕ_u/ϕ_y versus axial load ratio P/P_o, strap thickness t, and clear spacing between straps, s'; and plots of moment ratio, M_u/M_y versus strap thickness t were generated using different combinations of variable parameters discussed previously. The variables shown on the plots are: P = axial load; P_0 = ultimate axial load of unconfined column; ϕ_y = curvature at first yield of tension steel; ϕ_u = curvature at failure of confined column; M_y = moment at first yield of tension steel of unconfined column; M_u = moment at failure of confined column; t = thickness of composite strap; s' = clear spacing between straps. In these plots, each column is identified with an acronym, where the first symbol stands for the cross-section type, i.e., C = circular column and R = rectangular column; the second symbol stands for compressive strength of concrete, i.e., 3 indicates 20.67 MPa (3000 psi), 4 indicates 27.56 MPa (4,000 psi), and 5 indicates 34.45 MPa (5000 psi); the third symbol stands for thickness of strap, i.e., T5 indicates t = 5 mm (0.197), T10 indicates t = 10 mm (0.394 in.), and T15 indicates t = 15 mm (0.591 in.); and the fourth symbol stands for clear spacing between straps, i.e., S′0 indicates s' = 0.0 mm (0.0 in.), S′18 indicates s' = 152.4 mm (6 in.), and S′12 indicates s' = 305 mm (12.0 in.).

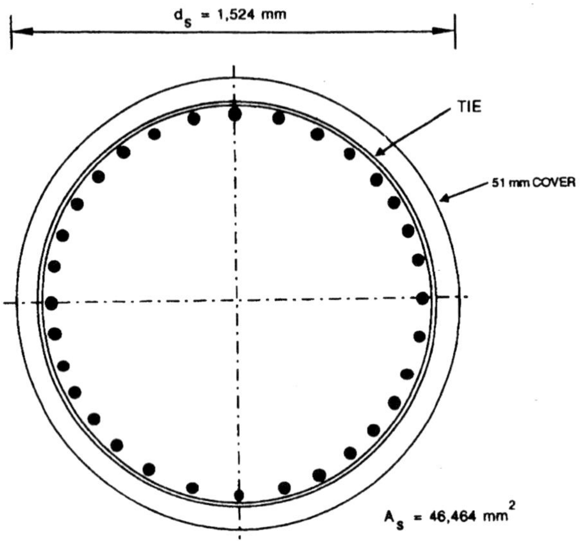

Figure 6. Cross-section and reinforcement details of circular column used in parametric study.

4.5.1. Circular Columns

A prototype bridge column was used in the parametric study.[1] Figure 6 shows the reinforcement details of the column. A 152 mm (6 in.) wide strap was used for retrofitting. Grade 60 steel reinforcing bar was used for the longitudinal reinforcement.[14]

Figures 7(a) through 7(c) show the plots of the interaction diagrams and curvature ductility for the column for three different concrete compressive strengths. The interaction diagrams and curvature ductility curves for the same section without strap are shown with dotted lines in each graph. From the interaction diagrams in the figures, it can be seen that the ultimate axial load is increased by 103, 92, and 82 percent for the column strengthened with E-glass strap, and 171, 162, and 151 percent for the column strengthened with carbon fiber strap, respectively, compared with the strength of unretrofitted column. The maximum moment capacity is increased by 53, 48, and 45 percent for the column strengthened with E-glass strap, and 87, 83, and 79 percent for the column strengthened with carbon fiber strap, respectively, compared with the moment capacity of unconfined column. The increase in maximum moment capacity is less than that in the ultimate axial load.

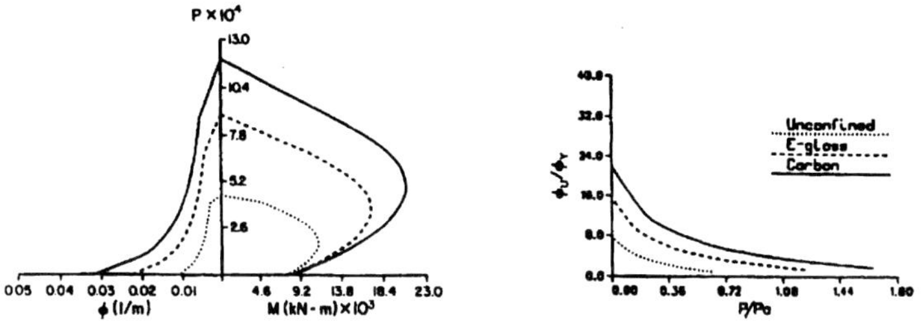

a) Column C3T5S′O

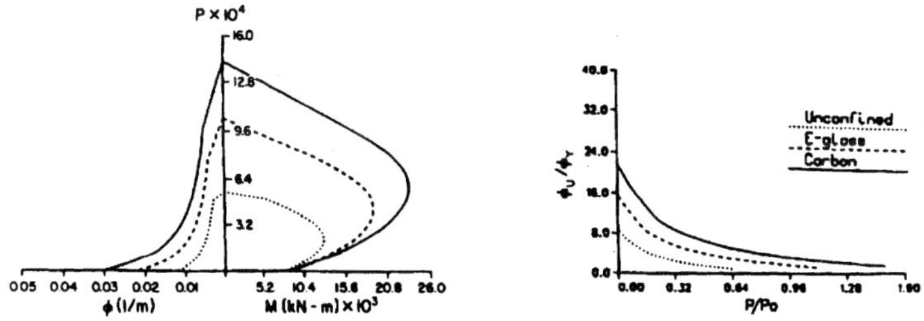

b) Column C4T5S′O

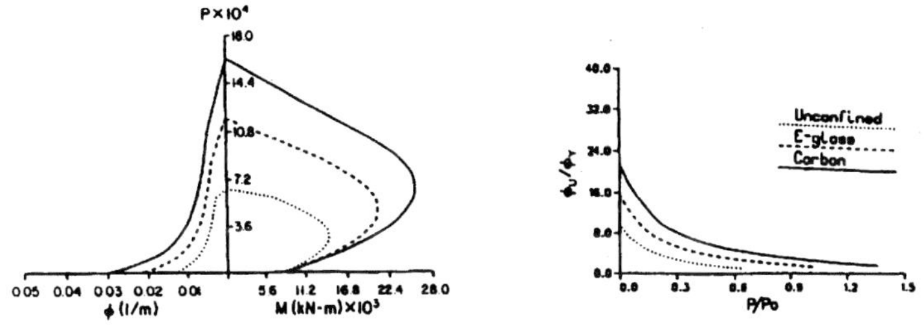

c) Column C5T5S′O

Figure 7. Interaction diagram and ductility factor of circular column for three different concrete compressive strengths.

This is not a shortcoming of this strengthening technique, since the majority of columns requiring strengthening lack adequate ductility, not flexural capacity. In fact, it is desirable to increase the ductility without increasing the moment capacity to prevent a brittle failure. From the plots of curvature ductility, it can be seen that the ductility factor ϕ_u/ϕ_y increases significantly as a result of confinement provided by the composite strap. However, this value does not change appreciably for the three different concrete compressive strengths.

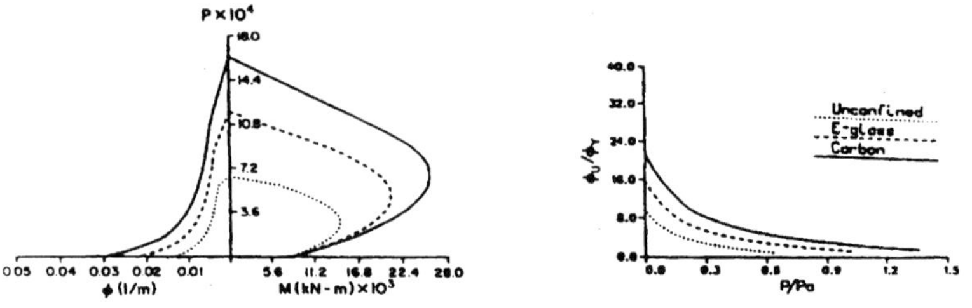

a) Column C5T5S'O

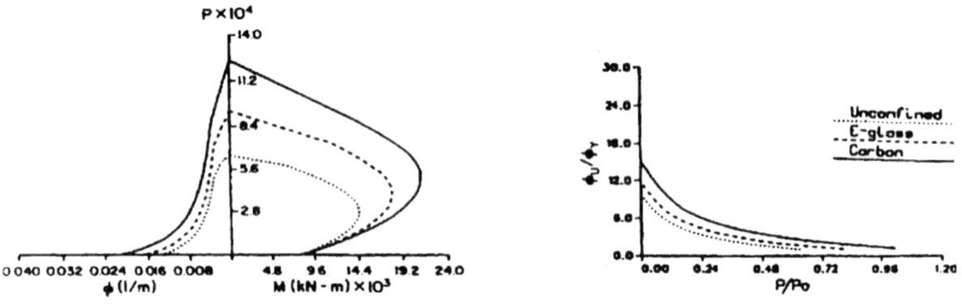

b) Column C5T5S'6

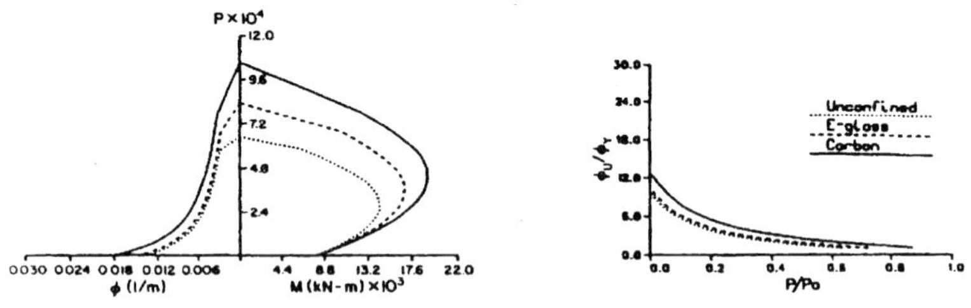

c) Column C5T5S'12

Figure 8. Interaction diagram and ductility factor of circular column for three different spacings between straps.

Figures 8(a) through 8(c) show the plots of the interaction diagrams and curvature ductility for the three different values of clear spacing between straps. A 152 mm (6 in.) wide strap was used for these figures. As can be seen from the figures, the ultimate axial load is increased by 82, 44, and 29 percent for strengthening with E-glass strap, and 151, 92, and 63 percent for strengthening with carbon fiber strap, respectively, compared with the values prior to strengthening. The maximum moment capacity is increased by 45, 25, and 17 percent for strengthening with E-glass strap, and 79, 47, and 33 percent for strengthening with carbon fiber strap, respectively.

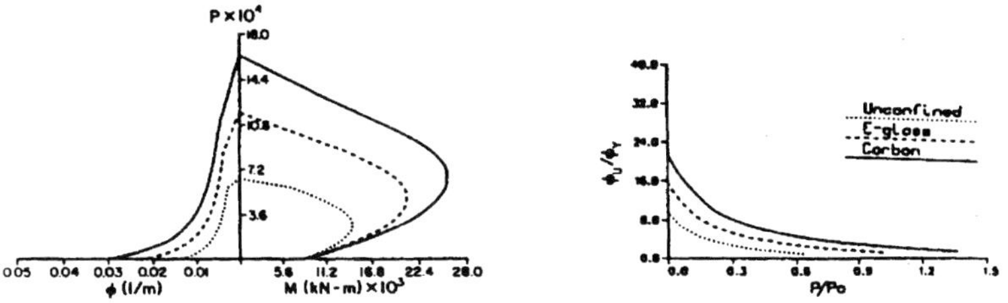

a) Column C5T5S′0

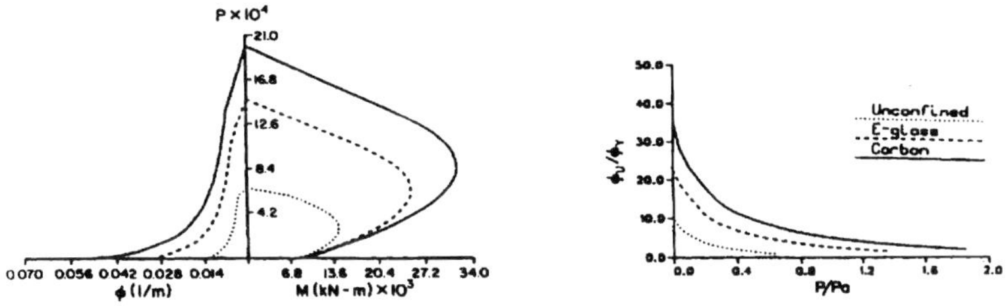

b) Column C5T10S′O

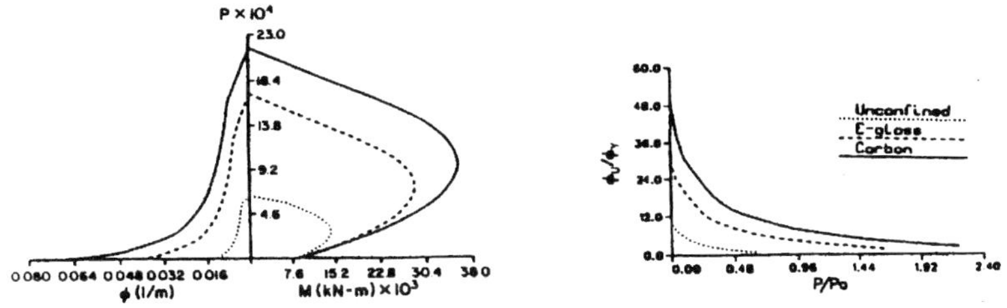

c) Column C5T15S′O

Figure 9. Interaction diagram and ductility factor of circular column for three different strap thicknesses.

As before, the increase in the maximum moment capacity is less than that in the ultimate axial load. From the plots of curvature ductility, it can be seen that larger strap spacings decrease the ductility factor significantly. In Figure 8(c), when a clear spacing of $s' = 304.8$ mm (12.0 in.) is used, the ductility factor of the section strengthened with E-glass is almost the same as that before strengthening.

Figures 9(a) through 9(c) show the plots of interaction diagrams and curvature ductility for the section for different thicknesses of the E-glass and carbon fiber straps. The figures show that the ultimate axial load is increased by 82, 130, and 163 percent for the section strengthened with

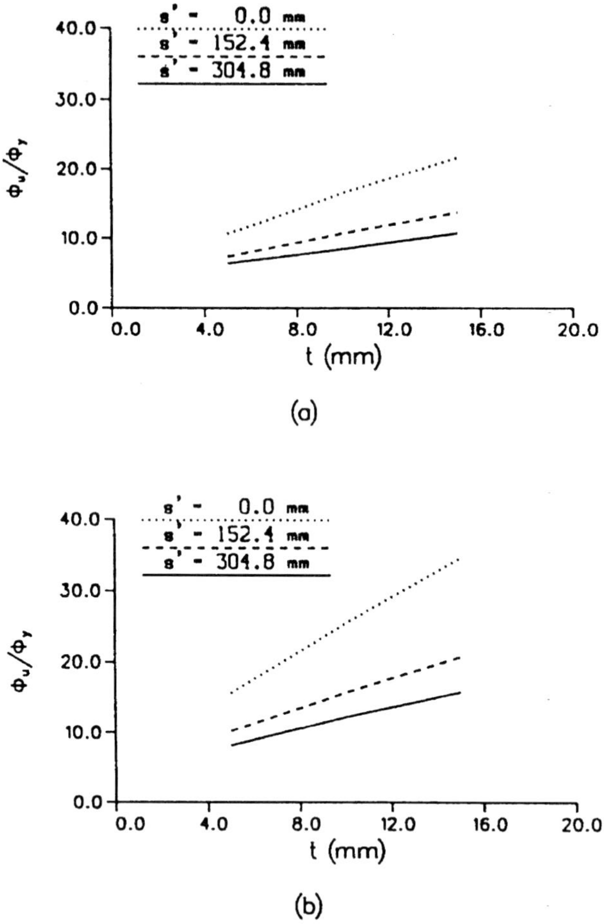

Figure 10. Ductility versus thickness for circular column [f'_{co} = 20.67 MPa (3000 psi), $P/P_o = 0.1$]: (a) E-glass; (b) carbon fiber.

E-glass strap, and 151, 208, and 235 percent for the same section strengthened with carbon fiber strap, respectively, compared to the ultimate axial load before strengthening. The maximum moment capacity is increased by 45, 75, and 98 percent for E-glass strap, and 79, 120, and 148 percent for the carbon fiber strap, respectively. From the plots of the interaction diagrams and curvature ductility, it can be seen that the ductility factor increases significantly as the thickness of strap increases.

Figure 10 shows the relationship between ductility factor ϕ_u/ϕ_y and strap thickness t for E-glass and carbon fiber straps for a typical retrofitted column.

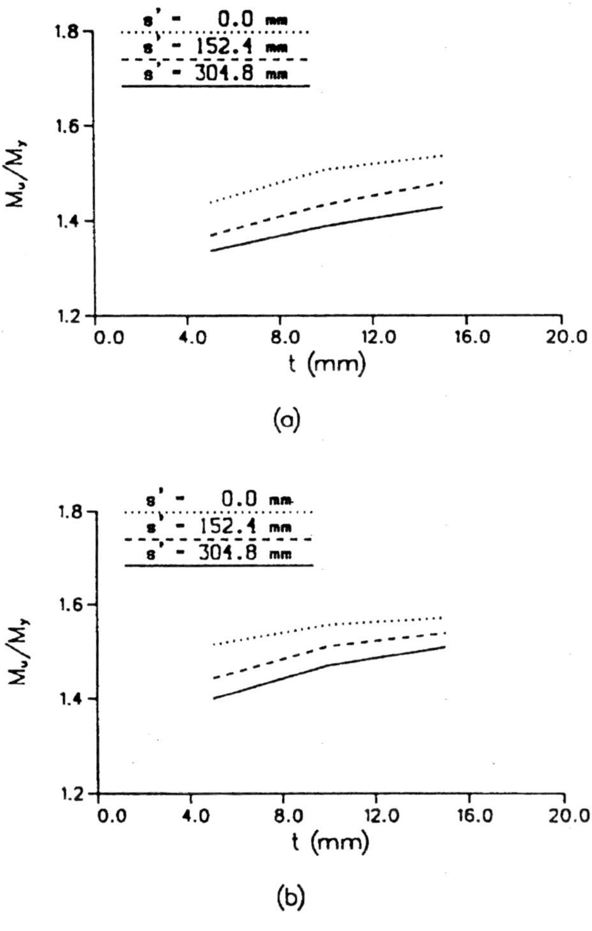

Figure 11. Moment versus thickness for circular column [f'_{co} = 20.67 MPa (3000 psi, $P/P_o = 0.2$]: (a) E-glass; (b) carbon fiber.

As can be seen from these graphs, the ductility factor is increased almost linearly with increasing thickness of the strap. However, the rate of increase of ductility factor decreases with an increase in the clear spacing between straps. A similar trend was also observed for other columns with different f'_{co} and P/P_o; however, the ductility ratio ϕ_u/ϕ_y decreases for higher concrete compressive strengths.

Figure 11 shows the relationship between the moment ratio M_u/M_y and strap thickness t where M_y is the moment of unconfined column at first yield

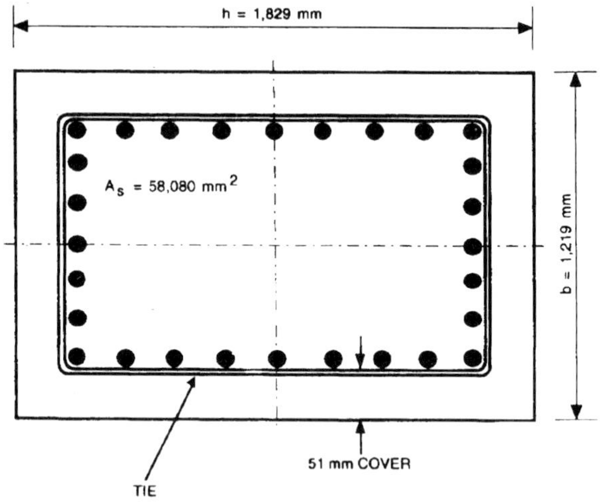

Figure 12. Cross-section and reinforcement details of rectangular column used in parametric study.

of tension steel, and M_u is the ultimate moment of confined column. From this graph, it can be seen that, as the thickness of the strap increases, the moment ratio is also increased. The rate of increase in the moment ratio, or the slope of the curve, decreases slightly as the strap thickness increases.

4.5.2. Rectangular Columns

Figure 12 shows the cross-section of the rectangular column used in the parametric study. Grade 60 longitudinal steel reinforcement was used in the analysis of this column. Figures 13(a) through (c) show the interaction diagrams and curvature ductility curves for a typical retrofitted column. Figure 14 shows the relationship between the ductility factor ϕ_u/ϕ_y and thickness t.[14] As can be seen from these figures, the same benefits as those discussed for circular columns can be observed for rectangular columns strengthened with high-strength composite straps. The plots of the curvature ductility versus strap thickness also indicates that the ductility increases as the strap thickness increases, although, however, at a slower rate for wider spacings of the strap, as indicated by the smaller slopes as spacing increases.

To compare the effectiveness of E-glass and carbon fiber straps for retrofitting of concrete columns, the column shown in Figure 6 was analyzed

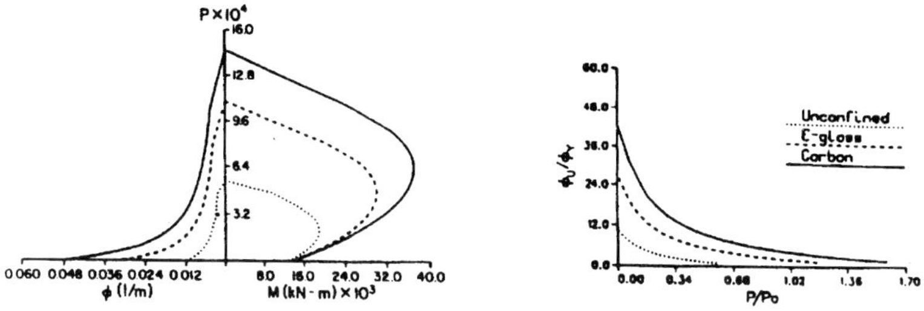

a) Column R3T5S´O

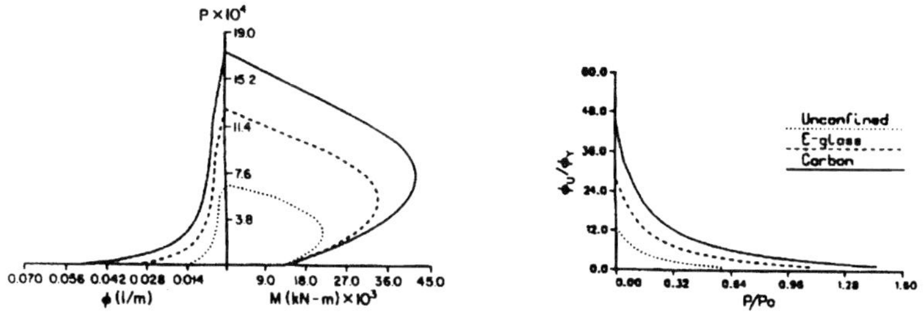

b) Column R4T5S´O

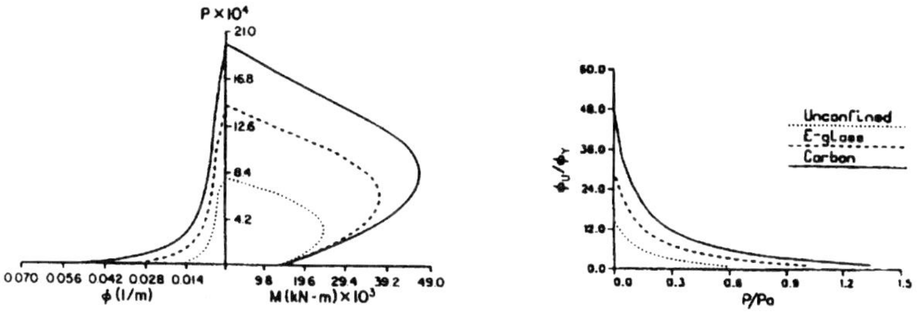

c) Column R5T5S´O

Figure 13. Interaction diagram and ductility factor of rectangular column for three different concrete compressive strengths.

for a case where the confining force of the strap was kept constant. The strap was 152 mm (6 in.) wide and fully confined the column, that is, the clear spacing between straps was zero. Due to the higher strength of the carbon fiber strap, and to keep the confining force constant, a larger thickness was used for the E-glass strap. The thickness of the E-glass strap was 25 mm (1 in.), and the thickness of the carbon fiber strap was 10 mm (0.4 in.).

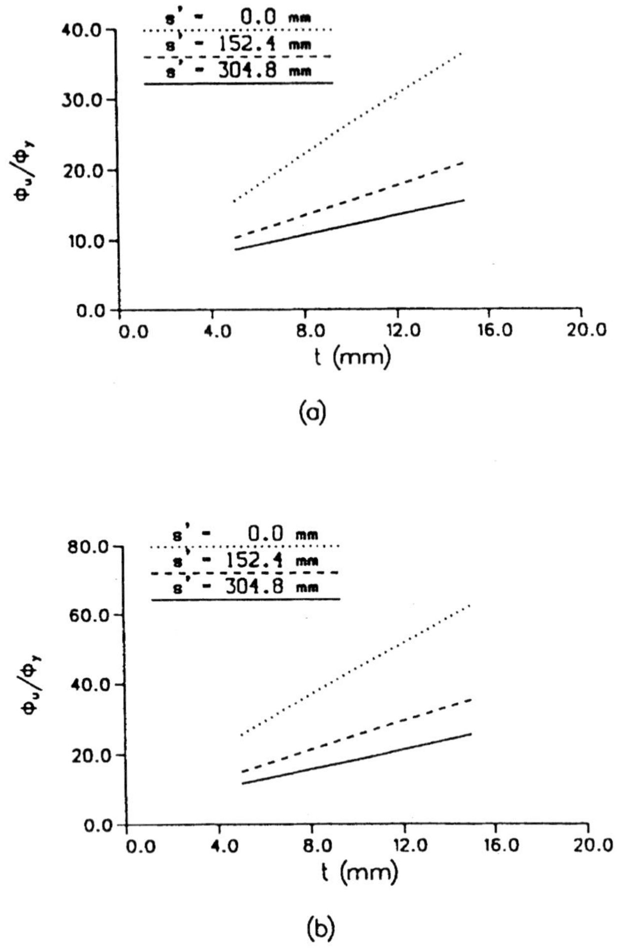

Figure 14. Ductility versus thickness for rectangular column [$f'_{co} = 20.67$ MPa (3000 psi), $P/P_o = 0.1$]: (a) E-glass; (b) carbon fiber.

Figure 15 shows, with the dashed line, the interaction diagram and curvature ductility curve of the column confined with E-glass strap. The solid and dotted lines represent curves belonging to the column confined with carbon fiber strap and the column prior to strengthening, respectively. Comparing the solid and dashed lines, it can be seen that confinement with E-glass results in a slight increase in both strength and ductility beyond those

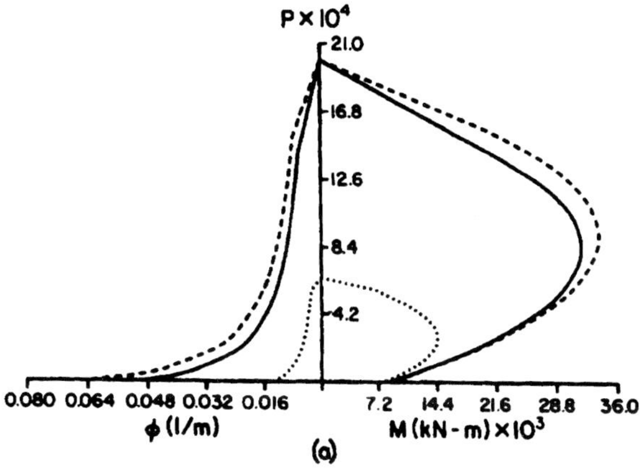

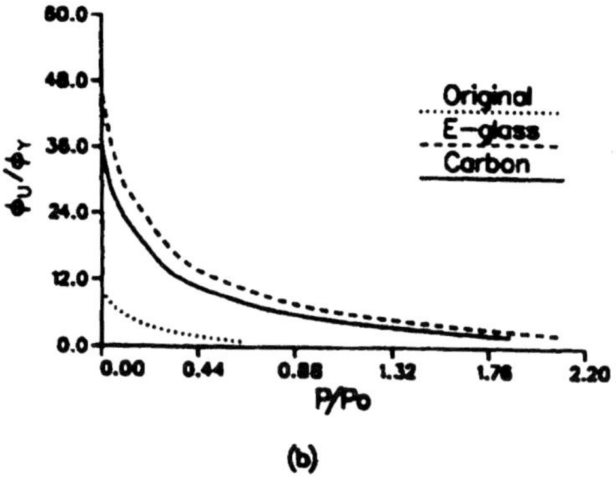

Figure 15. Interaction diagram and ductility factor of column C5 × S′O for E-glass and carbon fiber straps with the same tensile strength.

for confinement with carbon fiber strap. Considering that the cost of carbon fiber is significantly more than that of E-glass, well in excess of the thickness ratio of 2.5, E-glass straps appear more promising for field application of this technique.

4.6. EXPERIMENTAL STUDY OF CIRCULAR COLUMNS

4.6.1. Test Specimens

The test results for five scaled-down, circular reinforced concrete bridge column footing assemblages are reported here.[20] The test specimens were

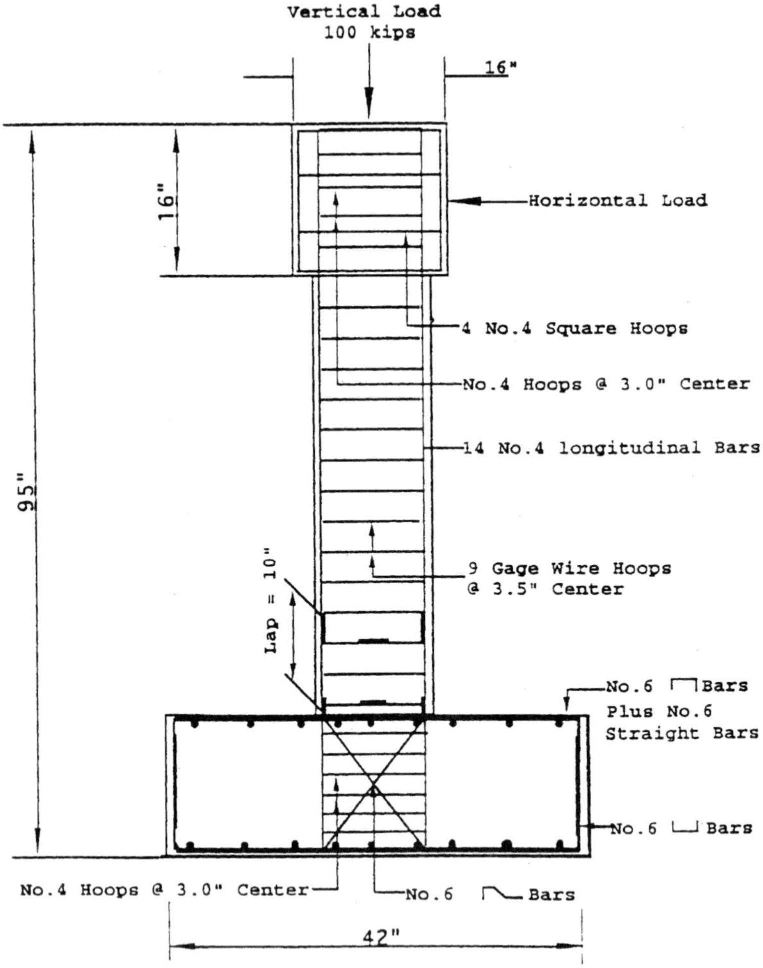

Figure 16. Geometric details of column specimens.

designed to approximately model typical pre-1971 design of existing highway bridge columns in a zone of high seismic risk. Each specimen consisted of a single column bent with strong footing details, as shown in Figure 16. The composite strap was applied only in the potential plastic hinge region; i.e., in the 635 mm (25 in.) long portion of the column above the top face of the footing.[21] Figure 17 shows a typical column specimen wrapped with the composite straps in the plastic hinge region prior to the beginning of the test.

Figure 17. Column wrapped with composite straps in plastic hinge region. (See Color Plate I)

Both active and passive retrofit methods were tested in the laboratory to show the effectiveness of different retrofitting schemes for enhancing the shear and flexural behaviors. For the passive retrofit scheme, the composite straps with fiber orientation in the circumferential direction were directly wrapped on to the column in the region of the reinforcement lap splices and/or the potential plastic hinge zone, as shown in Figure 18(a). In this case, as the concrete expands outward, tensile stresses are gradually developed

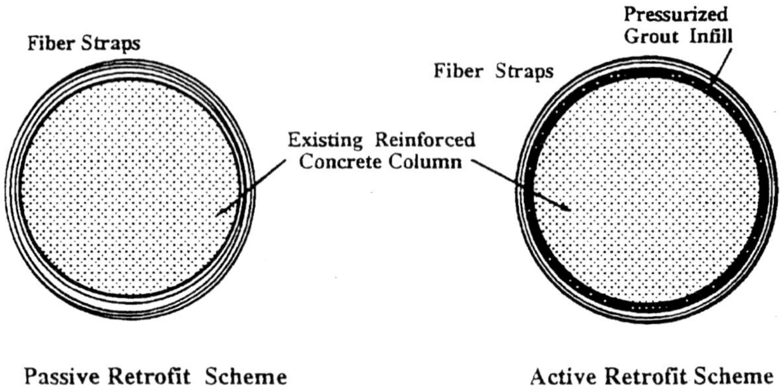

Figure 18. Active and passive retrofit schemes.

in the composite straps which will resist this expansion; thus, the system is referred to as passive confinement. For the active retrofit scheme, the composite straps were slightly oversized for the column and the resulting gap between the column and the straps was initially injected with pressurized epoxy resin, as shown in Figure 18(b). Spacers were provided between the column and the straps to maintain a uniform gap. This scheme of retrofitting induces initial tensile stresses in the straps. As a result, an active pressure is created around the column. This pressure reduces the amount of radial dilation and cracking of the core concrete.

The materials used in the construction of the column specimens included: concrete with $f_c' = 34.5$ MPa (5000 psi) and Grade 40 steel. Although the specific concrete compression strength for the prototype bridge column was 21 MPa (3000 psi), ready-mix concrete providing $f_c' = 34.5$ MPa (5000 psi) was used to include the effect of expected overstrength resulting from normal conservative mix design and strength gain with concrete aging.[21] The actual compression strength of the concrete for each specimen is given in Table 1.

The diameter of the columns was 305 mm (12 in.). These columns were reinforced longitudinally with 14 No. 4 bars [diameter = 13 mm ($\frac{1}{2}$ in.)], resulting in a longitudinal reinforcement ratio of 2.48%. The measured yield strength for these bars was 358 MPa (53 ksi). Transverse confinement was provided by 9-gage steel wire [diameter = 3.5 mm (0.135 in.)] hoops spaced at 89 mm (3.5 in.) on center throughout the entire height of the column. The average yield stress for these wires was 301 MPa (43.7 ksi). The only difference among specimens was in the detail of anchorage of

Table 1. Details of column specimens

Specimen	Measured concrete strength (MPa)	Longitudinal steel detail	Retrofit
C-1	36.5	Starter Bars	None
C-2	38.3	Starter Bars	Passive
C-3	38.5	Starter Bars	Active
C-4	36.6	Continuous Bars	None
C-5	36.5	Continuous Bars	Passive

column longitudinal steel reinforcement and the type of confinement, i.e., passive or active (Table 1). For specimens C-1, C-2 and C-3, the longitudinal reinforcement of the column was extended into the footing using starter bars which were lapped with the main longitudinal reinforcement of the column over a length of 20 bar diameters, i.e., 254 mm (10 in.). Columns C-2 and C-3 were strengthened using the passive and active retrofit schemes, respectively. For specimens C-4 and C-5, the column reinforcement extended into the footing and was anchored with a standard 90° hook. Column C-4 was used as the control specimen, while column C-5 was retrofitted with the passive scheme.

The composite straps constructed for this project were 0.8 mm (0.03 in.) thick and had a tensile strength and modulus of elasticity of 532 MPa (77 ksi) and 18.6 GPa (2700 ksi), respectively. The stress-strain relationship of the composite strap was linear elastic to failure. For both active and passive retrofit schemes, the columns were wrapped with six-layers of composite straps, resulting in a total thickness of 5 mm (0.2 in.) within the potential plastic hinge zone of the column, i.e., from the top of the footing to 635 mm (25 in.) above it. As the strap was wrapped around the column, an epoxy was applied to its surface and the multiple layers of the strap were adhered together to form a single composite wrap with the desired thickness. The epoxy was provided to ensure that the multiple layers of the straps act as one unit; with no interlaminar slippage.

4.6.2. Test Set-Up and Instrumentation

The test setup was designed for testing column-footing assemblages subjected to combined axial and lateral loadings.[20] The specimens were tested in a steel reaction frame as shown in Figure 19. Two independent loading systems were used to apply the load to the specimens. First, the axial load of 445 kN (100 kips) was applied to the column by prestressing a pair of 25 mm (1 in.) diameter high-strength steel rods against the base beams of the test

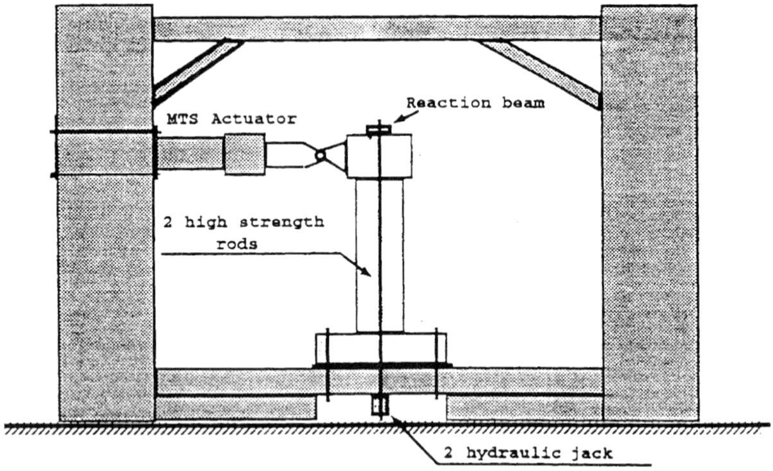

Figure 19. Test setup.

frame, which was bolted to the 915 mm (3 ft) thick concrete floor. This load was applied to simulate the dead load on the columns. Next, the reversing lateral forces were applied to the column by an MTS 489 kN (110 kips) hydraulic actuator mounted on the reaction frame. The actuator was capable of moving the top of the specimen 127 mm (5 in.) in both positive and negative directions. A displacement of 127 mm (5 in.) corresponds to a drift of approximately 7%. Each column was instrumented to monitor the applied displacement and corresponding loads, strains and deformations. Four types of instruments were used to measure the various quantities. These included:

(a) Load cell and displacement transducer of the actuator;
(b) Electrical inclinometers mounted on the potential plastic hinge region of the column to measure rotation;
(c) Displacement transducers installed on the steel reference frame;
(d) Electrical-resistance strain gages bonded to reinforcing bars; and
(e) Photographic records of the column cracking and failure modes.

The effect of an earthquake on the column specimen was simulated by reversed cyclic loading. The hydraulic actuator in the test set up was used to displace the top of the column to achieve a pre-determined load level or displacement. Then, the loading direction was reversed to achieve the same load or displacement level in the opposite direction. The load and

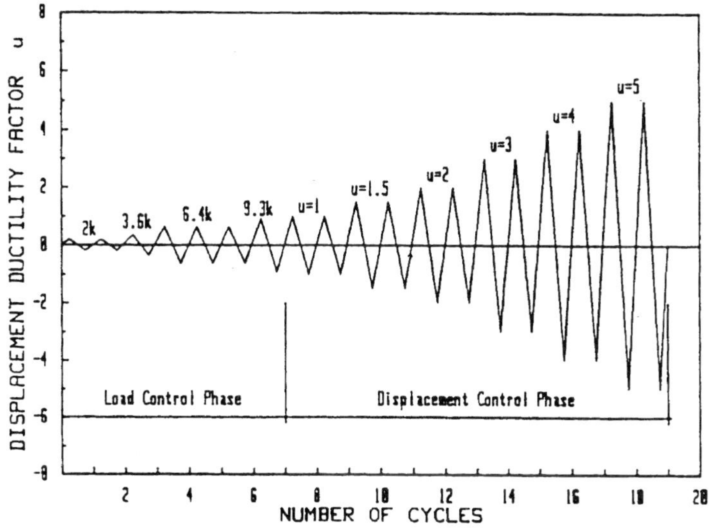

Figure 20. Loading sequence.

displacement input history was broken into two phases. At the initial stage, the test was in a load control mode. Upon yielding of the longitudinal rebars, a displacement control mode of loading was utilized. Figure 20 shows the loading sequence. The letter "u" in this figure indicates the displacement ductility factor defined as the ratio of the applied displacement over the displacement at first yielding of the longitudinal rebars. The hysteresis loops of applied lateral loads vs. the column free end displacement at the point of application of the load were continuously plotted and updated on the computer screen during the test.

Ten electrical inclinometers were distributed at 127 mm (5 in.) spacing, over both opposite faces of the column within the plastic hinge region, from the top of the footing up to a height of 635 mm (25 in.), as shown in Figure 21. The data from these inclinometers were used to measure the plastic hinge rotations at critical sections of the column and to plot moment-curvature relationship of these sections. Along the height of the column, four displacement transducers were used to monitor the column deflection, as shown in Figure 21.

The strains in the reinforcing bars at the locations within the plastic hinge zones of the column were measured by means of electrical-resistance strain gages bonded to the steel bars. Twelve strain gages were bonded to the column bars and hoops. In addition, for each retrofitted column, a total of twelve strain

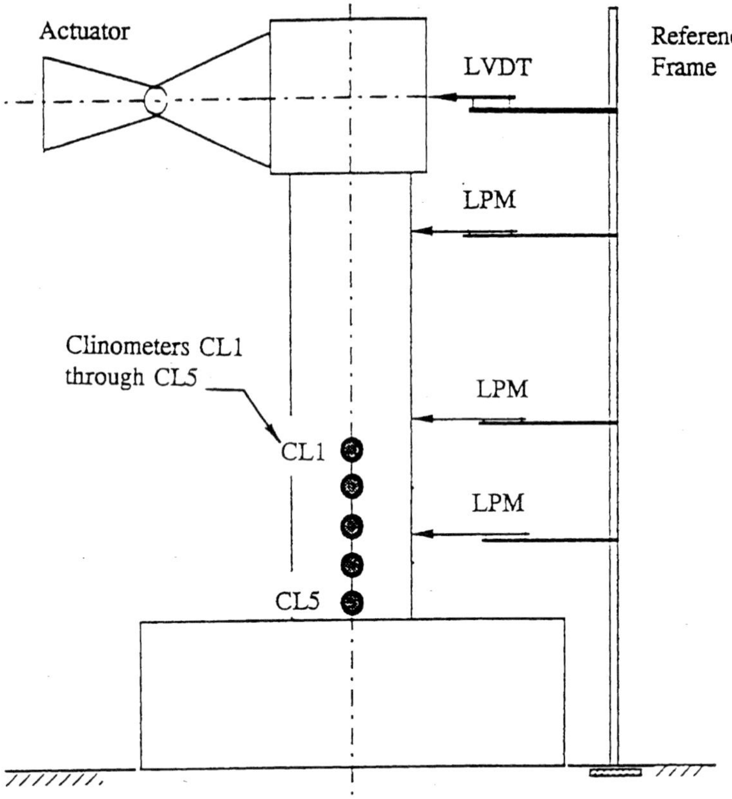

Figure 21. Clinometer and transducer arrangement.

gages were attached to the composite straps to measure the hoop strains in the straps.

During each loading cycle, testing was stopped for a few seconds at several points while all data were scanned through an HP 3947 data acquisition system.

4.6.3. Load vs. Displacement Response

Plots of the lateral load versus displacement for all five specimens are shown in Figures 22 through 26.[20] It should be noted that these figures have been plotted to the same scale. Because the columns were symmetrically reinforced

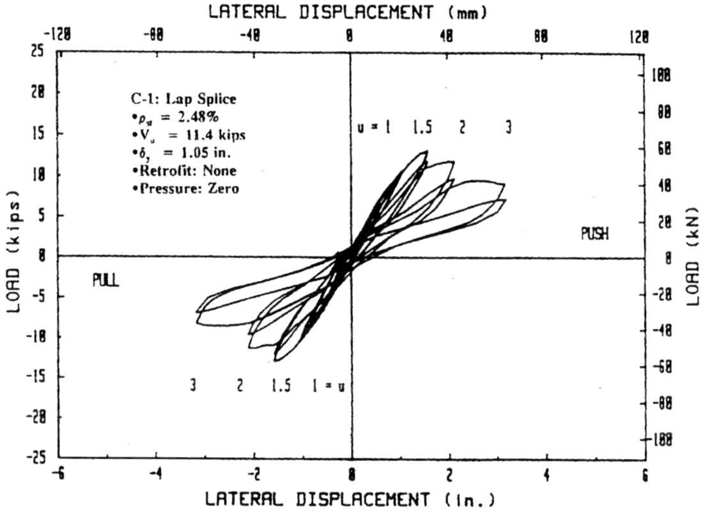

Figure 22. Load vs. displacement response of column C-1.

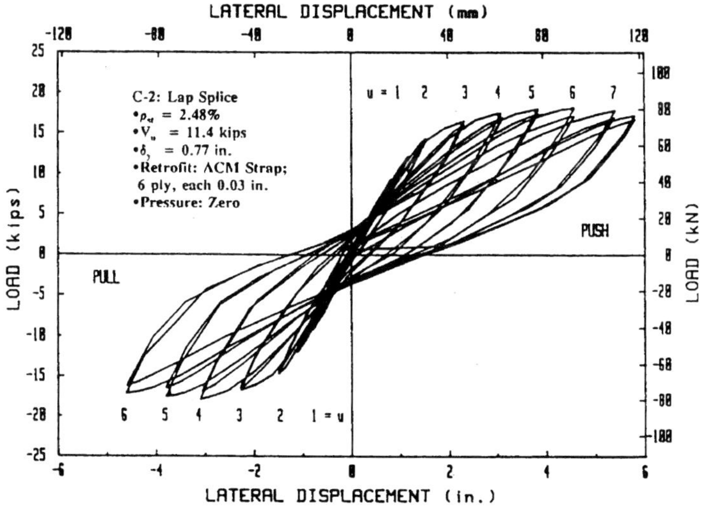

Figure 23. Load vs. displacement response of column C-2.

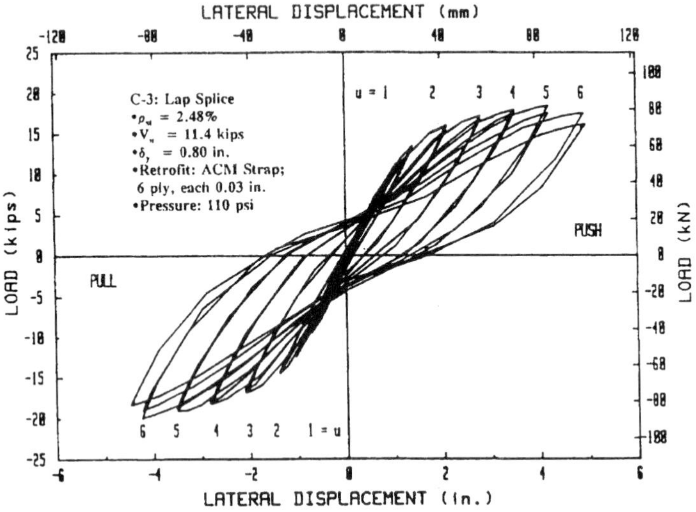

Figure 24. Load vs. displacement response of column C-3.

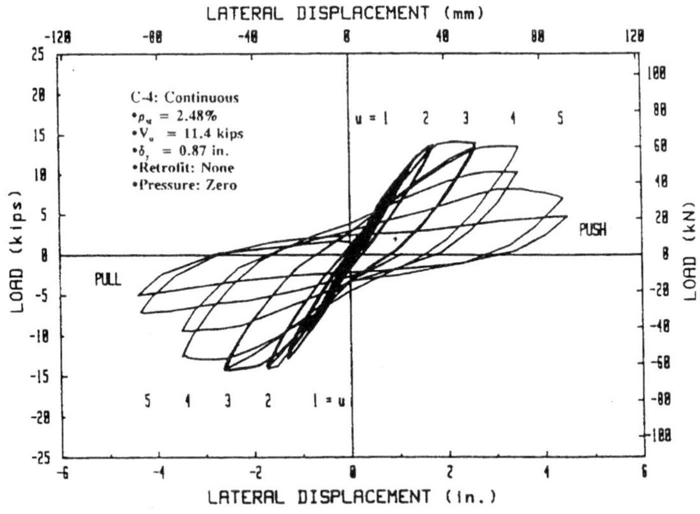

Figure 25. Load vs. displacement response of column C-4.

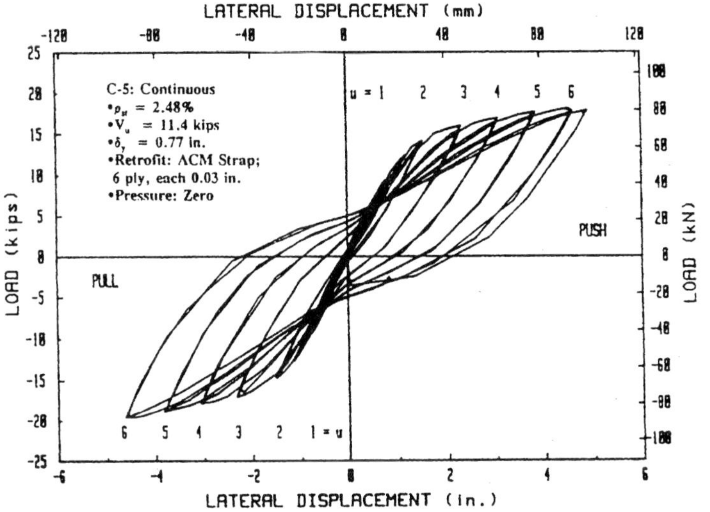

Figure 26. Load vs. displacement response of column C-5.

with respect to the positive and the negative sides for each specimen, the resulting positive and negative portions of each specimen's hysteresis loops should be symmetrical and identical in values. However, due to a limitation of the hydraulic actuator, the maximum strokes reached in the positive and the negative directions were slightly different for some of the test specimens.[21] This difference affected the symmetries of the hysteresis loops.

In Figures 22 through 26, V_u is the lateral force corresponding to the theoretical flexural capacity of the unconfined column section, δ_y is the yield displacement which was used as a reference value to determine each subsequent displacement ductility level that was applied to the specimen during the test, and ρ_{sl} is the longitudinal reinforcement ratio of the column. The measured and calculated maximum strengths for the retrofitted and unretrofitted columns are summarized in Table 2. The calculated strength was determined using the procedures outlined in the ACI Code and taking into consideration the actual strain hardening properties of the reinforcing steel. The percent increase in strength was calculated with reference to the measured value for the control specimen.

Figure 22 shows that the hysteresis loops for the unretrofitted circular column C-1 with lap splice reinforcement degrade rapidly after the first cycle to $u = 1.5$, due to the failure of the lapped reinforcement. The maximum lateral load of 58.3 kN (13.1 kips) was recorded during the push cycle to

Table 2. Measured and Calculated Strength of the Columns

Specimens	Calculated lateral load using ACI (kN) (unretrofitted)	Measured maximum lateral load (kN) (retrofitted)	Increase in strength resulting from retrofitting
C-1	50.7	58.3	Control
C-2	50.7	81.4	40%
C-3	50.7	89.4	53%
C-4	50.7	71.6	Control
C-5	50.7	87.2	22%

$u = 1.5$. The lateral responses of circular columns C-2 and C-3, strengthened with six plies of composite straps, applied with the passive and active retrofit schemes, respectively, show a significant improvement with stable hysteresis loops up to the displacement ductility level of $u = 6$, as shown in Figures 23 and 24. There was no sign of structural degradation of retrofitted columns associated with the bond failure of lap-spliced bars. The maximum strength of 81.4 kN (18.3 kips) was noted for column C-2. This value was approximately 40 percent higher than that of the reference column C-1. The increase in the maximum lateral load for column C-3 was higher than that for column C-2 perhaps due to the active confinement pressure.

Lateral load-displacement hysteresis loops for the two circular columns with continuous reinforcement are plotted in Figures 25 and 26. The loops for the control specimen, C-4, in Figure 25, show that the lateral strength did not decay until displacement ductility level of $u = 4$. The maximum lateral load-carrying capacity was 71.6 kN (16.1 kips) at the displacement ductility level of $u = 3$. However, the specimen showed rapid deterioration in its strength at $u = 5$, due to concrete failure and longitudinal bar buckling. Figure 26 shows hysteresis curves for column C-5 with the passive retrofit scheme. These curves are very similar to those of columns C-2 and C-3, with significant improvement in strength and excellent energy absorption and dissipation characteristics.

4.6.4. Curvature vs. Height

Ten electrical inclinometers were mounted along the centerline of each column on opposite faces (East and West), as shown in Figure 21. The plastic rotation of specific sections of the column was then obtained from the readings of each inclinometer. An average curvature at the section was calculated from[20]

$$\phi = \frac{\frac{(\theta_E + \theta_W)}{2}}{h_v} \tag{22}$$

where θ_E and θ_W are the East and West side rotations of the cross-section with respect to adjacent inclinometers, and h_v is the vertical distance between the adjacent inclinometers.

The vertical distributions of curvature within the plastic hinge region for the three circular columns with lapped starter bars are shown in Figures 27 through 29. At the location of each of the inclinometers CL1 through CL5, shown in Figure 21, the curvature was calculated and the corresponding points were connected by straight lines to show an approximate variation of the curvature within the plastic hinge region. The reduction in curvature from CL3 to CL4 is due to the fact that the reinforcement ratio is doubled within the splice region, resulting in a stiffer cross-section. By comparing Figures 27 and 28, one can see the increase in rotational capacity of the cross-section resulting from the confinement provided by the composite strap. However, comparing Figures 28 and 29, does not reveal measurable gain by actively confining the column. This could be partly due to the loss of the active pressure caused by seepage, creep, etc. The results from this one test are not conclusive and additional tests are required to confirm the benefits of active retrofitting. Figures 30 and 31 show the curvature distribution within the confined portions of columns C-4 and C-5, respectively. Comparison of these figures confirms the previous results indicating the enhancement of ductility and rotational capacity of retrofitted columns.

4.6.5. Load vs. Strain

4.6.5.1. Longitudinal reinforcement

The measured maximum strain in longitudinal bars recorded at a location immediately above the top face of the footing on the extreme tension side of the columns for C-1, C-3, C-4 and C-5 were 21, 40, 35 and 53×10^{-4}, respectively.[20] Due to the failure of the gages in Column C-2, the corresponding value for this column is not presented.

Generally, in the unretrofitted columns, the longitudinal reinforcements were less strained as compared to retrofitted columns. The factors contributing to this result were the insufficient confinement, bond failure in the lapped starter bars and/or buckling of continuous longitudinal bars in unretrofitted columns. Confinement by composite straps allowed higher strains in longitudinal reinforcements before failure, resulting in higher overall ductility and energy absorption capacity of the retrofitted columns.

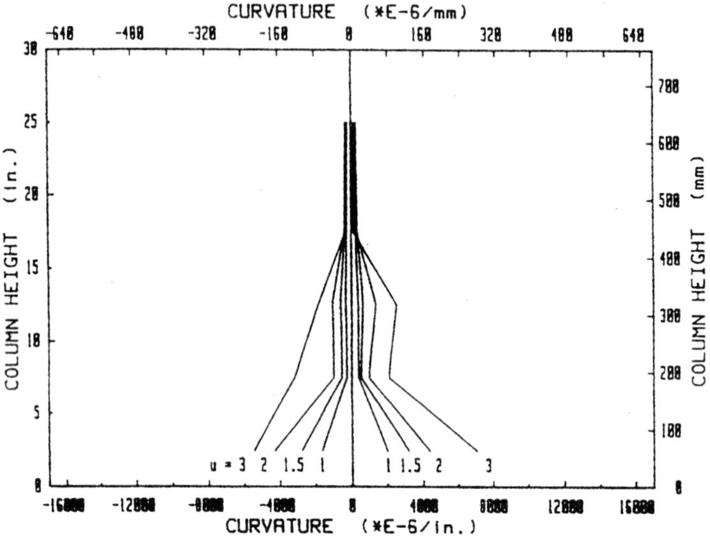

Figure 27. Curvature vs. height of column C-1.

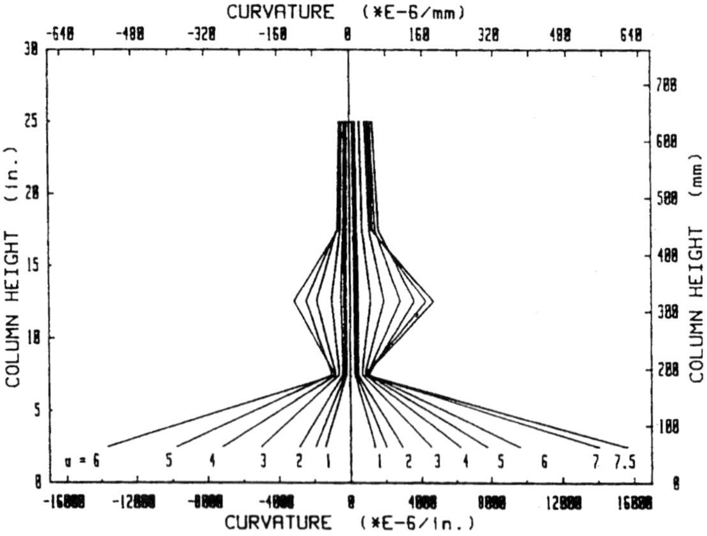

Figure 28. Curvature vs. height of column C-2.

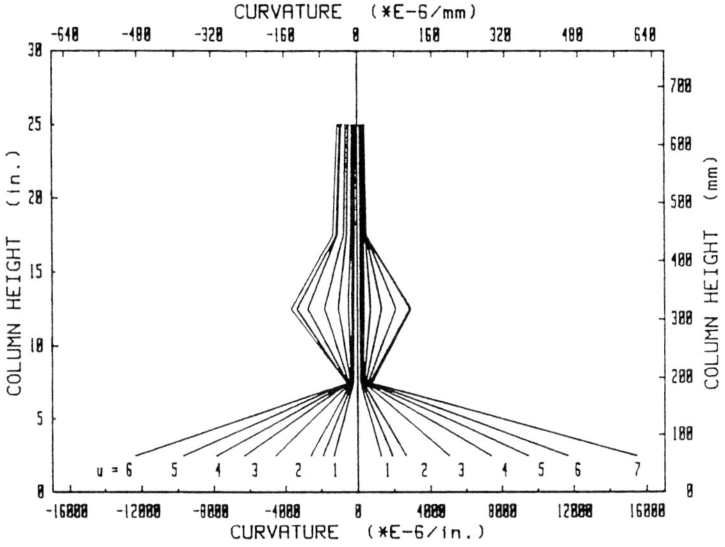

Figure 29. Curvature vs. height of column C-3.

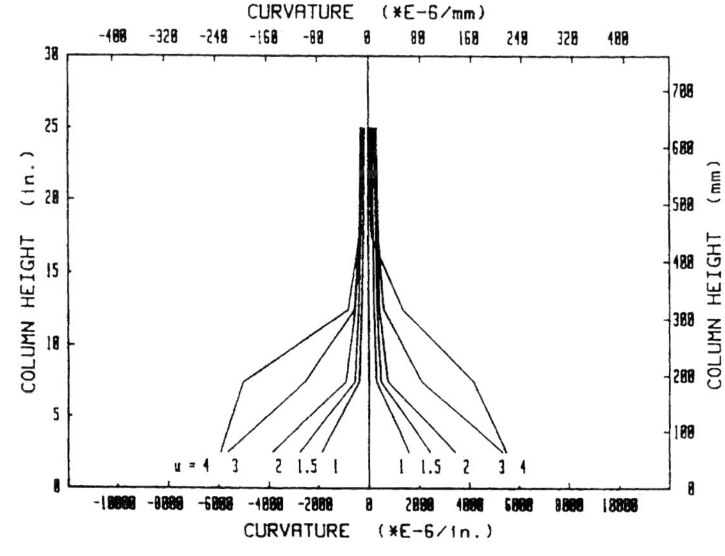

Figure 30. Curvature vs. height of column C-4.

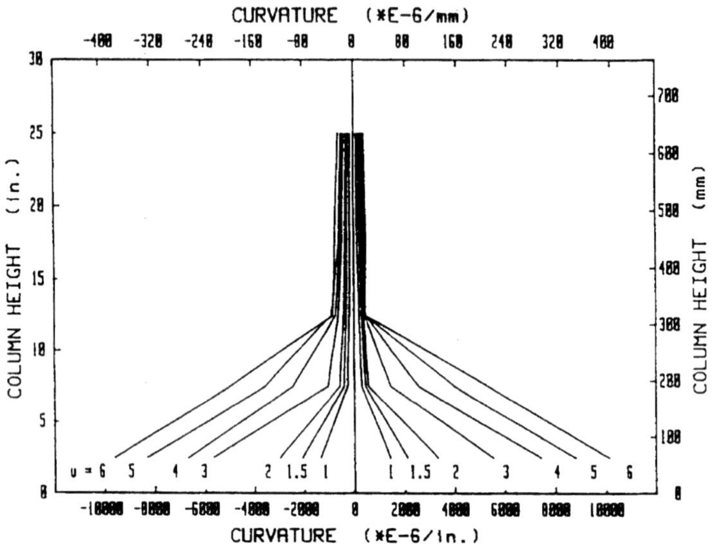

Figure 31. Curvature vs. height of column C-5.

4.6.5.2. Transverse reinforcements

Lack of adequate lateral reinforcement will result in premature yielding of hoops and therefore loss of confinement and rapid deterioration of the plastic zone region. The strains in the column transverse reinforcements were recorded using special electric resistance strain gages mounted on opposite sides of the hoop wires. In order to compare the behavior of typical hoops in the retrofitted and unretrofitted columns, the results of strain measurements at the same location for the first hoop above the footing of the columns C-4 and C-5 are presented here. This hoop was approximately 30 mm (1.2 in.) above the top face of the footing. Figures 32 and 33 show the load vs. strain in the hoop for columns C-4 and C-5, respectively. Normally, the hoops are in tension under the action of the applied loads. The slight compression strains shown in Figures 32 and 33 are due to local effects at the location of the strain gages. In column C-4, at the lateral load of approximately 62 kN (14 kips), the strain in the hoop reached 15×10^{-4}. However, at the same lateral load level, the strain in the retrofitted column C-5 was only 3×10^{-4}, indicating the effectiveness of the composite strap in sharing the load, and confining and reducing dilation of the core concrete.

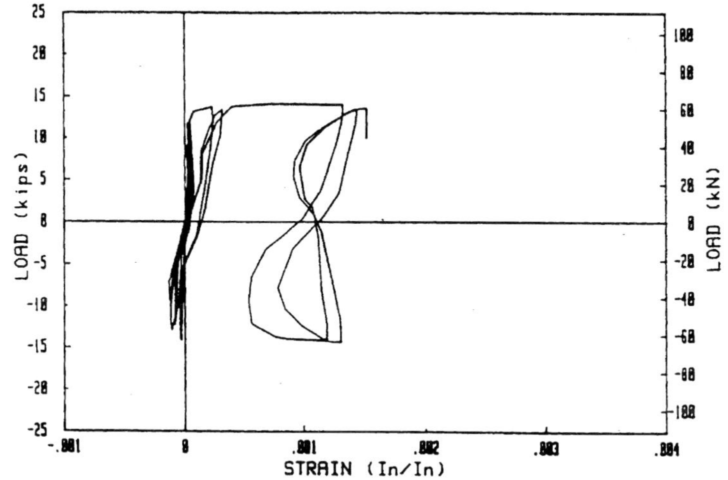

Figure 32. Load vs. strain in the hoop of column C-4.

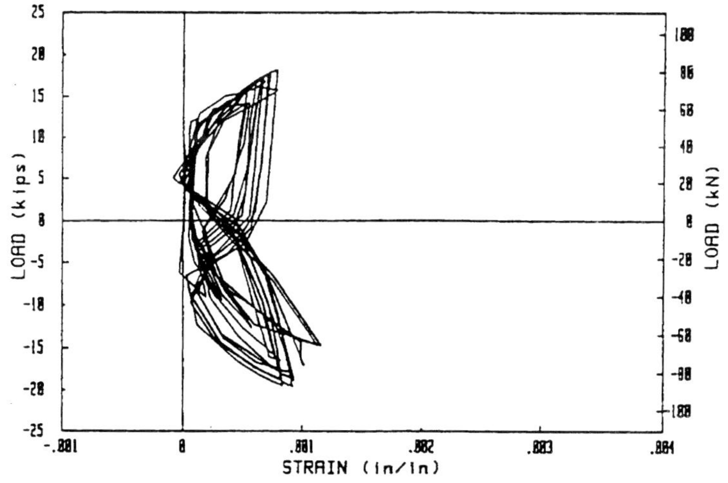

Figure 33. Load vs. strain in the hoop of column C-5 at Gage S8.

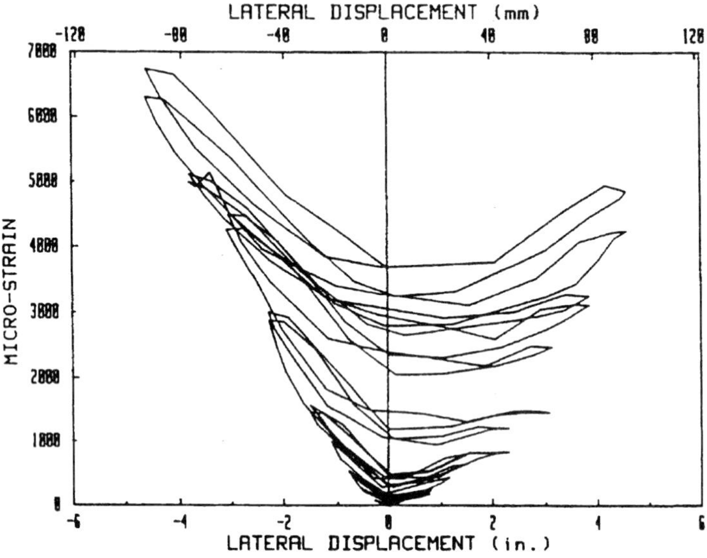

Figure 34. Deflection vs. strap strain of column C-2.

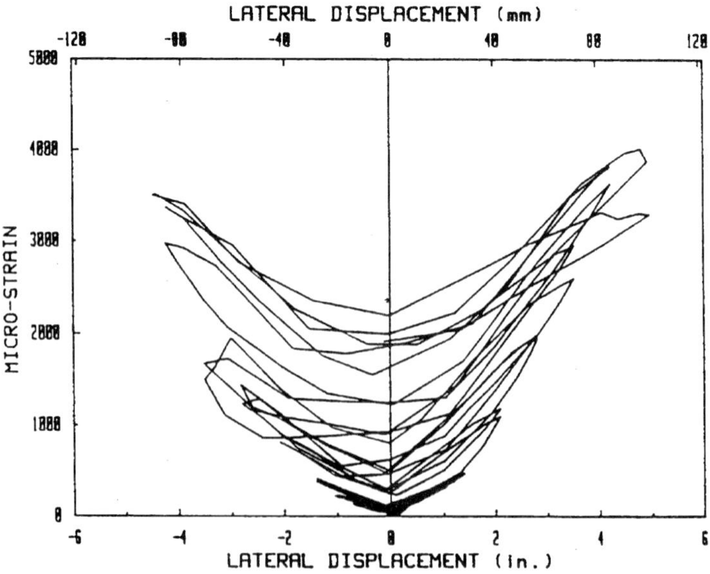

Figure 35. Deflection vs. strap strain of column C-3.

4.6.5.3. *Composite strap*

Strain gages were placed in the circumferential direction on the composite strap 572 mm (22.5 in.), 318 mm (12.5 in.) and immediately above the top face of the footing on the extreme tension and compression sides of the cross section.[20] Figures 34 and 35 show the circumferential (hoop) strain in the composite strap measured with the strain gages immediately above the top face of the footing for columns C-2 and C-3 vs. applied lateral displacement. From these figures, it can be seen that the strains were increasing in their magnitudes as the displacement was increasing, even though the lateral loads were essentially constant after the displacement ductility reached the level of $u = 3$, which, on the average, for all these columns, approximately corresponded to a drift of 2.8 percent. The maximum strain reached in column C-3 (active retrofit), not including the initial prestressing strain, is less than that of column C-2 (passive retrofit), indicating that the initial pressure reduced the dilation of concrete. However, as was observed from the load vs. displacement and load vs. curvature responses, this reduced dilation indicated by the smaller strain in the composite strap with active retrofitting did not translate into measurable improvement in the overall ductility of the column.

4.6.5.4. *Cracking and failure mechanisms*

Since all of the specimens for this study were designed with a strong footing detail, most of the cracking damage in the control specimens was concentrated in the column, especially within the plastic hinge region.[20] The primary change of the behavior in column C-1 occurred in the plastic hinge region at the displacement ductility level of $u = 1.5$. At this stage of the test, the cover concrete spalled and the longitudinal reinforcement bars started to debond in the lap-spliced region as shown in Figure 36. Column C-4 with continuous reinforcement started to fail at a displacement ductility level of $u = 4$. At this point, the longitudinal reinforcement started to buckle and the lateral load-carrying capacity reduced rapidly until complete failure of the column as shown in Figure 37.

4.7. EXPERIMENTAL STUDY OF RECTANGULAR COLUMNS

Five rectangular, reinforced concrete bridge column footing assemblages were designed with a scale factor 1/5 that of the prototype bridge columns.[20,22] The test specimens were designed to model typical pre-1971 design of existing highway bridge columns in a zone of high seismic risk. Each specimen consisted of a single column bent with a strong footing details.

Figure 36. Bond failure of column C-1 (See Color Plate II).

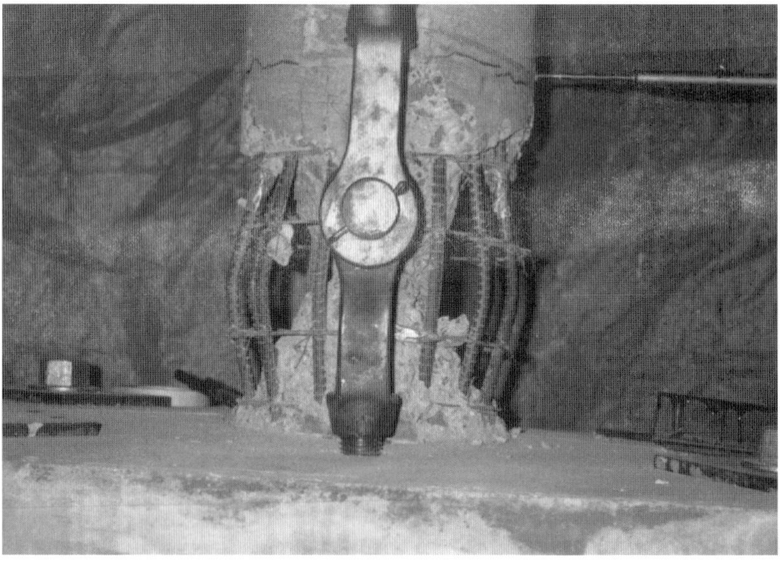

Figure 37. Longitudinal bar buckling in column C-4 (See Color Plate III).

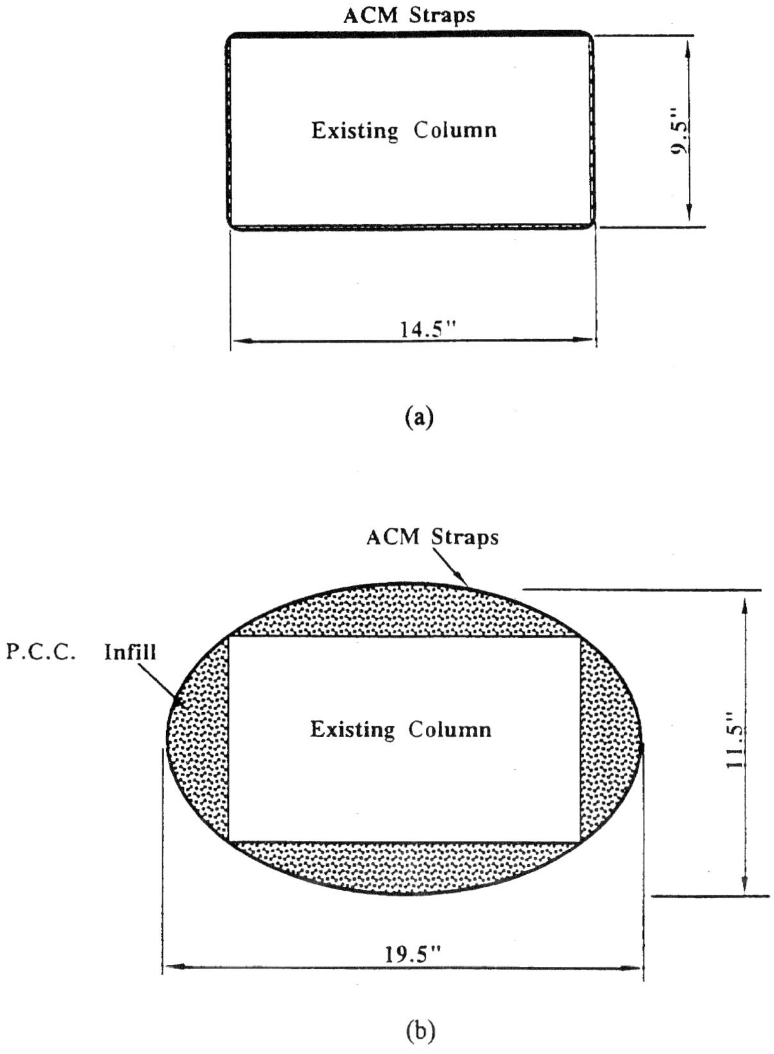

Figure 38. Retrofit with (a) Rectangular and (b) Oval-shaped composite straps.

The composite strap was applied only in the potential plastic hinge region; i.e., in the 635 mm (25 in.) long portion of the column above the top face of the footing. Two types of strap configurations were designed and fabricated in the laboratory, as shown in Figure 38. In addition to rectangular strap configuration, an oval shape (Figure 38b) was used to examine the

Figure 39. Rectangular column modified into an oval section in the plastic hinge region and wrapped with composite straps (See Color Plate IV).

effectiveness of the confining action of these two configurations. Figure 39 shows a typical column with its cross-section modified into an oval shape and wrapped with the composite straps in the plastic hinge region. The behavior of the columns with these strap configurations will be discussed in subsequent sections.

4.7.1. Materials

The materials used in the construction of specimens included: concrete with $f_c' = 34.5$ MPa (5000 psi) and Grade 40 steel rebars [nominal yield stress = 276 MPa (40 ksi)]. Although the specific concrete compression strength for the prototype bridge column was 21 MPa (3000 psi), ready-mix concrete providing $f_c' = 34.5$ MPa (5000 psi) was used to include the effect of expected overstrength resulting from normal conservative mix design and strength gain with concrete aging. All columns had a 240 × 368 mm (9.5 × 14.5 in.) rectangular cross-section. Columns R-1 and R-2 were reinforced with 12 No. 5 Grade 40 longitudinal bars resulting in a reinforcement ratio of 2.7%. Columns R-3, R-4 and R-5 were reinforced with 10 No. 5 and 22 No. 4 Grade 40 longitudinal bars resulting in a reinforcement ratio of 5.45%. The measured yield strength of the Grade 40 bars used for longitudinal reinforcement was 358 MPa (52 ksi). The transverse reinforcement for all five columns consisted of 9-gage steel wire ties placed 89 mm (3.5 in.) on center along the entire height of the column. The measured yield strength of the transverse ties was 301 MPa (43.7 ksi). Composite straps with fiber volume ratio of 50.2% were used for all retrofitted specimens. The measured stress-strain relationship of this type of strap is shown in Figure 40. Other design details of the specimens are summarized in Table 3. Active and passive retrofit schemes in the table refer to, respectively, pressure in the gap between the strap and column face, or no pressure. A test setup similar to that shown in Figure 19 for circular columns was also used here.

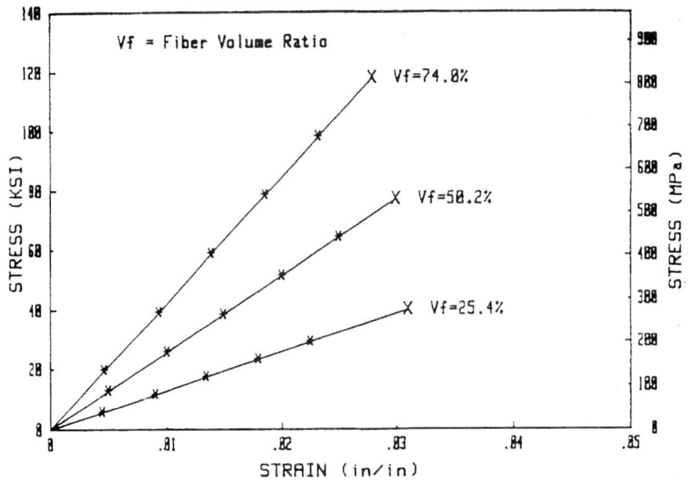

Figure 40. Typical stress-strain curves of composite strap.

Table 3. Details of column specimens

Specimen	Retrofit scheme	Longitudinal steel detail	Confining strap configuration
R-1	Control	Starter Bars	None
R-2	Active	Starter Bars	Rectangular
R-3	Control	Continuous Bars	None
R-4	Passive	Continuous Bars	Rectangular
R-5	Passive	Continuous Bars	Oval

4.7.2. Simulated Seismic Loading

The effect of an earthquake on the column specimen was simulated by reversed cyclic loading. The hydraulic actuator in the test setup was used to displace the top of the column to achieve a pre-determined load or displacement level.[20] Then, the actuator was reversed to achieve the same load or displacement level in the opposite direction. The load and displacement input history was broken into two phases. At the initial stage, the test was in a load control mode. Upon yielding of the longitudinal rebars, a displacement control mode of loading was utilized. Figure 41 shows typical loading sequences of the columns.

4.7.3. Lateral Load-displacement Response

The lateral load-displacement curves for all five specimens are shown in Figures 42 through 46. Due to the limitation of the hydraulic actuator, the maximum strokes reached by the actuator in the positive and negative directions were slightly different for some of the test specimens resulting in unsymmetric hysteresis loops for these symmetrically reinforced columns.[22] The measured and calculated maximum strengths for the retrofitted and unretrofitted columns are summarized in Table 4. The calculated strength was determined using the procedure outlined in the ACI Code and taking into consideration the actual strain hardening properties of the reinforcing steel.[23] The percent increase in strength was calculated with reference to the measured value for the control specimen.

The load-displacement curves of the two columns with lapped starter bars are shown in Figures 42 and 43. The hysteresis curves of the control specimen, R-1, shown in Figure 42, indicate that rapid degradation in strength occurred early and with very narrow energy dissipation loops. The lateral strength in column R-1 started to drop quickly at the first push cycle to $u = 1.5$. Because of bond failure in the lapped-spliced bars within the

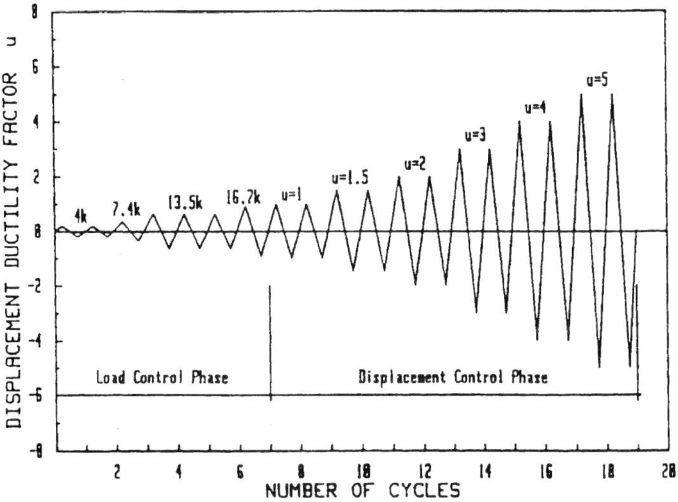

(a)

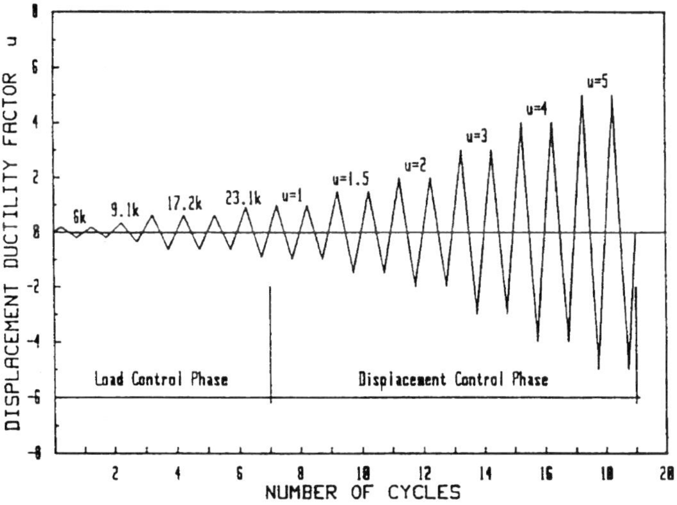

(b)

Figure 41. Loading sequence of test specimens; (a) columns with lapped starter bars, and (b) columns with continuous reinforcement.

Table 4. Measured and calculated lateral strength of columns

Specimens	Calculated lateral strength using ACI (kN) (unretrofitted)	Measured maximum lateral load (kN) (retrofitted)	Increase in strength resulting from retrofitting
R-1	102.7	96.5	Control
R-2	102.7	138.8	44%
R-3	132.5	161.5	Control
R-4	132.5	214	32%
R-5	132.5	226.8	40%

plastic hinge region of the column, the lateral load-carrying capacity was reduced approximately by 80 percent of the calculated value at the ductility level of $u = 4$. Column R-2 retrofitted with eight plies of composite straps resulting in a total strap thickness of 6 mm (0.25 in.) and an active scheme showed substantial improvement in the lateral load-displacement response as compared to column R-1, as shown in Figure 43. The lateral load exceeded the predicted capacity of the unretrofitted column at the first cycle to $u = 1.5$. The column continued to resist more lateral loads up to the ductility level of $u = 5$ with very stable hysteresis loops. The maximum strength of 138.8 kN (31.2 kips) was recorded of the pull cycle at $u = 5$. A slight decrease in the column strength was noted at the ductility level of $u = 6$, which appeared to be due to small slippage between the main longitudinal reinforcement and the lapped starter bars.

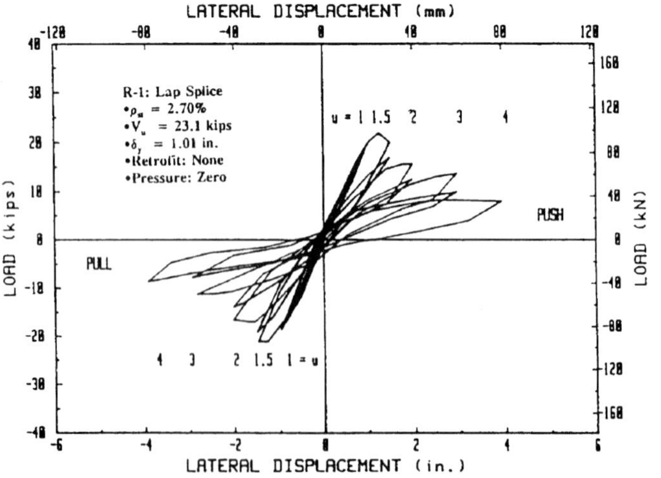

Figure 42. Load vs. displacement response of column R-1.

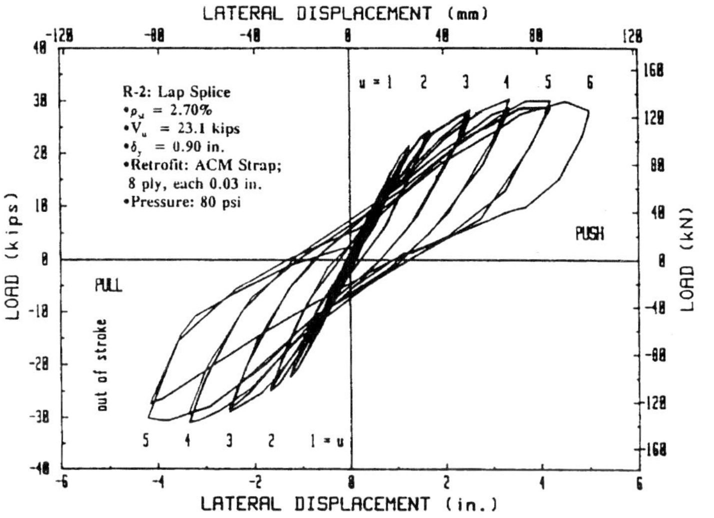

Figure 43. Load vs. displacement response of column R-2.

Figures 44 through 46 show the load vs. displacement responses of the three columns with continuous reinforcement and a higher longitudinal reinforcement ratio of 5.45 percent. Column R-3 was tested as the control specimen, and R-4 was retrofitted with rectangular shaped straps, while R-5 was with the oval shaped straps. It is believed that the oval-shaped composite straps are able to develop membrane confinement stresses, and hence, are expected to have better retrofit results, as compared with rectangular-shaped straps. Columns R-4 and R-5 were also wrapped with eight layers of composite strap. The load-displacement loops for column R-3 is shown in Figure 44. The predicted capacity of the column was reached at the displacement ductility level of $u = 1.5$. Some large flexural-shear cracks appeared within and beyond the plastic hinge zone at the ductility level of $u = 2$. On the way to the first push cycle to $u = 3$, extensive shear cracks along the whole height of the column were observed and longitudinal bars were suddenly separated from the core concrete due to the lack of adequate transverse reinforcement, resulting in a large drop of the lateral load-carrying capacity. Figure 47 shows Column R-3 near failure.

Compared to R-3, the load-displacement responses of R-4 and R-5 were greatly improved, as shown in Figures 45 and 46, respectively. For both columns R-4 and R-5, stable hysteresis loops developed up to $u = 6$, and the measured lateral strength showed no sign of structural degradation of

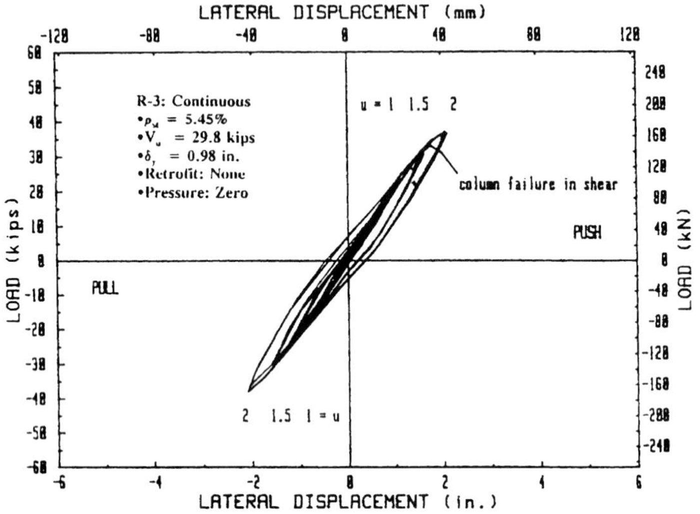

Figure 44. Load vs. displacement response of column R-3.

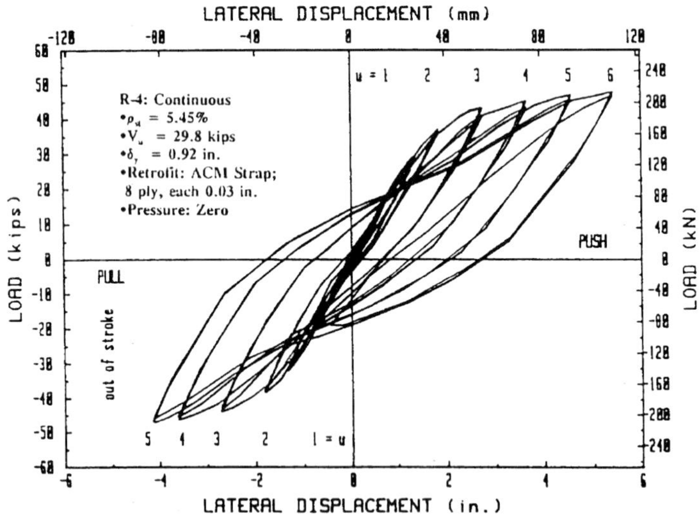

Figure 45. Load vs. displacement response of column R-4.

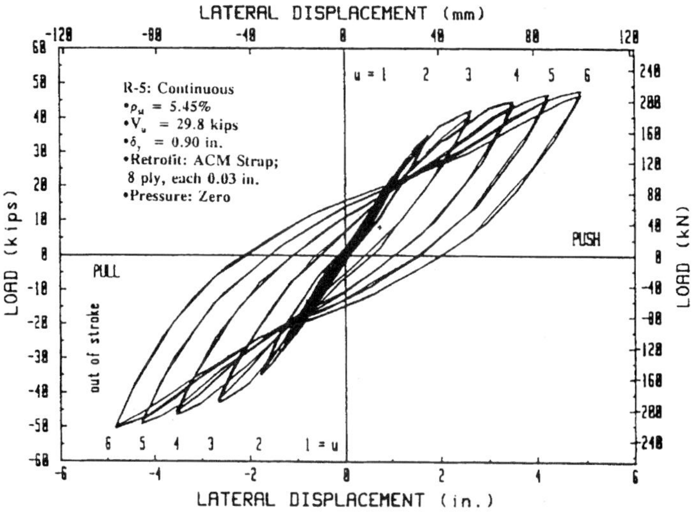

Figure 46. Load vs. displacement response of column R-5.

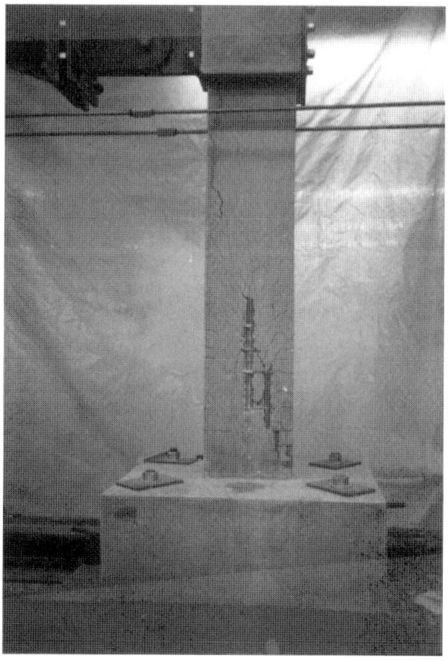

Figure 47. Shear cracks and failure of column R-3 (See Color Plate V).

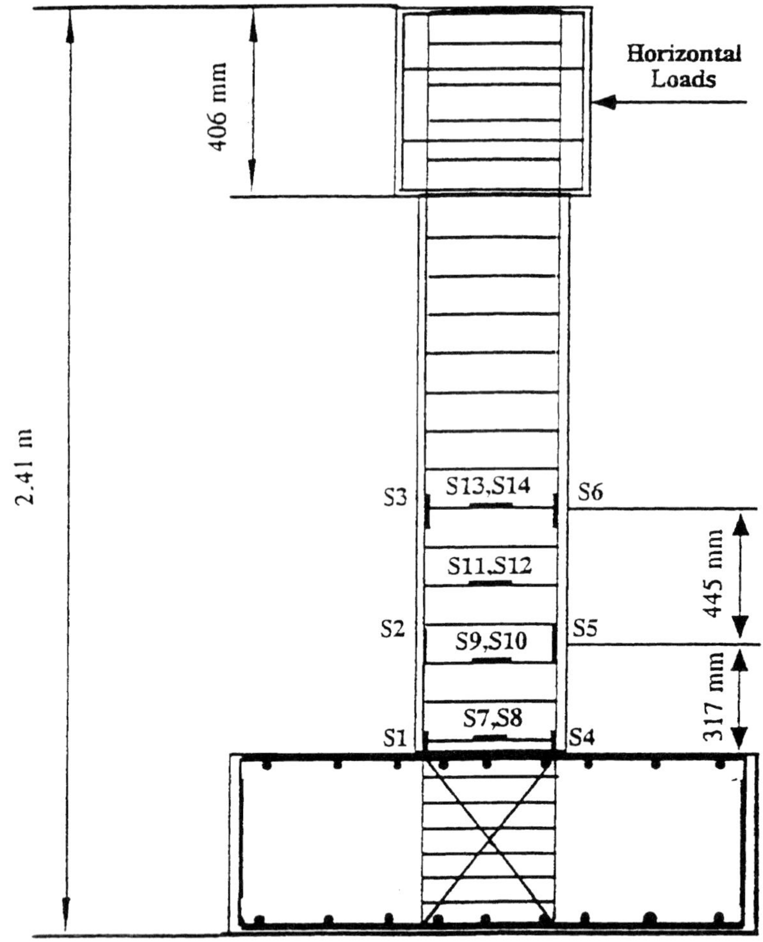

Figure 48. Locations of strain gages in columns.

the column. Unfortunately, the tests were terminated at a ductility level of $u = 6$, due to the limitation of the hydraulic actuator's stroke. The maximum increase in strength was measured at 214 kN (48.1 kips) for column R-4 and 224 kN (50.4 kips) for column R-5. The differences in the responses of the column with the rectangular shaped cross-section and the column with the oval shaped cross-section was not significant in this limited testing program. Additional tests must be conducted to establish the merits of modifying the cross-section.

An examination of the hysteresis loops revealed that all the specimens experienced some degree of pinching. More pinching in the loops were observed for the specimens with lapped starter bars than for the specimens with continuous reinforcement. Pinching is characterized by the narrow width of the hysteresis loops near mid-cycle, resulting in less energy dissipation per cycle as compared to a loop without pinching. Pinching is mainly caused by the slippage of the longitudinal bars and closing of flexural cracks in the plastic hinge zone of the columns.

4.7.4. Load vs. Strain

Figure 48 shows the location of various strain gages on the longitudinal reinforcements.

The following describes strain measurements in different elements within the specimens.[20]

4.7.5. Longitudinal Reinforcement

Generally, the strains in the longitudinal reinforcements of the retrofitted columns were higher than those in unretrofitted columns, indicating the effectiveness of the composite straps in confining the concrete and improving the bond, particularly at higher load levels. From the results of the limited study conducted here to examine the effect of the cross-section modification for improvement in the confining action of the straps, it was not possible to establish the merits of modifying the cross-section from the longitudinal bar strains. For example, there was not a clear difference between the strains of the longitudinal reinforcements in columns R-4 and R-5. Figures 49 and 50 show typical load vs. longitudinal bar strains at gage location S3 for columns R-4 and R-5, respectively. As can be seen from these figures, the strains were very close in both columns.

4.7.6. Transverse Reinforcements

The composite straps were very effective in confining the core concrete and sharing the hoop stresses as can be seen from typical load vs. hoop strains diagrams for unretrofitted and retrofitted columns shown in Figures 51 and 52.[20] These figures show the load vs. strain at gage locations S11 and S12 for columns R-1 and R-2, respectively. Comparing these two figures show that the strains in the hoop of the retrofitted column R-2 are smaller than the

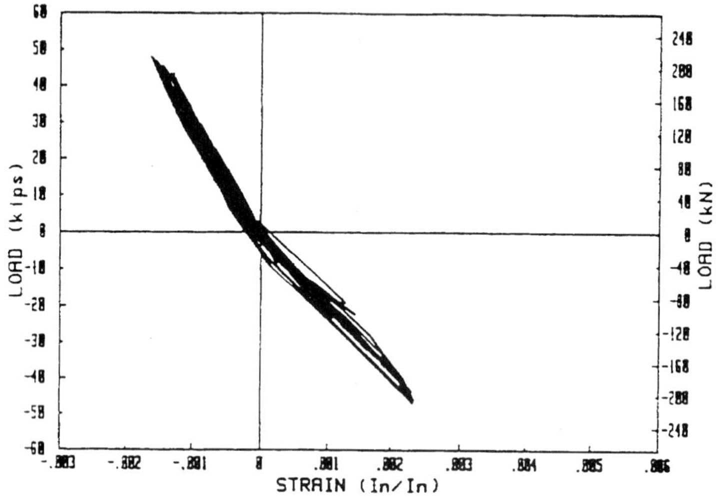

Figure 49. Load vs. strain in longitudinal bar of column R-4 at gage S3.

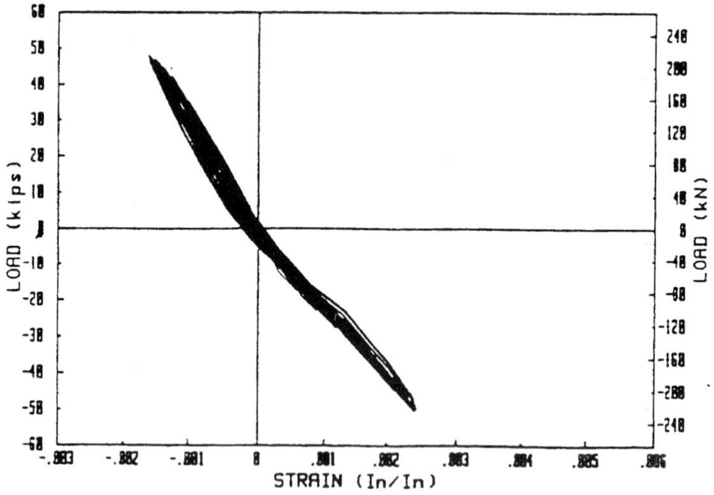

Figure 50. Load vs. strain in longitudinal bar of column R-5 at gage S3.

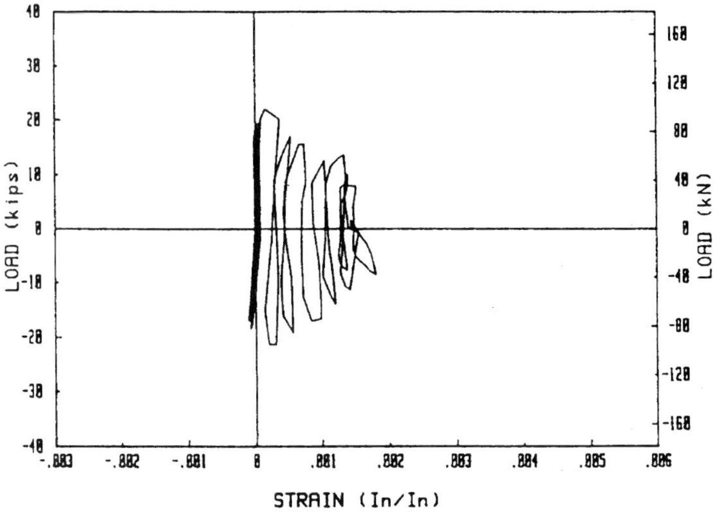

Figure 51. Load vs. strain in the hoop of column R-1 at gage S12.

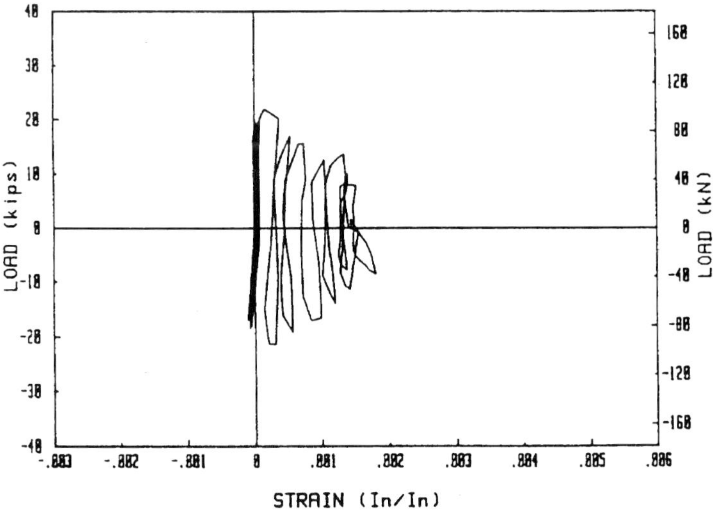

Figure 52. Load vs. strain in the hoop of column R-2 at gage S11.

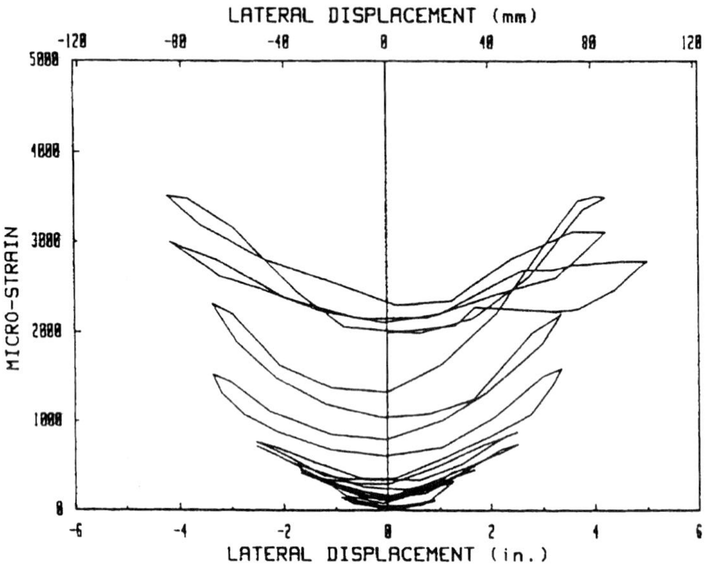

Figure 53. Deflection vs. average strap strain in column R-2 at gages SH3 & SH4.

strains in the same hoop in the unretrofitted column R-1, even under higher lateral loads. This indicates that the confinement provided by the composite strap prevented the dilation of the core concrete in the retrofitted column, resulting in smaller hoop stresses in the transverse reinforcement.

4.7.7. Composite Straps

The effectiveness of the composite strap in confining the concrete core can also be verified through the column load or lateral displacement vs. strap strain.[22] Figure 53 shows the lateral displacement at the top of the column R-2 vs. hoop strain in the strap at a location 318 mm (12.5 in.) above the top face of the footing. As can be seen from this figure, the strain in the strap increases successively with increasing lateral displacement, indicating resistance developed in the strap to prevent the lateral expansion of the concrete core. Figures 54 and 55 show the lateral displacement at the top of the column vs. the hoop strain in the strap at the same location; that is, 318 mm (12.5 in.) above the top of the footing for columns R-4 and R-5, respectively. The cross-section of column R-5 was modified into an oval

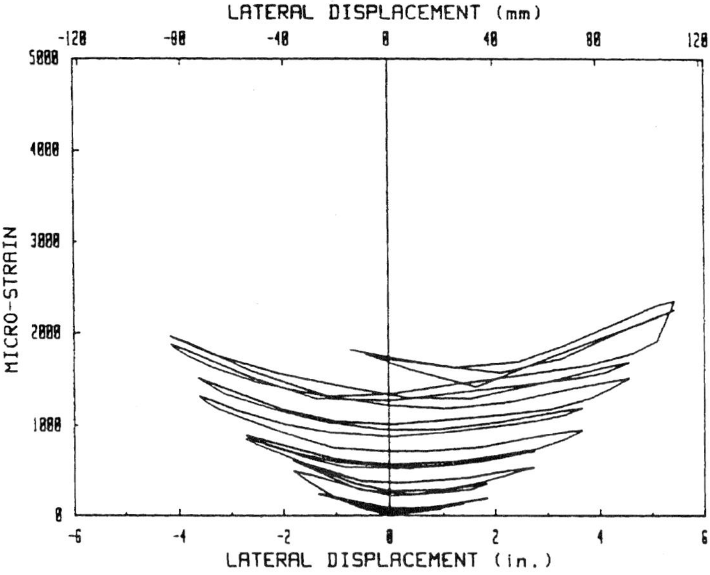

Figure 54. Deflection vs. average strap strain in column R-4 at gages SH3 & SH4.

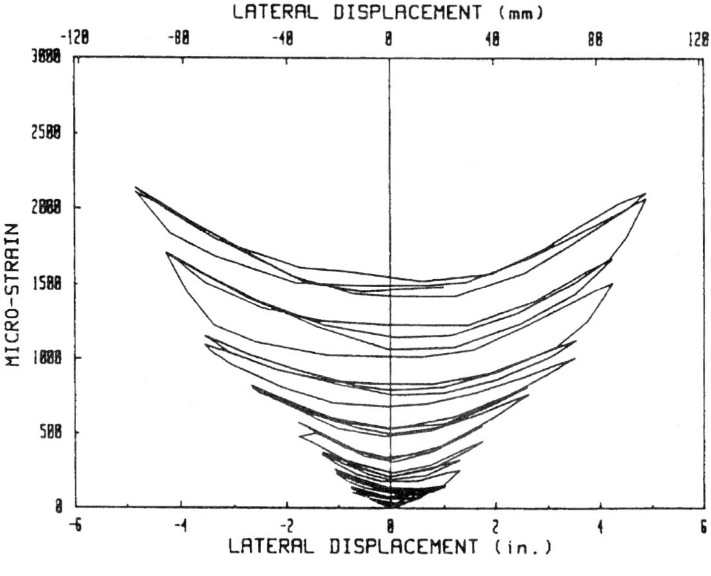

Figure 55. Deflection vs. average strap strain in column R-5 at gages SH3 & SH4.

shape. The strain records in the composite strap for the two columns were too close to discern conclusively the effectiveness of this shape modification in improving the confining action of the straps.

4.7.7. Column Cracking and Failure Mechanism

Because all specimens for this study were designed with a strong footing detail, most of the cracking damage was concentrated in the column, especially within the plastic hinge region.[20] For the retrofitted columns, no major column cracking was observed during any of the load cycles. Two types of cracks, namely, flexural and shear, were observed during the test.

4.7.8. Flexural Cracks

In the unretrofitted columns, the flexural cracks spread from the bottom of the column up to about the mid-height of each specimen. The intervals of the flexural cracks were nearly the same as that of the transverse reinforcement spacing. The primary change of the behavior in R-1 occurred in the plastic hinge region of the column at the displacement ductility level of $u = 1.5$. At this stage of the test, the cover concrete spalled and the longitudinal reinforcement bars started to debond in the lapped-spliced region. As a result, lateral load-carrying capacity reduced rapidly at this displacement level. Subsequently, the columns failed at the displacement ductility level of about $u = 3$, as a result of debonding of the longitudinal reinforcement in the lapped region, as shown in Figure 56. Column R-3, with the higher reinforcement ratio and continuous reinforcement, failed in shear at displacement ductility level of $u = 2$.

4.7.9. Shear Cracks

Large shear forces in column R-3, resulting from higher longitudinal reinforcement ratio of 5.45 percent, caused extensive diagonal shear cracks.[20] Diagonal cracking has a marked effect on the failure mechanism of bridge columns, and can develop well outside the potential plastic hinge zone, particularly in columns with inadequate transverse reinforcement, as was shown for column R-3 in Figure 47. When the lateral load reached 147 kN (33 kips), the formation of diagonal shear cracks occurred at the mid-height of the column. At the push cycle of $u = 2.0$, corresponding to a lateral load of 160 kN (36 kips), the column longitudinal bars were suddenly separated

Figure 56. Debonding of longitudinal reinforcement (See Color Plate VI).

from the core concrete, and the lateral load dropped significantly, and the test was terminated.

4.8. CONCLUSIONS

The results of the present study on concrete columns strengthened with composite straps indicate that this strengthening method can be used to increase effectively the strength and ductility of seismically deficient concrete columns. The following conclusions are drawn from the results of the study.

1. The stress-strain models for concrete confined with composite straps indicate significant increases in compressive strength and strain at failure when compared with the stress-strain behavior of unconfined concrete.
2. Although E-glass has a larger elongation at failure than carbon fiber, carbon fiber has a larger energy-absorbing capacity, indicated by its larger area under the stress-strain curve. Based on an energy balance approach, this results in an increase in ultimate axial load and ductility for strengthening with carbon fiber that is larger than that for strengthening with E-glass, if the volumes of straps were equal.
3. The increase in the maximum moment capacity is less than that in the ultimate axial load and ductility factor. This behavior is desirable in seismic strengthening of concrete columns because it results in a ductile flexural failure mode rather than a brittle shear mode of failure.
4. The rate of increase in the ultimate axial load, ductility, and maximum moment capacity decreases for increasing concrete compressive strength.
5. The ductility factor increases linearly with increase in strap thickness; however, the rate of increase in ductility factor decreases as strap spacing increases.
6. Reinforced concrete bridge columns, designed before the new seismic design provisions were in place, and with lap-spliced longitudinal reinforcement in the potential plastic hinge zone, appear to fail at low ductility levels of $u = 1.2$ to 1.5. This is due to the debonding of lapped starter bars, resulting from the lack of transverse reinforcement and insufficient development length of the longitudinal bars. The failure modes are brittle because strength deterioration is very rapid following debonding of longitudinal reinforcement.
7. For circular columns, the use of continuous reinforcement through the plastic hinge region improves moderately the lateral-displacement hysteresis loops. The structural degradation is likely to be delayed until ductility levels of $u = 4$ are reached. The column failure is caused by longitudinal reinforcement buckling within the hinge region due to the lack of lateral confinement.
8. Concrete columns externally wrapped with the composite straps in the potential plastic hinge region showed a significant improvement in both strength and displacement ductility. The retrofitted columns developed very stable load-displacement hysteresis loops up to displacement ductility level of $u = 6$, without evidence of significant structural deterioration associated with the bond failure of lapped started bars or longitudinal reinforcement buckling.

9. Both active and passive retrofit schemes provided additional confinement to existing core concrete, and were highly effective in preventing the columns from bond failure or longitudinal bar buckling, and hence, greatly increased the earthquake resistance of the column. The improvements resulting from the active retrofit scheme, as compared with the passive scheme, do not seem to justify the additional cost associated with the active retrofit scheme. This conclusion is, however, based on a limited number of tests. Additional studies are necessary to further investigate the benefits of active confinement.
10. For rectangular columns designed with high longitudinal reinforcement ratio ($\rho_{sl} \approx 5\%$) and continuous reinforcement, brittle shear failure occurs due to insufficient transverse reinforcement. In this type of column, moment capacity is high.
11. No major differences were observed in the test results between the oval-shaped straps and the rectangular-shaped straps within the range of the retrofit design parameters examined in this study. However, the results are not conclusive and additional studies must be undertaken to further investigate the behavior of rectangular columns retrofitted with oval-shaped composite straps.

ACKNOWLEDGEMENTS

The research reported in this paper was sponsored by the National Science Foundation (NSF) through grants MSS 9022667 and MSS 9257344, Dr. John B. Scalzi, Program Director. The support of the NSF is greatly appreciated. The experimental work was conducted in the Structural Laboratories of the University of Arizona. Assistance from the technical staff of the laboratories is gratefully acknowledged.

References

1. Chai, Y.H., M.J.N. Priestley and F. Seible, 1991, *ACI Structural Journal*, **88**(5), 565–584.
2. Saadatmanesh, H. and M.R. Ehsani, 1990, ACI Concrete International, *Design & Construction*, **12**(3), 65–71.
3. Saadatmanesh, H. and M.R. Ehsani, 1991, American Society of Civil Engineers, *J. Struct. Eng.*, **117**(11), 3417–3433.
4. An, W., H. Saadatmanesh and M.R. Ehsani, 1991, American Society of Civil Engineers, *J. of Struct. Eng.*, **117**(11), 3434–3455.
5. Saadatmanesh, H., M.R. Ehsani and M.W. Li, 1994, American Concrete Institute, *Struct. J.*, **91**(4), 434–447.
6. Katsumata, H., Y. Kobatake and T. Takeda, 1988, *Proceedings of the Ninth World Conference on Earthquake Engineering*, Tokyo, Japan, **7** (Aug. 2–9, 1988): 517–522.
7. N. Zamichow, 1991, *Los Angeles Times*, Sec. A:1, 22.

8. Hoskin, C.B. and A.A. Baker, 1986, *AIAA Education Series*, American Institute of Aeronautics and Astronautics, Inc., New York.
9. Pleiman, L.G., 1987, *Vega Technologies*, Marshal, Arkansas.
10. *FIBERITE Materials Handbook*, 1989, Imperial Chemical Company (ICI), p.v.b-16.
11. *FIBERITE Materials Handbook*, 1989, Imperial Chemical Company (ICI), p.v.b-53.
12. Mander, J.B., M.J.N. Priestley, R. and Park, 1988, ASCE, *J. Struct. Eng.*, **114**(8), 1804–1826.
13. Popovics, S., 1973, *Cement and Concrete Research*, **3**(5), 583–599.
14. Li, M.W., 1992, *Strengthening of Concrete Columns with Fiber Composites*, M.S. thesis, Dept. Of Civil Eng. University of Arizona.
15. Mander, J.B., M.J.N. Priestley and R. Park, 1984, *Research Report No. 84-2*, University of Canterbury, New Zealand.
16. Richart, F.E., A. Brandtzaeg and R.L. Brown, 1928, "A Study of the Failure of Concrete Under Combined Compressive Stresses," Bulletin 185, University of Illinois Engineering Experiment Station, Champaign, IL.
17. Schickert, G. and H. Winkler, 1979, *Deutscher Ausschuss for Stahlbeton*, Heft 277, Berlin, Germany.
18. Scott, B.D., R. Park, and M.J.N. Priestley, 1982, *American Concrete Institute Journal*, **79**(1), 13–27.
19. Sheikh, S.A. and S.M. Uzumeri, 1980, ASCE, *Journal of Structural Division*, **106**(5), 1079–1102.
20. Jin, L., 1995, *Seismic Retrofit and Design Recommendations for Reinforced Concrete Bridge Columns*, Doctoral Dissertation, Dept. Of Civil Eng., University of Arizona.
21. Saadatmanesh, H., M.R. Ehsani and Jin L., 1996, American Concrete Institute, *Struct. J.*, **93**(6), 639–647.
22. Saadatmanesh, H., M.R. Ehsani and L. Jin, 1997, *J. Earthquake Eng. Research Inst.*, **13**(2), 281–304.
23. ACI Committee 318, 1995, American Concrete Institute, pp. 351.

5 RELIABILITY ASPECTS IN DYNAMIC AND STRUCTURAL OPTIMIZATION

JIAN-JUN CHEN and BAO-YAN DUAN

Xidian University, Xi'an 710071, China

Preface

This special subject was developed based on our research work in recent years. The purpose of the subject is to introduce the main and important concepts, numerical methods, and possible applications of structural dynamic reliability and the reliability based optimization. Much work has been done and many studies have been published on this subject in recent years. It is an attempt to collect together the selected topics of this subject and to present them in a unified approch. In addition to the common points of this area, some special applications, for instance, antenna structures, are introduced too. The literature was attempted to be organised as far as possible in order to meet the appetite of different readers.

The emphasis throughout the literature is on problem formulation, relative merit of various method, and possible difficulties in applications. It is shown how alternative formulations and methods can be combined to treat with different problems. To stress that certain methods are applicable in various problems and to show that more than one method can be utilized in the solution of a specific problem, the methods of solution are developed separately from the presentation of design applications. The methods are discussed in general terms and their application to the engineering is illustrated by a variety of examples.

Part 1, composed of 13 sections, deals with the structural dynamic reliability. The role of main problem of it in the overall reliability analysis process is outlined and the general problem is mathematically formulated. Terminology used throughout the literature is defined and alternative approaches to structural dynamic reliability analysis are discussed.

Section 1–5 contain the basic concepts formulation, and treating method on structural dynamic analysis. Some useful examples are presented to emphasis the understanding and application of the methods.

In sections 6 and 7, both problems of level crossing and peak distribution for the random process are discussed.

Sections 8–10 develop analysis method — Poisson process algorithm — for the first passage of the system with single degree of freedom.

Section 11 is devoted to the problem of fatigue life of injuring — accumulative damage problem of structural system.

Sections 12–13 deal with the dynamic reliability analysis of the structure with multi-degree-of-freedoms.

Part 2 deals with the reliability based optimization of structures. Some general topics are discussed and some new optimization methods developed recently are introduced. Throughout the text, the general problem is formulated mathemalically. Typical methodology and alternative approaches to the reliability based potimum design are considered systematically.

The former 5 sections present the optimization with static reliability constraints for three cases such as the truss structures with element reliability, the frame structures with failure model probability, and general structures with the system reliability.

Section 6 introduces the optimum criterion method for structural reliability based optimization.

Section 7 shows how the reliability based optimization is applied to antenna structures, because antenna structures have some specific characteristics compared with the general structures, which is one of the main subjects of us.

Section 8 is devoted to optimization of structures with dynamic reliability constraints. Alternative formulations and solutions by different solving strategies are provided.

Section 9 presents an optimization method dealing with the problem by displaying the implicit reliability constraints. Thus this kind of the problem can be solved by the traditional optimization methods.

Section 10 introduces a new method for random programming problem by means of the equivalent determination strategy.

Finally, the authors wish to submit that only those directly utilized references for preparing this literature are listed in the reference, some other works, may be of great influence and importance, but have not been included because they are not directly relevant.

PART ONE: RELIABILITY ASPECTS IN STRUCTURAL DYNAMIC

5.1. STRUCTURAL RELIABILITY AND ITS MODEL

Structural reliability (or structural safety), considered from the viewpoint of probability, means the probability of implementing the required performance function under the conditions of fixed time interval and environment. Suppose that R and S address the random variables (or random process) of structural strength (resistance) and stress (load), and that the structural performance function is defined in terms of $Z = g(R, S) = R - S$, the reliability P_r and the probability of structural failure P_f can be expressed by the performance function in the following form,

$$P_r = P_{\text{rob}}\{Z > 0\} \tag{1-1}$$

$$P_f = P_{\text{rob}}\{Z \leqslant 0\} \tag{1-2}$$

It is easy to notice that $P_f + P_r = 1$. In the above formulae, $Z = R - S = 0$ is called the limit-state (or the safety-state) of structure, which divides the design space into the safety region Ω_r and the failure region Ω_f.

In structural reliability analysis, there are three basic models involved in the different situations of performance function Z.

(1) The model of a random variable, i.e., the performance function $Z = R - S$ is a random variable.
(2) The model of a semi-random process, i.e., the performance function $Z(t) = R - S(t)$ or $Z(t) = R(t) - S$ is a random process. In other words, one of R and S is the random process.
(3) The model of a random process, i.e., the performance function $Z(t) = R(t) - S(t)$ is a random process. That is to say both R and S are random processes.

The first model is that of a static situation, in which the time factor t is neglected. The performance function Z (i.e., R and S) is considered as a random variable that is independent of the time factor. Clearly, it is an unrealistic model. The research here has already matured to a certain extent in both the theory and the methodology. On the other hand, cases two and three belong to the dynamic model, in which the performance function Z (i.e., R and S) is considered as the time dependent parameter. This is a rather general reliability analysis model, the static analysis model is just a special case of it. The characteristic description and the analysis computation of reliability for the dynamic model are far more complicated than that for the static model. It is a main research area of structural reliability analysis.

For the dynamic model of structural reliability, the different reliability target can be defined according to the different circumstances. However, the most general situations are the following two: instantaneous reliability of the structure and the structural reliability in the valid serving life.

For the arbitrary time t, the probability

$$P_r(t) = P_{\text{rob}}\{Z(t) > 0\} \tag{1-3}$$

is called the reliability of structure at the instantaneous time point t. Obviously, the instantaneous reliability will change with time t.

If the random process $Z(t)$ of structure within the time interval $[0, T]$ can be expressed by the state variables (random variables) $Z(t_1), Z(t_2), \ldots, Z(t_n)$, the probability

$$P_r(T) = P\{\cap_{t_i \in T} Z(t_i) > 0\} \tag{1-4}$$

will be the structural reliability in the valid serving time T.

In addition, the equivalent method can be employed to deal with the semi-random process model. The concrete corresponding method is that the minimizing or maximizing method is applied to the performance function, thus the problem of solving a dynamic model can be transformed into the problem of solving an equivalent static model. The details are discussed as follows:

If the minimizing deformation is applied to the performance function $Z(t) = R - S(t)$ in the time interval $[0, T]$, i.e.,

$$Z_{\min} = \min_{t \in T} Z(t) = \min_{t \in T}[R - S(t)] = R - \max_{t \in T} S(t) \tag{1-5}$$

where, let $S_{\max} = \max_{t \in T} S(t)$, the dynamic reliability of the structure within the time interval can be expressed by means of reliability in the following form,

$$P_r(T) = P\{Z(t), 0 \leqslant t \leqslant T\} = P\{Z_{\min} > 0\} = P\{R - S_{\max} > 0\} \tag{1-6}$$

Likewise, applying the maximizing deformation to the performance function $Z(t) = R(t) - S$ in $[0, T]$ yields the following static reliability,

$$P_r(T) = P\{R_{\min} - S > 0\} \qquad (1\text{-}7))$$

in which $R_{\min} = \min_{t \in T} R(t)$.

So far as the problem where the performance function is a random process is concerned, the method of semi-random processes is usually employed in the most practical computation cases due to the complexity of the problem. It means that what is really to be done in practice is the computation of random variables instead of the random process itself.

5.2. THE RELIABILITY OF VIBRATION FOR THE SYSTEM WITH A SINGLE DEGREE OF FREEDOM UNDER THE EXCITATION OF THE SIMPLE HARMONIC FORCE

Suppose X_m denotes the maximum vibration amplitude of a single-degree-of-freedom system under the excitation of a simple harmonic force, X^* is the allowable value of the vibration amplitude, and both are random variables independent of each other, the safety criterion (performance function) and the limit state equation of the system can be described as,

$$X^* - X_m > 0 \tag{2-1}$$

and

$$X^* - X_m = 0 \tag{2-2}$$

In order to simplify the above expression, let $\delta^* = X^*/X_s$ and $\delta_m = X_m/X_s$, where X_s is the static displacement of structure under the maximum amplitude of the simple harmonic force, thus the formulae (2-1) and (2-2) can be written as a random variable as,

$$\delta^* - \delta_m > 0 \tag{2-3}$$

and

$$\delta^* - \delta_m = 0 \tag{2-4}$$

According to the steady-state solution of structural system with a single degree of freedom under the excitation of the simple harmonic force, the random variable δ_m can be further described in the following form,

$$\delta_m = \frac{X_m}{X_s} = \frac{1}{[(1-\lambda^2)^2 + (2\xi\lambda)^2]^{1/2}} \tag{2-5}$$

in which, $\lambda = \omega/\omega_0$; ω and ω_0 are the frequencies of the simple harmonic force and the natural frequency of the system, respectively. It should be noted that ω and ω_0 are independent random variables. ξ is the damping ratio of the vibration mode of system.

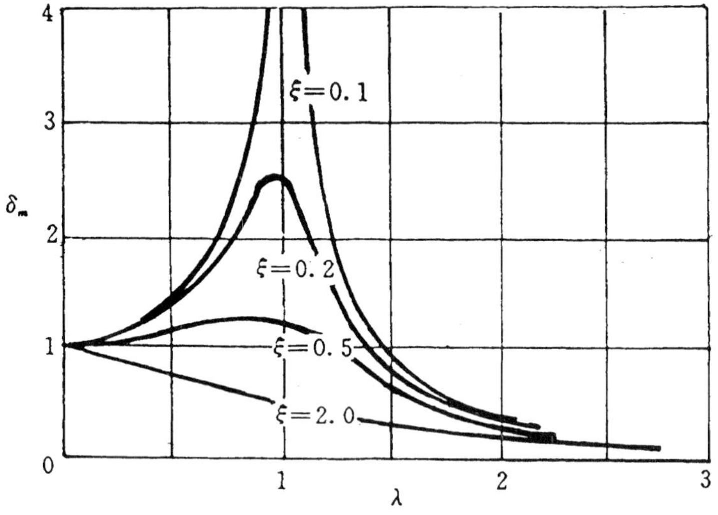

Figure 1. The vibration curve of δ_m with λ and ξ.

Further more, the relationship between δ_m and both λ and ξ can also be described in terms of the following chart Figure 1.

Inspecting Figure 1 leads to the following two conclusions:

(1) For the fixed ξ, there is a maximum value $\delta_{m,\max}$ of δ_m with respect to variable λ;

(2) If δ^* is less than or equal to $\delta_{m,\max}$, there exist the upper and the lower limits λ_u and λ_l of variable λ corresponding to δ^*.

Therefore, if the relation $\{(\lambda < \lambda_l) \cup (\lambda > \lambda_u)\}$ holds, the safety criterion $\delta^* - \delta_m > 0$ can be confirmed. Synthetically considering the above discussion will give the limit state equations of the system vibration as follows[1],

$$\lambda_l - \lambda = 0 \quad \text{when } \lambda < 1 \tag{2-6}$$

It also means

$$\lambda_l \omega_0 - \omega = 0 \quad \text{when } \lambda < 1 \tag{2 - 6a}$$

$$\lambda - \lambda_u = 0 \quad \text{when } \lambda > 1 \tag{2-7}$$

i.e.,

$$\omega - \lambda_u \omega_0 = 0 \quad \text{when } \lambda > 1 \tag{2-7a}$$

The upper and the lower bounds λ_u and λ_l can be found through the following procedure.

From formula (2-5), thc following equation results

$$\frac{1}{\delta_m^2} = \lambda^4 + (4\xi^2 - 2)\lambda^2 + 1 \tag{2-8}$$

which is a quadratic polynomial with respect to λ^2. There is obviously a minimum value $B_{\min}$ of $1/\delta_m^2$ for the given ξ and $B_{\min}$ can be written as,

$$B_{\min} = 4\xi^2(1 - \xi^2) \tag{2-9}$$

In contrast, there is a maximum value $\delta_{m,\max}$ of δ_m corresponding to $B_{\min}$ and it is

$$\delta_{m,\max} = \frac{1}{\sqrt{B_{\min}}} = \frac{1}{2\xi^2\sqrt{1 - \xi^2}} \tag{2-10}$$

The above formula (2-10) tells us that the variable $\delta_{m,\max}$ increases with ξ's decreasing. The safety criterion (2-3) can be satisfied whatever the value of λ is as long as $\delta_{m,\max}$ is less than δ^*. However, the upper and the lower limits λ_u and λ_l will not exist any more in this case. On the other hand, if $\delta_{m,\max}$ is greater than δ^*, for the sake of satisfying the limit state equation (2-4), the value of δ_m must be equal to the value δ^*. If the condition of $\delta_m = \delta^*$ is substituted into formula (2-8), the following equation results,

$$\lambda^4 + (4\xi^2 - 2)\lambda^2 + 1 - \frac{1}{\delta^{*2}} = 0 \tag{2-11}$$

It is clear that the above formula is a quadratic algebraic equation of λ^2, and λ can be solved by

$$\lambda_{1,2} = \left\{(1 - 2\xi^2) \mp \left[(1 - 2\xi^2)^2 + \left(\frac{1}{\delta^{*2}} - 1\right)\right]^{1/2}\right\}^{1/2} \tag{2-12}$$

As a result of this, the upper and the lower limits of λ can be easily found as,

$$\lambda_l = \lambda_l = \left\{(1 - 2\xi^2) - \left[(1 - 2\xi^2)^2 + \left(\frac{1}{\delta^{*2}} - 1\right)\right]^{1/2}\right\}^{1/2} \tag{2-13}$$

and

$$\lambda_u = \lambda_2 = \left\{(1 - 2\xi^2) + \left[(1 - 2\xi^2)^2 + \left(\frac{1}{\delta^{*2}} - 1\right)\right]^{1/2}\right\}^{1/2} \tag{2-14}$$

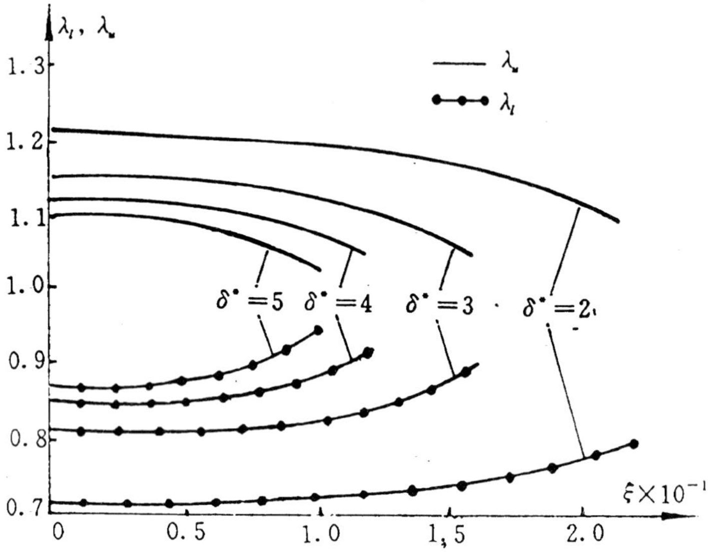

Figure 2. The shifting curve of λ_u and λ_l with respect to ξ under the different value of δ^*.

It is obvious that both λ_u and λ_l are the nonlinear functions of ξ and δ^*. The relationship that λ_u and λ_l are dependent upon ξ for different values of δ^* is described in Figure 2.

If the frequency ω of the simple harmonic force and the natural frequency ω_0 of system are all subjected to the normal distribution, i.e.,

$$\omega \sim \mathcal{N}(\mu_\omega, \sigma_\omega^2), \omega_0 \sim \mathcal{N}(\mu_{\omega_0}, \sigma_{\omega_0}^2)$$

the reliability index β and the reliability P_r of system's vibration can be deduced from the limit state equations (2-6a) and (2-7a) as,

$$\begin{cases} \beta = (\lambda_l \mu_{\omega_0} - \mu_\omega)[\lambda_l^2 \sigma_{\omega_0}^2 + \sigma_\omega^2]^{-1/2} \\ P_r = \Theta(\beta) \end{cases} \quad \text{if } \mu_\lambda = \frac{\mu_\omega}{\mu_{\omega_0}} < 1 \qquad (2\text{-}15)$$

and

$$\begin{cases} \beta = (\mu_\omega - \lambda_u \mu_{\omega_0})[\sigma_\omega^2 + \lambda_u^2 \sigma_{\omega_0}^2]^{-1/2} \\ P_r = \Phi(\beta) \end{cases} \quad \text{if } \mu_\lambda = \frac{\mu_\omega}{\mu_{\omega_0}} > 1 \qquad (2\text{-}16)$$

where, $\Phi(\cdot)$ denotes the standard normalized distribution function. Inspecting formulae (2-13)–(2-16) can result in the following remarks such as: (1) If the damping ratio ξ and the allowable value δ^* of the vibration amplitude are increased, the reliability index β will rise and thus the reliability P_r of the system will also increase correspondingly; (2) If the variance σ_ω^2 and $\sigma_{\omega_0}^2$ of the exciting force frequency and the natural frequency increase, the reliability index β will decrease and therefore the reliability P_r of system will be reduced correspondingly. It is obvious that in order to have a structural system with high reliability, the damping ratio ξ should be increased as much as possible. Meanwhile, the variance of the exciting force frequency ω and natural frequency ω_0 should be decreased as little as possible.

5.3. THE RELIABILITY OF THE VIBRATION RESPONSE OF A STRUCTURAL SYSTEM

If a linear elastic structure is subjected to a stationary random load, the dynamic equation of the system can be established by the finite element method as follows,

$$[M]\{[\ddot{X}(t)\} + [C]\{\dot{X}(t)\} + [K]\{X(t)\} = \{F(t)\} \tag{3-1}$$

in which, $[M]$, $[C]$ and $[K]$ are the mass matrix, damped matrix and stiffness matrix respectively; $\{X(t)\}$, $\{\dot{X}(t)\}$ and $\{\ddot{X}(t)\}$ are the displacement vector, velocity vector and acceleration vector separately; $\{F(t)\}$ is the random load vector.

Generally speaking, Equation (3-1) is a set of differential equations with multidegree of freedoms coupled with each other. By means of the model decomposing technique and the Duhamel integral, its formal solution of displacement response of the structural system can be obtained as follows,

$$\{X(t)\} = \int_0^t [\Phi][h(\tau)][\Phi]^T\{F(t-\tau)\}d\tau \qquad (3\text{-}2))$$

where, $[\Phi]$ is a normal model matrix of system; $[h(t)] = \text{diag}[h_j(t)]$ is the impulse response function matrix of the system.

Starting from formula (3-2), the average value vector and the mean square value matrix of the displacement response can be obtained, they are

$$\{\mu_X\} = E(\{X(t)\}) \tag{3-3}$$

$$[\psi_X^2] = \int_0^\infty [\Phi][H(\omega)][\Phi]^T[S_F(\omega)][\Phi][H^*(\omega)][\Phi]^T d\omega \tag{3-4}$$

in which, $[H(\omega)] = \mathrm{diag}(H_j(\omega))$ is the frequency response function matrix of system; $[H^*(\omega)]$ is the conjugate matrix of $[H(\omega)]$; $[S_F(\omega)]$ is the power spectrum density function matrix of the random load $\{F(t)\}$.

To go a step further, the stress response of the eth element in the system can be described, by means of the relationship between the nodal displacement and the element stress of the finite element method, in the following form,

$$\{S^{(e)}\} = [D^{(e)}][T^{(e)}]\{X^{(e)}\} \quad (e = 1, 2, \ldots, n_e) \tag{3-5}$$

where, $\{S^{(e)}\}$ is the stress response vector of element (e) in the local coordinate system; $[D^{(e)}]$ is the related matrix between stress and displacement of element (e); $[T^{(e)}]$ is the coordinate transforming matrix of element (e) from the local coordinate system to the global coordinate system; $\{X^{(e)}\}$ is the nodal displacement vector of element (e) in the global coordinate system; n_e is the total number of elements in the system.

Next, the average value and the mean square value matrix of the eth element's stress response can be easily obtained,

$$\{\mu_S^{(e)}\} = E(\{S^{(e)}\}) = [D^{(e)}][T^{(e)}]E(\{X^{(e)}\}) = [D^{(e)}][T^{(e)}]\{\mu_X^{(e)}\} \tag{3-6}$$

and

$$[\psi_S^{2(e)}] = [D^{(e)}][T^{(e)}][\psi_X^{2(e)}][T^{(e)}]^T[D^{(e)}]^T \tag{3-7}$$

Then the variance of the ith response component (displacement or stress) $y_i(t)$ can be deduced from the relationship between the mean square value and the variance as

$$\sigma_{y_i}^2 = \psi_{y_i}^2 - \mu_{y_i}^2 \tag{3-8}$$

Clearly, if the exciting force $\{F(t)\}$ is a zero average process, i.e., $\{\mu_F\} = \{0\}$, the response of structural displacement and element stress are both a zero average process, i.e., $\{\mu_X\} = \{0\}$, $\{\mu_S^{(e)}\} = \{0\}(e = 1, 2, \ldots, n_e)$. Therefore, the mean square value, i.e., the variance, of the response becomes,

$$\psi_{y_i}^2 = \sigma_{y_i}^2 \tag{3-9}$$

If $\{F(t)\}$ is a stationary Gaussian process, the ith response component y_i is also subjected to a Gaussian process according to that the linear structural system is not changed with respect to Gaussian excitation, i.e.,

$$y_i \sim \mathcal{N}(\mu_{y_i}, \sigma_{y_i}^2)$$

If the allowable value of the ith response component y_i is also submitted to Gaussian distribution, i.e., $y_i^* \sim \mathcal{N}(\mu_{y_i^*}, \sigma_{y_i^*}^2)$, then the performance function Z_i and the reliability P_{ri} of the ith response component can be mathematically written as,

$$Z_i = y_i^* - y_i \tag{3-10}$$

$$\begin{cases} P_{ri} = P\{Z_i > 0\} = \Phi(\beta_i) \\ \beta_i = (\mu_{y_i^*} - \mu_{y_i})(\sigma^2_{y_i^*} + \sigma^2_{y_i})^{-1/2} \end{cases} \tag{3-11}$$

Finally the reliability P_r of the vibration response for a structural system can be obtained by the following formula,

$$\begin{aligned} P_r &= P\{(Z_1 > 0) \cap (Z_2 > 0) \cap \cdots \cap (Z_i > 0) \cap \cdots\} \\ &= P\{\cap_i (Z_i > 0)\} \end{aligned} \tag{3-12}$$

5.4. THE FATIGUE RELIABILITY OF A STRUCTURAL ELEMENT UNDER THE CONDITION OF HARMONIC STRESS

In most cases, the structural element is subjected to a random dynamic stress. When the value of dynamic stress is much less than that of the strength limit, fatigue failure will be a main factor.

The reliability of the element under the random dynamic stress will be discussed in Section 11. This section concerns only the case when the amplitude of the element's dynamic stress is a random variable — simple harmonic stress $S(t)$, i.e., $S(t)$ is a stationary and a narrow band random process, its mathematical expression is,

$$S(t) = S_a \cdot \sin \omega t \tag{4-1}$$

where, the amplitude S_a is a random variable; the frequency ω is a constant.

It is easy to determine from the peak distribution of the random process[2] that S_a obeys the Rayleigh distribution for a stationary and a narrow band process $S(t)$, i.e., $S_a \sim Ra(\sigma_{Sa})$. The corresponding probability density function is,

$$f_{Sa}(x) = \frac{x}{\sigma^2_{Sa}} \cdot \exp\left[-\frac{x^2}{2\sigma^2_{Sa}}\right] \quad (x \geqslant 0) \tag{4-2}$$

where, σ^2_{Sa} is the mean square value of stress peak.

The above simple harmonic stress $S(t)$, for which the amplitude modulation is considered as a random variable, can be classified into three types of models as wave type, pulse type and symmetric type. These three kinds of models are given in Figure 3.

In the Figure 3, $S_{\max}$ and $S_{\min}$ are the maximum and minimum values of stress respectively. In order to be able to describe the cyclic variation rules of the different types of stress, the cycle characteristic parameter C is specially defined as follows,

$$C = \frac{S_{\min}}{S_{\max}} \tag{4-3}$$

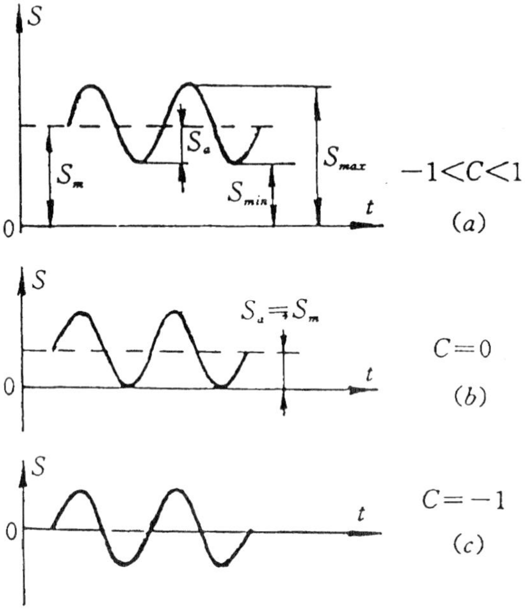

Figure 3. Simple harmonic stress.

Then the following types of stress can be identified by,

$S(t)$ is a wave type, if $-1 < C < 1$;
$S(t)$ is a pulse type, if $C = 0$;
$S(t)$ is a symmetric type if $C = -1$;
$S(t)$ is a constant type (i.e., static stress), if $C = 1$.

For each engineering material, its fatigue strength S_r' can be obtained by either the fatigue limit chart of standard test samples or the hand book of engineering materials provided the cycle characteristic parameter C and the cycle number N are known. The fatigue strength S_r' usually is a random variable. Once S_r' is found, the element's fatigue strength S_r can be obtained through the adjusted formula that is used in engineering usually as,

$$S_r = \frac{\varepsilon\beta}{k} S_r' \tag{4-4}$$

in which, ε, β and k are the size affecting factor, the surface quality affecting factor, and the concentrated stress affecting factor of the element respectively.

Since these factors are indetermined and disconcentrated, they should also be viewed as the random variables. In engineering, they are treated as the normal variables, which are independent of each other. That is

$$\varepsilon \sim \mathcal{N}(\mu_\varepsilon, \sigma_\varepsilon^2),\ \beta \sim \mathcal{N}(\mu_\beta, \sigma_\beta^2),\ k \sim \mathcal{N}(\mu_k, \sigma_k^2)$$

By using the formula of the function moment for a random variable, the average value and the variance of fatigue strength S_r of element can be determined as follows,

$$\mu_{S_r} = \frac{\mu_\varepsilon \mu_\beta}{\mu_k} \mu_{S'_r} \tag{4-5}$$

$$\sigma_{Sr}^2 = \left(\frac{\mu_\beta \mu_{S'_r}}{\mu_k}\right)^2 \cdot \sigma_\varepsilon^2 + \left(\frac{\mu_\varepsilon \mu_{S'_r}}{\mu_k}\right)^2 \cdot \sigma_\beta^2 + \left(\frac{\mu_\varepsilon \mu_\beta \mu_{S'_r}}{\mu_k^2}\right)^2 \cdot \sigma_k^2 + \left(\frac{\mu_\varepsilon \mu_\beta}{\mu_k}\right)^2 \cdot \sigma_{S'_r}^2 \tag{4-6}$$

If the probability density function $f_{Sa}(\cdot)$ of the stress amplitude modulation S_a and the probability density function $f_{Sr}(\cdot)$ of the fatigue strength S_r are known, the fatigue reliability can be found through the theory of stress-strength interference as follows,

$$\begin{aligned} P_r &= P_{\text{rob}}\{S_r - S_a > 0\} \\ &= \int_0^\infty f_{Sr}(S_r) \cdot F_{Sa}(S_r) dS_r \\ &\overset{\text{or}}{=} \int_0^\infty f_{Sa}(S_a) \cdot [1 - F_{Sr}(S_a)] dS_a \end{aligned} \tag{4-7}$$

in which, $F_{Sr}(\cdot)$ and $F_{Sa}(\cdot)$ are the probability distribution functions of variables S_r and S_a separately.

Compared with the case of static strength, it is easy to notice that both cases have the same form. The difference is that the stress and the strength in formula (4-7) are the time-variation working stress and fatigue strength.

From the stress $S_a \sim Ra(\sigma_{sa})$ and strength $S_r \sim \mathcal{N}(\mu_{Sr}, \sigma_{Sr}^2)$, the following formula can be deduced from formula (4-7) as,

$$P_r = \int_0^\infty \frac{S_a}{\sigma_{Sa}^2} \cdot \exp\left[-\frac{S_a^2}{2\sigma_{Sa}^2}\right] \cdot \left[1 - \Phi\left(\frac{S_a - \mu_{Sr}}{\sigma_{Sr}}\right)\right] dS_a \tag{4-8}$$

If the stress is $S_a \sim R_a(\sigma_{Sa})$ then the logarithm $y = \ln S_r$ of strength S_r satisfies $y \sim \mathcal{N}(\mu_y, \sigma_y^2)$, formula (4-7) gives,

$$P_r = \int_0^\infty \frac{S_a}{\sigma_{Sa}^2} \cdot \exp\left[-\frac{S_a^2}{2\sigma_{Sa}^2}\right] \cdot \Phi\left(\frac{\mu_y - \ln S_a}{\sigma_y}\right) dS_a \tag{4-9}$$

where, $\mu_y = \ln \mu_{Sa} - \frac{1}{2}\sigma_y^2$ and $\sigma_y^2 = \ln\left[\left(\frac{\sigma_{Sa}}{\mu_{Sa}}\right)^2 + 1\right]$ are the average value and the variance of variable $\ln S_a = y$ respectively.

In practical computation, the integration in the above formulae can be implemented by either numerical methods or the method of Mellin deformation based on the integrated function.

5.5. OUTLINE FOR THE DYNAMIC RELIABILITY OF STRUCTURES

The dynamic reliability of structures means the structural ability to implement the required function under environmental conditions, fixed time interval and random dynamic loads.

The theory of structural dynamic reliability is a new branch of structural dynamics, which concerns the study of structural reliability analysis as well as the corresponding computational methods under random dynamic loads through the use of combined methods of structural dynamics and probability theory. Since dynamic reliability theory is relatively recent and is of considerable difficulty, some fundamental problems about it have not been solved thus far.

In structural dynamic reliability analysis, there exist two main failure mechanisms. These are the first-passage and the fatigue.

5.5.1. Failure Mechanism of the First-passage*

In this kind of problem, the failure of a structure is due to the dynamic response $x(t)$ such that (displacement, stress or strain) the structure oversteps its limit value or safety limit. According to the different safety limits, the concept of the first-passage and the corresponding dynamic reliability can be classified into the following three types:

(1) Single limit (B limit)

In this case, the pregiven limit is a constant as $x = \lambda$. The safety region of the structural response $x(t)$ is $\Omega_r = \{x(t) < \lambda\}$. Then the dynamic reliability, i.e., the probability that structural dynamic response $x(t)$ in $[0, T]$ does not exceed the safety limit λ, will be,

$$P_r(\lambda) = P_{\text{rob}}\{x(t) \leqslant \lambda, 0 \leqslant t \leqslant T\} \tag{5-1}$$

*By First-passage is meant that the first time at which a structure's design stress limit is exceeded.

The above formula (5-1) can also be written as,

$$P_r(\lambda) = P_{\text{rob}}\{\max_{t\in T} x(t) \leqslant \lambda\} \tag{5-1a}$$

Suppose T_{f1} denotes the time when response $x(t)$ under the initial condition of $x(0) = 0$ and $\dot{x}(0) = 0$ exceeds the limit λ for the first time, it is obvious that T_{f1} is a random variable and its probability distribution function is

$$F_{T_{f1}}(t) = P\{T_{f1} \leqslant t, 0 \leqslant t \leqslant T\} \tag{5-2}$$

So the maximum value of the distribution function is the passage probability of the response $x(t)$ within the time interval T. The dynamic reliability in T can also be expressed by the distribution function $F_{T_{f1}}$ in the following form,

$$P_r(\lambda) = 1 - F_{T_{f1}}(T) \tag{5-1b}$$

(2) Double limit (D limit)

Assuming that the pregiven upper and lower safety limits are $x = \lambda_u$ and $x = -\lambda_l(\lambda_u, \lambda_l$ are the constants) separately and the safety region of the structural response $x(t)$ is $\Omega_r = \{-\lambda_l < x(t) < \lambda_u\}$, the dynamic reliability, i.e., the probability of the response $x(t)$ in $[0, T]$ is located in the interval $[-\lambda_l, \lambda_u]$, becomes,

$$P_r(-\lambda_l, \lambda_u) = P_{\text{rob}}\{-\lambda_l \leqslant x(t) \leqslant \lambda_u, 0 \leqslant t \leqslant T\} \tag{5-3}$$

It is clear that the double limit will become a single limit provided the lower limit $\lambda_l \to \infty$. If the same procedure in a single limit is applied to dynamic reliability (5-3), the reliability can be expressed by both the maximum and the minimum value of response and the probability distribution function at the first passage time as,

$$P_r(-\lambda_l, \lambda_u) = P_{\text{rob}}(\max_{t\in T} x(t) \leqslant \lambda_u) \cap (\min_{t\in T} x(t) \geqslant -\lambda_l) \tag{5-3a}$$

$$\overset{\text{or}}{=} 1 - F_{T_{f2}}(T) \tag{5-3b}$$

where, $F_{T_{f2}}(\cdot)$ is the probability distribution function of the time when the response $x(t)$ exceeds the double limits λ_u and $-\lambda_l$ for the first time.

If the upper and the lower limits are symmetrical with respect to zero, i.e., $\lambda_u = \lambda_l = \lambda$, formula (5-3a) will become,

$$P_r(-\lambda, \lambda) = P_{\text{rob}}\{\max_{t\in T} \leqslant \lambda\} \tag{5-4}$$

If we let $x_m = \max_{t \in T} |x(t)|$ and define the probability distribution function of x_m as

$$F_{x_m}(\lambda) = P\{x_m \leqslant \lambda\} \tag{5-5}$$

the dynamic reliability can be addressed, from formula (5-4), as follows,

$$P_r(-\lambda, \lambda) = F_{x_m}(\lambda) \tag{5-6}$$

(3) Envelope limit (E limit)

The envelope $A(t)$ of structural dynamic response $x(t)$ is generally defined as,

$$A(t) = \left[x^2(t) + \frac{\dot{x}^2(t)}{\omega_0^2}\right]^{1/2} \tag{5-7}$$

in which, $\dot{x}(t)$ is the derivative of $x(t)$; ω_0 is the dominant frequency of the response $x(t)$.

Since the term $\frac{1}{2}A^2(t)$ is the energy of the structure at time t, the envelope $A(t)$ defined by formula (5-7) is also called energy envelope.

If the pregiven safety limit is $A = \lambda$ (constant), the safety region of structural envelope response is,

$$\Omega_r = \{A(t) < \lambda\}$$

As a result of this, the dynamic reliability of A(t) in the time interval $[0, T]$ can be written as,

$$P_{Ar}(\lambda) = P\{A(t) \leqslant \lambda, 0 \leqslant t \leqslant T\} = P\{\max_{t \in T} A(t) \leqslant \lambda\} \tag{5-8a}$$

$$\overset{\text{or}}{=} 1 - F_{T_f}(T) \tag{5-8b}$$

where $F_{T_f}(\cdot)$ is the probability distribution function for the first-passage time of the envelope response $A(t)$.

The phase plane descriptions of the above three limits are described in Figure 4.

From the above dynamic reliability formulae, it can be seen that the key to find the first passage dynamic reliability is to determine the probability distribution function of the first-passage time, $F_{T_{f1}}(\cdot)$, $F_{T_{f2}}(\cdot)$ and $F_{T_f}(\cdot)$. However, the accurate solution of this classical problem has not been found yet except for the first order Markov process.[2] At present, several approximate solutions for the probability distribution of the first-passage time are all based on the second order statistical parameter of the original response process $x(t)$ and its derivative process $\dot{x}(t)$.

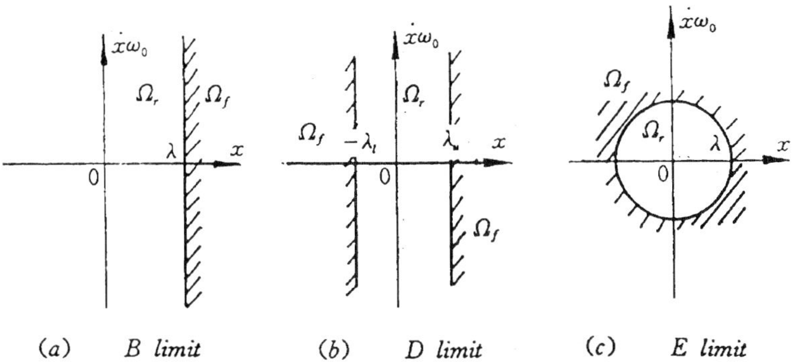

Figure 4. The phase plane description of three limits, (a) B-limit, (b) D-limit, (c) E-limit.

5.5.2. Failure Mechanism of the Fatigue

In this case, the failure of a structure is mainly due to the fact that the limit value of the structural strength is finally reached due to the stress response experiencing multiple loads under the lower ultimate stress amplitude of dynamic response, that is, the structural response's amplitude is less than or much less than its ultimate stress value. Generally speaking, the failure model of a structure can be classified into two categories as the accumulated stress and the crack expansion.

(1) The model of accumulated stress

This kind of model is based on both the theory of accumulated stress and the statistical relationship between stress amplitude S and the cycling number N. Thus the statistical parameters with respect to the accumulated stress or the fatigue life can be obtained from the above theory and the statistical relationship.

(2) The model of crack expansion

In this model, the fatigue procedure is classified into three stages such as the crack forming, expansion and fracture. The final fracture is considered due to the fact that the corresponding stress exceeds the strength capability at the first-passage.

The above two kinds of failure mechanisms, the first-passage and fatigue, are the two basic types of the damage models in reliability analysis of structural dynamics. In addition to that, there are two damage criteria, i.e., energy criteria and deformation-energy criteria. The energy damage criteria involve fatigue damage while the double damaging criterion of deformation-energy correspond to both damage resulting from first-passage or fatigue. When structural deformation is large enough, the damage will be the first-passage damage. In contrast, if the deformation is small, the damage will be fatigue damage. If structural deformation is neither large nor small, the damage mechanism will be the combination of both.

5.6. THE LEVEL CROSSINGS OF THE RANDOM PROCESS

Within the time interval $[0, T]$, the number of the system random process $x(t)$ response crossing the given limit level $x = \lambda$ is considered as a random variable. The study of such random processes was begun in 1944 when Rice published his paper.[3]

Suppose $x(t)$ is a random process with zero average, and the total number of $x(t)$ crossing of the limit $x = \lambda$ under the initial condition $x(0) = 0$ within $[0, T]$ is taken as a number recording process $n(\lambda, T)$, the statistical values of which include the expected value $N(\lambda, T)$ and the expected value $\upsilon_\lambda(t)$ within the unit time (i.e., the crossing rate of $x(t)$ through $x = \lambda$), the following formulae obviously holds,

$$N(\lambda, T) = E[n(\lambda, T)] \tag{6-1}$$

$$\overset{\text{or}}{=} \int_0^T \upsilon_\lambda(t)dt \tag{6-2}$$

If $\upsilon_\lambda(t)$ is independent of the time t, i.e., $\upsilon_\lambda(t) = \upsilon_\lambda = \text{constant}$, the following process results as,

$$N(\lambda, T) = \upsilon_\lambda \cdot T \tag{6-3}$$

In order to deduce the computation formulae of $N(\lambda, T)$ and $\upsilon_\lambda(t)$, the analysis method presented by Middleton[4] is employed. First of all, utilizing the following random process,

$$y(t) = U\{x(t) - \lambda\} \tag{6-4}$$

and the corresponding derivative

$$\dot{y}(t) = \dot{x}(t)\delta\{x(t) - \lambda\} \tag{6-5}$$

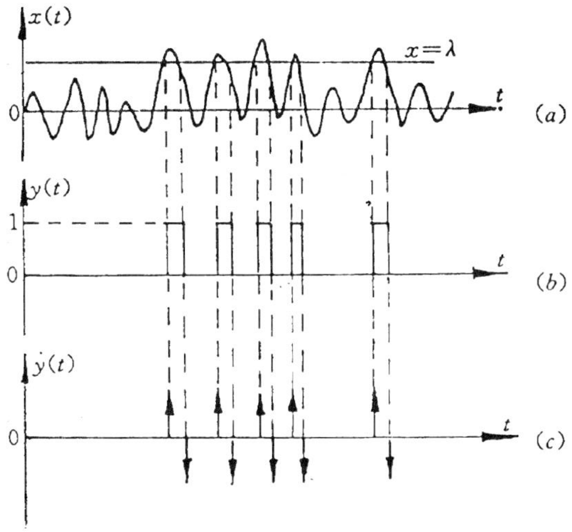

Figure 5. (a) $x(t)$ and $x = \lambda$, (b) $y(t)$, (c) $\dot{y}(t)$.

where $U\{\cdot\}$ is a unit jump function, i.e., Heaviside function; $\delta\{\cdot\}$ is Dirac-δ function.

One sample function of the process $x(t)$, the pregiven limit $x = \lambda$ as well as $y(t)$ and $\dot{y}(t)$ defined in formulae (6-4) and (6-5) are described in Figure 5.

Investigating Figure 5 indicates that the random process $y(t)$ is a number function whenever $x(t)$ crosses the limit $x = \lambda$ with a plus or minus derivative. And the value of the process $y(t)$ equals 1 or 0. In addition, the total number $n(\lambda, T)$, for which $x(t)$ in $[0, T]$ crosses the limit $x = \lambda$ with a plus or minus derivative, can be deduced by integrating formula (6-5), i.e.,

$$n(\lambda, T) = \int_0^T |\dot{x}(t)|\delta\{x(t) - \lambda\}dt \tag{6-6}$$

If the joint probability density function of $x(t)$ and $\dot{x}(t)$ is known as $f_{x\dot{x}}(x, \dot{x}, t)$, the expected value of $n(\lambda, T)$ is,

$$\begin{aligned} N(\lambda, T) &= E[n(\lambda, T)] \\ &= \int_0^T \int_{-\infty}^{\infty} \int_{-\infty}^{\infty} |\dot{x}(t)|\delta\{x(t) - \lambda\} \cdot f_{x\dot{x}}(x, \dot{x}, t)dxd\dot{x}dt \\ &= \int_0^T \int_{-\infty}^{\infty} |\dot{x}(t)| f_{x\dot{x}}(\lambda, \dot{x}, t)d\dot{x}dt \end{aligned} \tag{6-7}$$

Comparing the above formula with formula (6-2) gives the following famous Rice formula[3], that is the crossing rate

$$v_\lambda(t) = \int_{-\infty}^{\infty} |\dot{x}(t)| f_{x\dot{x}}(\lambda, \dot{x}, t) d\dot{x} \tag{6-8}$$

If only the case that $x(t)$ in $[0, T]$ crosses the limit $x = \lambda$ with a plus derivate ($\dot{x} > 0$) is taken into account, formulae (6-7) and (6-8) can also be separately written as,

$$N(\lambda^+, T) = \int_0^T \int_0^{\infty} \dot{x}(t) \cdot f_{x\dot{x}}(\lambda, \cdot x, t) d\dot{x} dt \tag{6-9}$$

and

$$v_\lambda^+(t) = \int_0^{\infty} \dot{x}(t) \cdot f_{x\dot{x}}(\lambda, \dot{x}, t) d\dot{x} \tag{6-10}$$

In contrast, if only the case that $x(t)$ in $[0, T]$ crosses the limit $x = \lambda$ with a negative derivative ($\dot{x} < 0$) is considered, formulae (6-7) and (6-8) can be respectively described as,

$$N(\lambda^-, T) = \int_0^T \int_{-\infty}^{0} |\dot{x}(t)| \cdot f_{x\dot{x}}(\lambda, \dot{x}, t) d\dot{x} dt \tag{6-11}$$

and

$$v_\lambda^-(t) = \int_{-\infty}^{0} |\dot{x}(t)| \cdot f_{x\dot{x}}(\lambda, \dot{x}, t) d\dot{x} \tag{6-12}$$

If the response $x(t)$ is a stationary random process, that is to say, the crossing rate $v_\lambda(t)$ independent of time, then,

$$v_\lambda(t) = v_\lambda = \text{const.} \tag{6-13}$$

In addition, if $f_{x\dot{x}}(x, \dot{x}, t)$ is symmetrical with respect to the coordinate origin, i.e.,

$$f_{x\dot{x}}(x, \dot{x}, t) = f_{x\dot{x}}(-x, -\dot{x}, t) \tag{6-14}$$

Equations (6-10) and (6-12) gives

$$v_\lambda^+(t) = v_\lambda^-(t) = \frac{1}{2} v_\lambda(t) \tag{6-15}$$

If the response $x(t)$ is a Guassian stationary process with zero average, $x(t)$ will be independent to $\dot{x}(t)$. Therefore, the joint probability density function is,

$$f_{x\dot{x}}(x, \dot{x}, t) = \frac{1}{2\pi\sigma_x\sigma_{\dot{x}}} \exp\left\{-\frac{1}{2}\left[\frac{x^2}{\sigma_x^2} + \frac{\dot{x}^2}{\sigma_{\dot{x}}^2}\right]\right\} \tag{6-16}$$

where, σ_x^2 and $\sigma_{\dot{x}}^2$ are the mean square values of the response $x(t)$ and its derivative $\dot{x}(t)$ respectively.

Substituting the above formula into equations (6-10), (6-12) and (6-15) leads to,

$$v_\lambda^+ = v_\lambda^- = \frac{1}{2}v_\lambda = \frac{1}{2\pi}\frac{\sigma_{\dot{x}}}{\sigma_x}\exp\left[-\frac{\lambda^2}{2\sigma_x^2}\right] \tag{6-17}$$

If the limit is taken to zero, i.e., $\lambda = 0$, the above formula (6-17) becomes,

$$v_0^+ = v_0^- = \frac{1}{2}v_0 = \frac{1}{2\pi}\frac{\sigma_{\dot{x}}}{\sigma_x} \tag{6-18}$$

where, $v_0 = v_0^+ + v_0^-$, v_0^+ and v_0^- are the crossing rate of $x(t)$ through the line $x = 0$ by plus and minus derivatives of $x(t)$ separately.

By employing formula (6-18), the formula (6-17) can also be expressed as,

$$v_\lambda^+ = v_\lambda^- = \frac{1}{2}v_\lambda = v_0^\pm \exp\left[-\frac{\lambda^2}{2\sigma_x^2}\right] \tag{6-19}$$

If the response $x(t)$ is a nonstationary Gaussian process with zero average, i.e., the crossing rate of $x(t)$ is the function of time t, the following formulae hold,

$$v_0^+(t) = v_0^-(t) = \frac{1}{2\pi}\frac{\sigma_{\dot{x}}(t)}{\sigma_x(t)} \tag{6-20}$$

$$v_\lambda^+(t) = v_0^+(t)\exp\left[-\frac{\lambda^2}{2\sigma_x^2(t)}\right] \tag{6-21}$$

$$v_\lambda^-(t) = v_0^-(t)\exp\left[-\frac{\lambda^2}{2\sigma_x^2(t)}\right] \tag{6-22}$$

5.7. THE PEAK DISTRIBUTION OF THE RANDOM PROCESS

In the reliability analysis of the structural dynamic problem, the peak (maximum value) distribution of structural response process $x(t)$, is usually involved.

Suppose that the process $x(t)$ in $[0, T]$ is a continuous random process with the second order derivative, the process $x(t)$ reaches the maximum or minimum value at time t_0 provided $\dot{x}(t_0) = 0$ and $\ddot{x}(t_0) < 0$ or $\dot{x}(t_0) = 0$

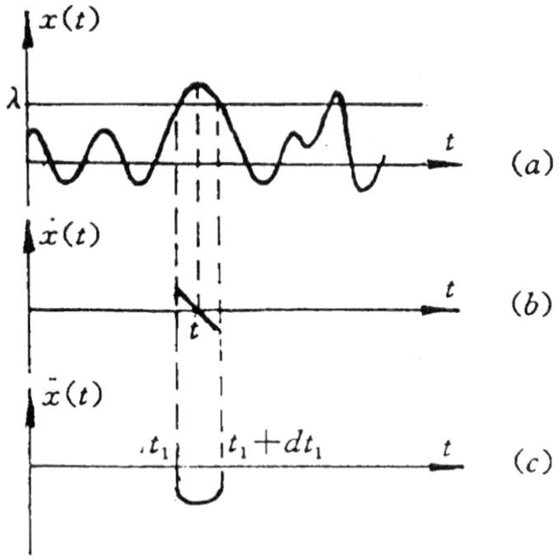

Figure 6. The sample function of $x(t)$, $\dot{x}(t)$ and $\ddot{x}(t)$.

and $\ddot{x}(t_0) > 0$ according to the necessary condition to which the extreme point is subjected. The sample function and the first and the second order derivative characteristics in the neighborhood of the extreme point of random process $x(t)$ are described in Figure 6. Obviously, if the process $x(t)$ at time $t \in [t_1, t_1 + dt_1]$ is considered at a peak above the fixed limit $x = \lambda$, the following conditions must be satisfied simultaneously,

$$x(t) > \lambda, \dot{x}(t) = 0, \ddot{x}(t) < 0 \tag{7-1}$$

Up to this point, the peak distribution of process $x(t)$ can be determined by the above extreme point condition.

With the same procedure as the study of crossing a specific level, Middleton[4] has deduced the total number $m(\lambda, T)$ for which the peak process $x(t)$ in $[0, T]$ is over the limit $x = \lambda$ by introducing the number function as follows,

$$m(\lambda, T) = \int_0^T \ddot{x}(t) \cdot \delta\{\dot{x}(t)\} \cdot U\{x(t) - \lambda\} dt \tag{7-2}$$

If the joint probability density function $f_{x\dot{x}\ddot{x}}(x, \dot{x}, \ddot{x}, t)$ of processes $x(t)$, $\dot{x}(t)$ and $\ddot{x}(t)$ are known, the expected value $M(\lambda, T)$ of the process $m(\lambda, T)$ is given by (7-2) as,

$$\begin{aligned} M(\lambda, T) &= E[m(\lambda, T)] \\ &= -\int_0^T dt \int_\lambda^\infty \int_{-\infty}^0 \ddot{x}(t) f_{x\dot{x}\ddot{x}}(x, 0, \ddot{x}, t) d\ddot{x} dx \end{aligned} \tag{7-3}$$

It can be noticed that the crossing rate $\mu(\lambda, t)$ of the peak of the process $x(t)$ with respect to the limit $x = \lambda$ is,

$$\mu(\lambda, t) = -\int_{\lambda}^{\infty}\int_{-\infty}^{0} \ddot{x}(t) f_{x\dot{x}\ddot{x}}(x, 0, \ddot{x}, t)d\ddot{x}dx \tag{7-4}$$

If the limit is $\lambda = -\infty$, formula (7-4) will become,

$$\mu(-\infty, t) = -\int_{-\infty}^{\infty}\int_{-\infty}^{0} \ddot{x}(t) f_{x\dot{x}\ddot{x}}(x, 0, \ddot{x}, t)d\ddot{x}dx \tag{7-5}$$

The formula (7-5) means that $\mu(-\infty, t)$ is the expected value of the total number for which the peak of process $x(t)$ in the unit time is greater than the given limit $\lambda = -\infty$.

According to the study by Huston and Skopinski[5], the probability distribution function $F_p(\lambda, t)$ of the peak of process $x(t)$ not exceeding the limit $x = \lambda$ can be expressed by the crossing rate of the peak as,

$$F_p(\lambda, t) = \frac{\mu(-\infty, t) - \mu(\lambda, t)}{\mu(-\infty, t)} = 1 - \frac{\mu(\lambda, t)}{\mu(-\infty, t)} \tag{7-6}$$

where $t \geqslant T_0 = \frac{2\pi}{\omega_0}$, ω_0 is the dominant frequency of the process $x(t)$.

Differentiating the above formula with respect to λ gives the probability density function of the peak of the process $x(t)$ as follows,

$$\begin{aligned} f_p(\lambda, t) &= \frac{-1}{\mu(-\infty, t)} \frac{\partial}{\partial\lambda}\mu(\lambda, t) \\ &= \frac{1}{\mu(-\infty, t)} \int_{-\infty}^{0} \ddot{x}(t) \cdot f_{x\dot{x}\ddot{x}}(\lambda, 0, \ddot{x}, t)d\ddot{x} \end{aligned} \tag{7-7}$$

If $x(t)$ is a Gaussian stationary process with zero average, the three processes $x(t), \dot{x}(t)$ and $\ddot{x}(t)$ are independent of each other, the joint probability density function $f_{x\dot{x}\ddot{x}}(x, \dot{x}, \ddot{x})$ at the point $(x, 0, \ddot{x})$ can be expressed in the following form,

$$f_{x\dot{x}\ddot{x}}(x, 0, \ddot{x}) = \frac{1}{(2\pi)^{3/2}|\mathbf{C}|^{1/2}} \exp\left[-\frac{1}{2|\mathbf{C}|}(\sigma_{\dot{x}}^2\sigma_{\ddot{x}}^2x^2 + 2\sigma_{\dot{x}}^4x\ddot{x} + \sigma_x^2\sigma_{\dot{x}}^2\ddot{x}^2)\right] \tag{7-8}$$

where,

$$\mathbf{C} = \begin{bmatrix} \sigma_x^2 & 0 & -\sigma_{\dot{x}}^2 \\ 0 & \sigma_{\dot{x}}^2 & 0 \\ -\sigma_{\dot{x}}^2 & 0 & \sigma_{\ddot{x}}^2 \end{bmatrix} \tag{7-9}$$

is the covariance matrix of $(x, \dot{x}, \ddot{x})$.

Substituting formula (7-8) into formula (7-7), we obtain the following probability density function for the peak of Gaussian stationary process $x(t)$ with zero average,

$$f_p(\lambda) = \frac{\varepsilon}{\sqrt{2\pi}\sigma_x} \exp\left[-\frac{\lambda^2}{2\sigma_x^2\varepsilon^2}\right] + \frac{\alpha\lambda}{2\sigma_x^2}\left\{1 + \text{erf}\left[\frac{\alpha\lambda}{\sqrt{2\sigma_x^2}\varepsilon}\right]\right\} \cdot \exp\left[-\frac{\lambda^2}{2\sigma_x^2}\right] \tag{7-10}$$

where, erf[·] is the error function and defined as

$$\text{erf}[x] = \int_0^x \frac{1}{\sqrt{2\pi}} \exp\left[-\frac{t^2}{2}\right] dt$$

$$\alpha = \frac{v_0^+(t)}{\mu(-\infty, t)} = \frac{v_0^+}{\mu(-\infty)} = \frac{\sigma_{\dot{x}}^2}{\sigma_x \sigma_{\ddot{x}}} \tag{7-11}$$

$$\varepsilon = (1 - \alpha^2)^{1/2} \tag{7-12}$$

and ε is called the band-width coefficient of the power spectrum of process $x(t)$, its value is $0 \leqslant \varepsilon \leqslant 1.0$.

If $x(t)$ is an ideal narrow-band Gaussian stationary process, ε will tend to 0. That is to say, $\varepsilon \to 0$. Consequently the density function of formula (7-10) become,

$$f_p(\lambda) = \frac{\lambda}{\sigma_x^2} \exp\left[-\frac{\lambda^2}{2\sigma_x^2}\right] \quad (\lambda \geqslant 0) \tag{7-13}$$

In this case, the peak of $x(t)$ is a Rayleigh distribution.

If $x(t)$ is an ideal wide-band Guassian stationary process (white noise process), i.e., $\varepsilon \to 1.0$, the density function of formula (7-10) is

$$f_p(\lambda) = \frac{1}{\sqrt{2\pi}\sigma_x} \exp\left[-\frac{\lambda^2}{2\sigma_x^2}\right] \quad (-\infty < \lambda < \infty) \tag{7-14}$$

In this circumstance, the peak of $x(t)$ is a Gaussian distribution.

If the band-width coefficient of Gaussian stationary random process takes values between 0 and 1.0, i.e., $0 < \varepsilon < 1.0$, the probability distribution of the peak of the process $x(t)$ will be situated between the Rayleigh and the Gaussian distributions. The curves of the probability density function of the peak for different values of ε are described in Figure 7.

If $x(t)$ is a nonstationary random process, the probability density function $f_p(\lambda, t)$ under the condition that the peak of $x(t)$ is not greater than the limit $x = \lambda$ can be obtained theoretically[6,7] by substituting the joint probability density function $f_{x\dot{x}\ddot{x}}$ into formula (7-7). If $x(t)$ is a nonstationary Gaussian narrow-band process, the probability density function of the peak can be approximately expressed in the following formula,

$$f_p(\lambda, t) = -\frac{1}{v_0^+(t)} \frac{\partial}{\partial\lambda} v_\lambda^+(t) \tag{7-15}$$

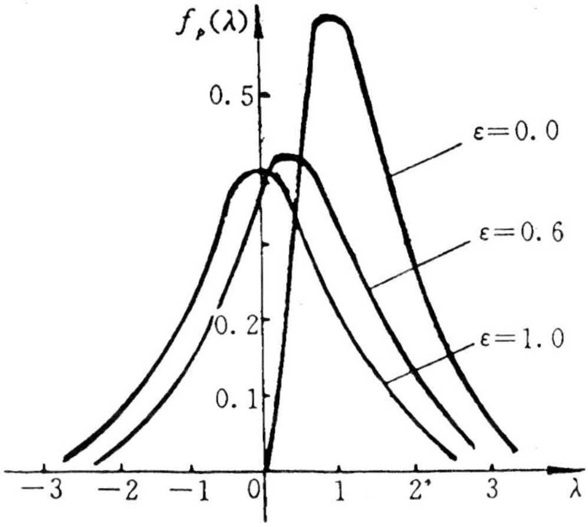

Figure 7. The curves of probability density function for the peak of process $x(t)$ corresponding to the different band-width coefficient ε.

in which $t \geqslant T_0 = \frac{2\pi}{\omega_0}$; ω_0 is the dominant frequency of the process $x(t)$, the crossing rate $v_0^+(t)$ and $v_0^-(t)$ can be found from equations (6-20) and (6-21).

5.8. POISSON PROCEDURE METHOD OF DYNAMIC RELIABILITY ANALYSIS FOR THE FIRST PASSAGE OF SYSTEM WITH A SINGLE DEGREE OF FREEDOM

As in the previous discussion, the key problem to determining the dynamic reliability of the first-passage*is to find the probability distribution function of process $x(t)$'s first passage time. Unfortunately, to date, the accurate solution of the probability distribution function has not been found yet for every process. Therefore, various kinds of approximate methods are used to solve the problem of dynamic reliability. Of all the approximate methods, the Poisson procedure method is a simple and an important one. What the Poisson method is based on is that the number of crossing of both the response process $x(t)$ and the pregiven limit is a Poisson process.

*To repeat, first passage means the first time the stress limit of a structure is exceeded.

If the limit λ (or λ_u, λ_l) is very large, the arbitrary two crossing events can be considered to be independent because the probability of the event that the process $x(t)$ crossed the limit λ is very small. It means that the crossing number $n(\lambda, T)$ of $x(t)$ and the limit λ in $[0, T]$ can be reasonably considered to obey the Poisson process.

Without compromising generality, we first consider the circumstance that $n(\lambda, T)$ is a nonstationary Poisson process and the limit type is a double limit (D limit). According to the Poisson distribution, the probability of $x(t)$ in $[0, T]$ crossing the upper limit $x = \lambda_u$ with a positive gradient at the ith time is,

$$P\{n(\lambda_u^+, T) = i\} = \frac{1}{i!}\left[\int_0^T v_{\lambda_u}^+(t)dt\right]^i \cdot \exp\left[-\int_0^T v_{\lambda_u}^+(t)dt\right] \quad (8\text{-}1)$$

Likewise, the probability of $x(t)$ in $[0, T]$ crossing the lower limit $x = -\lambda_l$ with a negative gradient at the ith time becomes,

$$P\{n(-\lambda_l^-, T) = i\} = \frac{1}{i!}\left[\int_0^T v_{-\lambda_l}^-(t)dt\right]^i \cdot \exp\left[-\int_0^T v_{-\lambda_l}^-(t)dt\right] \quad (8\text{-}2)$$

Since the double limit dynamic reliability is defined as the probability of response $x(t)$ in $[0, T]$ without crossing over the upper and the lower limit, the dynamic reliability of the system can be written as,

$$\begin{aligned} P_r(-\lambda_l, \lambda_u) &= P\{n(\lambda_u^+, T) = 0 \cap n(-\lambda_l^-, T) = 0\} \\ &= P\{n(\lambda_u^+, T) = 0\} \cdot P\{n(-\lambda_l^-, T) = 0\} \\ &= \exp\{-\int_0^T [v_{\lambda_u}^+(t) + v_{-\lambda_l}^-(t)]dt\} \end{aligned} \quad (8\text{-}3)$$

If the joint probability density function $f_{x\dot{x}}(x, \dot{x}, t)$ of the processes $x(t)$ and $\dot{x}(t)$ is symmetrical with respect to the origin, and the given upper and lower limits are symmetrical with respect to the zero line, i.e.,

$$f_{x\dot{x}}(x, \dot{x}, t) = f_{x\dot{x}}(-x, -\dot{x}, t),\ \lambda_u = \lambda_l = \lambda$$

the dynamic reliability of system can be further deduced as,

$$P_r(-\lambda, \lambda) = F_{x_m}(\lambda) = \exp\{-2\int_0^T v_\lambda^+(t)dt\} \quad (8\text{-}4)$$

As for the case of the single limit (B limit), the systematic dynamic reliability in $[0, T]$ is given by equation (8-3) by letting $\lambda_u = \lambda$ and $\lambda_l = \infty$,

$$P_r(\lambda) = 1 - F_{T_{f1}}(T) = \exp\{-\int_0^T v_\lambda^+(t)dt\} \quad (8\text{-}5)$$

If the response $x(t)$ is a stationary random process, the corresponding crossing number $n(\lambda, T)$ will be a stationary Poisson process. In this case, the crossing rate of $x(t)$ is independent of time, t. Consequently, the dynamic reliability formula (8-3) with double limits can be simplified as,

$$P_r(-\lambda_l, \lambda_u) = \exp\{-(v^+_{\lambda_u} + v^-_{-\lambda_l}) \cdot T\} \tag{8-6}$$

By going a step further, if $f_{x\dot{x}}(x, \dot{x}, t) = f_{x\dot{x}}(-x, -\dot{x}, t)$ and $\lambda_u = \lambda_l = \lambda$, the dynamic reliability formula (8-4) with symmetric double limits can be simplified as follows,

$$P_r(-\lambda, \lambda) = F_{x_m}(\lambda) = \exp\{-2v^+_{\lambda} T\} \tag{8-7}$$

Considering $\lambda_u = \lambda$ and $\lambda_l = \infty$ in formula (8-6) leads to the dynamic reliability of the system with a single limit,

$$P_r(\lambda) = 1 - F_{T_{f1}}(T) = \exp\{-v^+_{\lambda} T\} \tag{8-8}$$

It can be found from the above several formulae that finding the crossing rate is the key point to compute the systematic dynamic reliability.

If the process $x(t)$ is a stationary Gaussian process with zero average, the crossing rate is unrelated to the time t (see formulae (6-17) and (6-18). Substituting formulae (6-17) and (6-18) into formulae (8-6), (8-7) and (8-8), respectively, leads to the following dynamic reliability formulae of the system with double, symmetric double and single limits in the interval $[0, T]$,

$$P_r(-\lambda_l, \lambda_u) = \exp\left\{-\frac{1}{2\pi}\frac{\sigma_{\dot{x}}}{\sigma_x}T\left[\exp\left(-\frac{\lambda_u^2}{2\sigma_x^2}\right) + \exp\left(-\frac{\lambda_l^2}{2\sigma_x^2}\right)\right]\right\} \tag{8-9}$$

$$P_r(-\lambda, \lambda) = \exp\left\{-\frac{1}{\pi}\frac{\sigma_{\dot{x}}}{\sigma_x}T \cdot exp\left(-\frac{\lambda^2}{2\sigma_x^2}\right)\right\} \tag{8-10}$$

$$P_r(\lambda) = \exp\left\{-\frac{1}{2\pi}\frac{\sigma_{\dot{x}}}{\sigma_x}T \cdot \exp\left(-\frac{\lambda^2}{2\sigma_x^2}\right)\right\} \tag{8-11}$$

The above three formulae are the famous Poisson formulae by which the dynamic reliability of the system with the single degree-of-freedom can be found.

If $x(t)$ is a nonstationary Gaussian process with zero average, the crossing rate will change with time t, and the corresponding dynamic reliability of the system in $[0, T]$ with double, symmetric double and single limit can be determined by substituting formulae (6-20)–(6-22) into formulae (8-3)–(8-5) from the following three formulae,

$$P_r(-\lambda_l, \lambda_u) = \exp\left\{-\frac{1}{2\pi}\int_0^T \frac{\sigma_{\dot{x}}(t)}{\sigma_x(t)}\left[\exp\left(-\frac{\lambda_u^2}{2\sigma_x^2(t)}\right) + \exp\left(-\frac{\lambda_l^2}{2\sigma_x^2(t)}\right)\right]dt\right\} \tag{8-12}$$

$$P_r(-\lambda, \lambda) = \exp\left\{-\frac{1}{\pi}\int_0^T \frac{\sigma_{\dot{x}}(t)}{\sigma_x(t)}\exp\left(-\frac{\lambda^2}{2\sigma_x^2(t)}\right)dt\right\} \tag{8-13}$$

$$P_r(\lambda) = \exp\left\{-\frac{1}{2\pi}\int_0^T \frac{\sigma_{\dot{x}}(t)}{\sigma_x(t)}\exp\left(-\frac{\lambda^2}{2\sigma_x^2(t)}\right)dt\right\} \tag{8-14}$$

5.9. THE IMPROVED POISSON PROCEDURE METHOD

According to the study of Cramer[8], only when the response $x(t)$ is a stationary Gaussian process with zero average and the given limit λ (or λ_u, λ_l) $\to \infty$, the dynamic reliability by Poisson procedure method is the accurate solution of the problem. If the limit λ (or λ_u, λ_l) is not big enough, the dynamic reliability given by Poisson procedure method will be a conservative (rather safe) approximation due to the following reasons: The Poisson method is based on noting that the crossing events between the response process $x(t)$ and the given limit are independent of each other (i.e., the grant of Poisson process). However, this is usually not true for the narrow-band process owing to the fact that the sampling curve is a simple harmonic wave with the amplitude changing slowly. In this case, if the process, $x(t)$, exceeds the limit at this time, the probability that the process, $x(t)$, will exceed the limit at the next time is quite high. That is to say, the crossing events for the narrow-band process will tend to occur consecutively. Therefore, such limit exceeding for a Poisson process will not conform with reality.[9]

It is obvious that if the given limit λ (or λ_u, λ_l) is rather high (for instance, $\lambda > 3\sigma_x$), the hypothesis of a crossing event by a Poisson process will be acceptable either for a process with a narrow-band of frequency or for a process with a wide-band of frequency. Moreover, the higher the limit λ is, the more the hypothesis is realistic.

For the sake of overcoming the above defect of the Poisson procedure method, some researchers such as Vanmarcke[10] and Crandall[11] have presented several improved methods. Based on the hypothesis that the crossing event between the process and the limit is described by a Markov process (i.e., the crossing event of the next time is related to the crossing event of previous time only, there is no relation with the crossing event of time before*)Corotis[12] has presented an improvement for the dynamic reliability computing formula (8-9) of the Poisson procedure method. The improved expression is given as

$$P_r(-\lambda_l, \lambda_u) = \exp\left\{-\frac{1}{2\pi}\frac{\sigma_{\dot{x}}}{\sigma_x}T\left[\sum_{j=1}^{2}\exp\left(\frac{r_j^2}{2}\right)\alpha(r_j)\right]\right\} \tag{9-1}$$

where,

$$r_1 = \frac{\lambda_u}{\sigma_x}, r_2 = \frac{\lambda_l}{\sigma_x} \tag{9-2}$$

$$\alpha(r_j) = \frac{1 - \exp\left[-\left(\frac{\pi}{2}\right)^{1/2} r_j\varepsilon\right]}{1 - \exp\left[-\frac{r_j^2}{2}\right]} \quad (j = 1, 2) \tag{9-3}$$

$$\varepsilon = \left(1 - \frac{a_1^2}{a_0a_2}\right)^{1/2} \tag{9-4}$$

$$a_k = \int_0^\infty \omega^k S_x(\omega)d\omega \quad (k = 0, 1, 2) \tag{9-5}$$

in which, $S_x(\omega)$ is the single-side power spectral density function of the response $x(t)$; $a_k(k = 0, 1, 2)$ is the kth order spectral moment of $S_x(\omega)$; $\varepsilon(0 \leqslant \varepsilon \leqslant 1.0)$ is the band-width coefficient. If $\varepsilon \to 0$, $x(t)$ will tend to be a process with a narrow-band of frequency; If ε is located between 0.35 and 1.0, $x(t)$ will become a process with the wide-band of frequency.

For the symmetric double limit, i.e., $\lambda_u = \lambda_l = \lambda$, the improved formula (9-1) can be simplified as,

$$P_r(-\lambda, \lambda) = \exp\left\{-\frac{1}{\pi}\frac{\sigma_{\dot{x}}}{\sigma_x}T \cdot \exp\left(-\frac{r^2}{2}\right)\frac{1 - \exp\left[-\left(\frac{\pi}{2}\right)^{1/2} r\varepsilon\right]}{1 - \exp\left[-\frac{r^2}{2}\right]}\right\} \tag{9-6}$$

where,

$$r = \frac{\lambda}{\sigma_x} \tag{9-7}$$

*This is the definition of a first order Markou process.

It can be confirmed that if $x(t)$ is a process with a narrow band ($\varepsilon \to 0$) and λ is the lower limit ($r_j \leqslant 2$), the result of the improved formula will be better than that of the original Poisson formula. Further, with the limit λ (i.e., r_j) $\to \infty$, the result of the improved formula will tend to be an accurate solution of the problem.

Similarly, based on the hypothesis that the crossing of the process over the limit is subjected to a Markov process, an improvement through the use of (8-12)–(8-14) for the dynamic reliability of a nonstationary Gaussian process in Poisson procedure method results. The formula (8-12) of the double limit dynamic reliability could be modified to,

$$P_r(-\lambda_l, \lambda_u) = \exp\left\{-\frac{1}{2\pi}\int_0^T \frac{\sigma_{\dot{x}}(t)}{\sigma_x(t)}\left[\sum_{j=1}^{2}\exp\left(\frac{r_j^2(t)}{2}\right)\alpha(r_j, t)\right]dt\right\} \tag{9-8}$$

where,

$$r_1(t) = \frac{\lambda_u}{\sigma_x(t)}, \; r_2(t) = \frac{\lambda_l}{\sigma_x(t)} \tag{9-9}$$

$$\alpha(r_j, t) = \frac{1 - \exp\left[-\left(\frac{\pi}{2}\right)^{1/2} r_j(t)\varepsilon(t)\right]}{1 - \exp\left[-\frac{r_j^2(t)}{2}\right]} \tag{9-10}$$

$$\varepsilon(t) = \left[1 - \frac{\sigma_{\dot{x}}^2(t)}{\sigma_x(t)\sigma_{\ddot{x}}(t)}\right]^{1/2} \tag{9-11}$$

in which, $\sigma_x(t)$, $\sigma_{\dot{x}}(t)$ and $\sigma_{\ddot{x}}(t)$ are the root of mean square of the nonstationary response process $x(t)$ and its 1st-order and 2nd-order derivatives at time t respectively.

By applying the above formulae, Corotis[12] has computed the dynamic reliability for an equivalent single-degree-of-freedom system (with ξ = 0.001, 0.01 and 0.1 separately) under the excitation of the wide frequency band with zero average, and obtained the five dynamic reliability curves shown in Figures 8–10.

In the Figures 8–10, the curve PS is the result of the case wherein the process is a stationary Gaussian process and the crossing is subjected to a Poisson process; PN describes the result for the case wherein the process is a nonstationary Gaussian process and the crossing is subjected to a Poisson process: MS presents the result for the case with a stationary Gaussian process with the crossing subjected to a Markov process; MN shows the result of the solution with a nonstationary Gaussian process and the crossing subjected to a Markov process; ME gives the result for σ_x in MS, which is the function of time t.

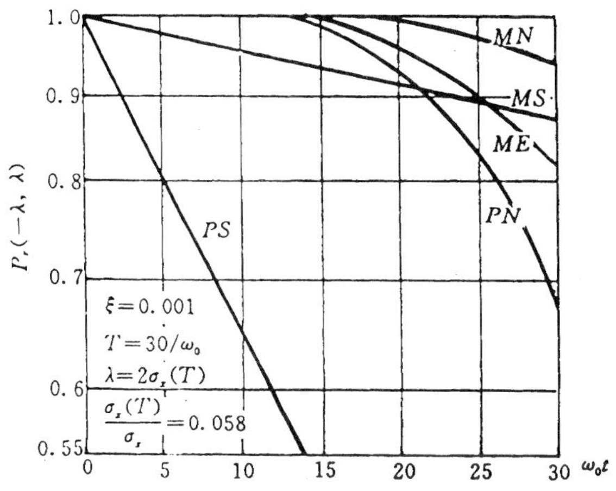

Figure 8. The computation result of dynamic reliability for the single-degree-of-freedom system with $\xi = 0.001$.

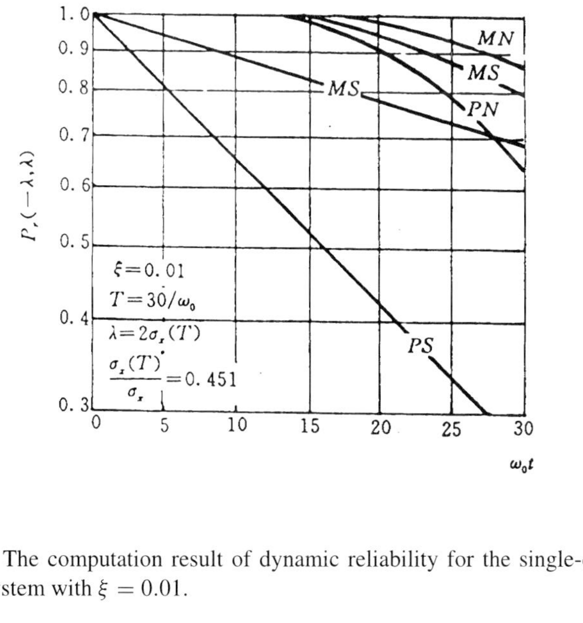

Figure 9. The computation result of dynamic reliability for the single-degree-of-freedom system with $\xi = 0.01$.

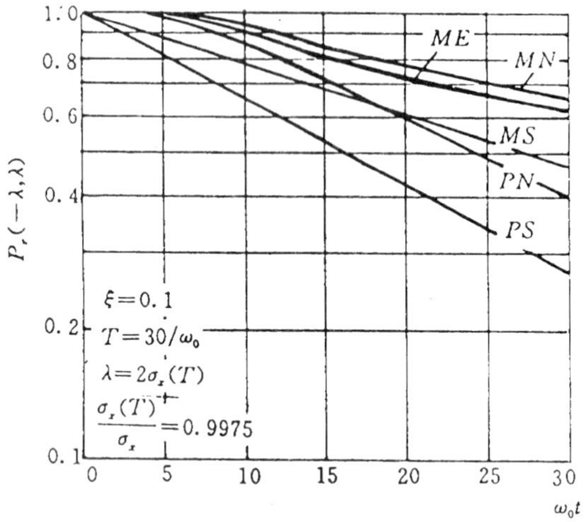

Figure 10. The computation result of dynamic reliability for the single-degree-of-freedom system with $\xi = 0.1$.

It can be seen from the results that the PS's situation is the most conservative, particularly for the small damped situation. This is because of the hypothesis of independent crossing and a stationary response process. Since the influence of a nonstationary response is considered in the result of PN, the result of PN is improved compared with that of PS, but it is still relatively conservative. Owing to the consideration of the influence of clump in MS, its result is better than that of PS. However, the influence of nonstationary response process's rising is not concerned in the MS's result, it is not suitable to the problem within the short time interval. While in the result MN, both influence of the clump and the nonstationary response process's rising are all considered, it is the most approached to the accurate solution.

5.10. APPROXIMATING SIMPLIFICATION TO THE POISSON FORMULA

The Poisson formulae (8-9)–(8-11) are the most important and useful formulae for computing the dynamic reliability of the single-degree-of-freedom system so far. These formulae are based on the well known mean square value σ_x^2 and $\sigma_{\dot{x}}^2$ of the response process $x(t)$ and its derivative $\dot{x}(t)$. For the case of a process $x(t)$ being a narrow-band process, Chen[13] has presented the following simple approximation for the Poisson formula.

According to the theory of random vibration, the ratio term $\frac{\sigma_{\dot{x}}}{\sigma_x}$ in the Poisson formula can be expressed by the integrated form of the power spectral density function $S_x(\omega)$ in the following form,

$$\frac{\sigma_{\dot{x}}}{\sigma_x} = \left[\frac{\int_0^{\omega_c} \omega^2 S_x(\omega) d\omega}{\int_0^{\omega_c} S_x(\omega) d\omega} \right]^{1/2} \tag{10-1}$$

where, ω_c is the closure frequency of $S_x(\omega)$.

By considering the continuity of function ω^2 in the interval $[0, \omega_c]$ and the characteristics of a nonnegative and integrable function $S_x(\omega)$, the above formula (10-1) can be written by employing the theorem of the mean in the integration,

$$\frac{\sigma_{\dot{x}}}{\sigma_x} = \left[\frac{\omega_m^2 \int_0^{\omega_c} S_x(\omega) d\omega}{\int_0^{\omega_c} S_x(\omega) d\omega} \right]^{1/2} = \omega_m \quad (0 < \omega_m < \omega_c) \tag{10-2}$$

in which, ω_m is the integrated middle point.

Even though the existence of ω_m is correct in theory, it is very difficult to determine ω_m in practice. As an approximation, the upper bound of the integrated interval is approximately taken as the middle point in the paper[13], i.e., let $\omega_m = \omega_c$. Consequently, the Poisson formula (8-9) can be further approximately expressed as,

$$P_r(-\lambda_l, \lambda_u) = \exp\left\{ -\frac{\omega_c T}{2\pi} \left[\exp\left(-\frac{\lambda_u^2}{2\sigma_x^2} \right) + \exp\left(-\frac{\lambda_l^2}{2\sigma_x^2} \right) \right] \right\} \tag{10-3}$$

Because the process of solving the root of mean square $\sigma_{\dot{x}}$ for the response derivative process $\dot{x}(t)$ can be avoided in formula (10-3), it will be convenient for appreciation in engineering. Further more, as a result of properties of exponential functions, the result of formula (10-3) is relatively safe if ω_c is taken as the middle point, and that the approximating extent of formula (10-3) to the accurate solution will be increased with the subtraction $(\omega_c - \omega_m)$. Therefore, the formula (10-3) is more appropriate if the response is that of a narrow band. Most structural system responses are narrow-band processes even under the excitation of a wide-band processes, there is a certain practical value to utilizing (10-3) as a conservative and a simple evaluation formula of the Poisson formula in the computation of dynamic reliability.

5.11. THE FATIGUE LIFE OF STRUCTURES—ACCUMULATIVE DAMAGE

In random vibration, the damage of some structural elements (or mechanical elements and mechanical parts) is due to the fatigue of the accumulative utilization. Since the mechanism of fatigue damage is far more complicated than that for the first-passage, so far, there is little developed theory and few experimental results about it. In this section, the evaluation method of fatigue life under the condition of random stress is discussed by employing the Miner's theory of linear accumulative fatigue.[14]

Suppose that the stress response $S(t)$ of an element subjected to random dynamic loads is a random process. According to the theory of linear accumulative fatigue by Miner–Palmgren, the related fatigue measure D_s, when the peak value of stress is S, will be,

$$D_s = \frac{n_s}{N_s} \tag{11-1}$$

where, n_s and N_s are the actual cycle number and the cycle number at which the element is destroyed if the stress peak value equals S.

When the sum of the related fatigue accumulation reaches one, the element will be destroyed due to fatigue. That is to say,

$$\int_0^{\infty} D_s ds = 1 \tag{11-2}$$

If N_f denotes the total cycle number of stresses (i.e., fatigue life) when the element is damaged, the probability density function $f_p(s)$ at the point where the stress peak is S can be expressed approximately by,

$$f_p(s) = \frac{n_s}{N_f} \tag{11-3}$$

Substituting the above formula into formula (11-1) gives

$$D_s = f_p(s)\frac{N_f}{N_s} \tag{11-4}$$

By considering the fatigue curve ($S \sim N$ curve) where the element is subjected to the constant amplitude harmonic stress S, i.e.,

$$N_s = C.S^{-\beta} \tag{11-5}$$

where, c, β are the element dominated behaviour constant, substituting formula (11-5) for the formulae (11-4) and (11-2) yields the total number of cycles at the time when a structural element's damage occurs as follows,

$$N_f = \frac{C}{\int_0^{\infty} f_p(s) S^{\beta} ds} \tag{11-6}$$

It can be observed from the above formula that the key problem of finding the total number of cycles (fatigue life) is to find the probability density function $f_p(s)$ of the stress process $S(t)$'s peak.

If $S(t)$ is a stationary Gaussian process with the narrow frequency band, the peak of $S(t)$ will be subjected to a Rayleigh distribution. Substituting the probability density function $f_p(s)$ (formula (7-13)) into the above formula results in the following expression of N_f,

$$N_f = \frac{C}{(\sqrt{2})^{\beta} \cdot \Gamma\left(1 + \frac{\beta}{2}\right)} \cdot \sigma_S^{-\beta} \tag{11-7}$$

in which, $\Gamma(\cdot)$ is the Gamma function; σ_s is the r.m.s. value of the stress process $S(t)$.

Because the total number of cycles, N_f, when the stress $S(t)$ is a stationary process, can be described as follows by the crossing rate v_0^+ of $S(t)$ with respect to the zero line and the valid life time T (i.e., fatigue life),

$$N_f = v_0^+ T \tag{11-8}$$

the fatigue life T of the element can be expressed as,

$$T = \frac{N_f}{v_0^+} = \frac{1}{v_0^+} \frac{C}{(\sqrt{2})^{\beta} \cdot \Gamma\left(1 + \frac{\beta}{2}\right)} \cdot \sigma_s^{-\beta} \tag{11-9}$$

If stress $S(t)$ is that of a process with wide frequency band, Wirshing[15] has presented an improved formula,

$$T' = \frac{T}{\delta} = \frac{C}{\delta v_0^+ \left(\sqrt{2}\right)^{\beta} \Gamma\left(1 + \frac{\beta}{2}\right)} \sigma_s^{-\beta} \tag{11-10}$$

in which, T' is the improved fatigue life of element; δ is the adjustment factor, the value of which depends on the behavior constant β of element and the band width coefficient ε of the stress process. δ can be determined by the following formula,

$$\delta = a + (1 - a)(1 - \varepsilon)^b \tag{11-11}$$

where, a and b are the β-related coefficients that need to be found.

It is obvious in the above formula that $\delta = 1.0$ (the ideal narrow band) if $\varepsilon = 0.0$; that $\frac{d\delta}{d\varepsilon} = 0$ (white noise) if $\varepsilon = 1.0$. These are two extreme situations. According to the statistic result given in the paper[16], by employing the least square method, the value of a and b can be determined as,

$$\begin{cases} a = 0.926 - 0.033\beta \\ b = 1.587\beta - 2.3323 \end{cases} \tag{11-12}$$

To demonstrate the method, the computation result of an engineering example[16] is provided below.

The material of an automobile direction controller's axis is a 20 Cr steel pipe. According to statistics, the axis's fracture rate is 0.2%. Detailed analysis of the cranny section reveals that the damage owing to fatigue is invalid. The fatigue life of a rotating axis can be evaluated as follows.

According to the experimental result of rotation fatigue about 20 Cr steel, and through some improvement for the influence of concentrated stress, surface quality as well as size, the S-N curve equation, when the survival rate of the axis is 99.8%, can be deduced as,

$$N_s = CS^{-\beta} = 10^{14.18} S^{-3.03}$$

The related parameters of stress process $S(t)$ at the point where the rotation axis is crannied are given by,

$$\sigma_s = 52.32 \text{ kg/cm}^2,\ \upsilon_0^+ = 8.978 \text{ Hz},\ \varepsilon = 0.769$$

so the value of δ can be calculated to be 0.826 by means of the formulae (11-12) and (11-11).

If the above values are substituted into formula (11-10), the fatigue life of the rotation axis can be determined as $T' = 3.278 \times 10^7 (s) = 9.1 \times 10^3 (h)$. Assuming the automobile serves 300 days per year and 12 hours each day, its life will be $T' = 2.53$ (year).

If the result is not adequate, the fatigue life evaluated by the narrow band process formula (11-9) is 2.09 (year). However, the fatigue life evaluated by means of the method of the peak value region is 2.49 (year). The statistics of the experimental result has shown that the fatigue life of this kind of rotation axis is 2.5 (year). It is obvious that the result from the wide band improvement is rather close to the real result.

5.12. THE POISSON PROCEDURE METHOD FOR DYNAMIC RELIABILITY ANALYSIS OF A STRUCTURAL SYSTEM WITH MULTIDEGREE OF FREEDOM

Generally speaking, every engineering structure will become a system with multidegree of freedom through the method of finite element discritezation. The dynamic reliability analysis of a multidegree-of-freedom system is related to several factors, such as the correlation between the dynamic responses of all degree of freedoms in the system, and the determination of a high dimensional joint probability density function, as well as the computation of a high dimensional integration and so on. The above computation is obviously much more difficult to implement than that of a single degree-of-freedom system. There is not a method that could be used generally. Nevertheless, this computation is necessary no matter whether it is considered from the viewpoint of either applications or research. Consequently, some approximation or reasonable simplification has to be introduced in practical applications.

What will be done in this section is to introduce the Poisson procedure method to the dynamic reliability analysis of the linear structural system with multidegree-of-freedoms. The basic fundamentals of the method is that the crossing between the response and the limit is subjected to the hypothesis of a Poisson process.

The main point of the Poisson procedure method is to find the crossing rate between the response process and the given limit. However, the dynamic response of multidegree-of-freedom systems will be the vector random process, the corresponding safety boundary and region will be a high-D curve and a high-D space. Therefore, the solution of the crossing rate of the system will be considerably complicated. Only the simple situation can be treated by Poisson procedure method.

Suppose the system is composed of n degrees of freedoms, under the excitation of the random dynamic load, system response $\mathbf{X}(t) = (x_1(t), x_2(t), \ldots, x_n(t))^T$ will be a $n - D$ vector random process. It means that the state of the vector process at any time is the $n - D$ random vector in space R^n. Moreover, if the system has m limit state equations (curves) $g_i(\mathbf{X}) = \lambda_i (i = 1, 2, \ldots, m)$, the safety boundary B of the system can be mathematically stated as,

$$B = \{\mathbf{X} | g_i(\mathbf{X}) = \lambda_i, i = 1, 2, \cdots, m\} \qquad (12\text{-}1)$$

i.e., boundary B is comprised of m curved planes. Going a step further, the safety region Ω_r of the system can be described in the following form,

$$\Omega_r = \{\mathbf{X} | g_i(\mathbf{X}) \leqslant \lambda_i, (i = 1, 2, \cdots m)\} \qquad (12\text{-}2)$$

Consequently, the first passage damage of a system can be written as

$$\begin{cases} \mathbf{X}(t_0) \in B \\ \dot{y}_n(t_0) = \mathbf{N}^T(t_0) \cdot \dot{X}(t_0) > 0 \end{cases} \qquad (12\text{-}3)$$

where, $\mathbf{X}(t_0)$ and $\dot{\mathbf{X}}(t_0)$ are the value and the derivative of any sample in the response of the vector process $\mathbf{X}(t)$ at time t_0, respectively; $\mathbf{N}(t_0)$ is the unit normal vector of boundary B at point $\mathbf{X}(t_0)$, and the vector is directed from the safe region to the damage region.

As for the crossing rate $v_B(t)$ of a multidegree-of-freedom system, since a formula similar to the Rice formula for the single degree-of-freedom system is impossible to be deduced, Bolotin[17] has established the extensive Rice formula for systems with multidegree-of-freedom according to the extensive Rice formula for the system with a single degree-of-freedom,

$$\begin{aligned} v_B &= \int_B d\mathbf{X} \int_0^{\infty} \dot{y}_n \cdot f_{\mathbf{X}\dot{y}_n}(\mathbf{X}, \dot{y}_n) d\dot{y}_n \\ &= \int_B E_0^{\infty}[\dot{y}_n | \mathbf{X} = \mathbf{x}] \cdot f_{\mathbf{X}}(\mathbf{X}) d\mathbf{X} \end{aligned} \quad (12\text{-}4)$$

in which,

$$E_0^{\infty}[\dot{y}_n | \mathbf{X} = \mathbf{x}] = \int_0^{\infty} \dot{y}_n f_{\dot{y}_n|\mathbf{X}}(\dot{y}_n | \mathbf{x}) d\dot{y}_n \quad (12\text{-}5)$$

means the expectation of variable $\dot{y}_n$ in $[0, \infty]$ under the condition of $\mathbf{X} = \mathbf{x}$.

If $\dot{y}_n$ is independent to $\mathbf{X}$, the conditional expectation (12-5) can be written as,

$$E_0^{\infty}[\dot{y}_n | \mathbf{X} = \mathbf{x}] = E_0^{\infty}[\dot{y}_n] \quad (12\text{-}6)$$

If

$$f(B) = \int_B f_{\mathbf{X}}(\mathbf{X}) d\mathbf{X} \quad (12\text{-}7)$$

the formula (12-4) will assume the following simple form,

$$v_B = E_0^{\infty}[\dot{y}_n] \cdot f(B) \quad (12\text{-}8)$$

It should be noted that the extensive Rice formula (12-4) is valid for either a stationary vector process or a nonstationary vector process. What will be provided below is the computation for the crossing rate v_B with two cases of specific safety boundary when $\mathbf{X}(t)$ is a standard Gaussian vector process.

Suppose the response of a multidegree-of-freedom structural system is a Gaussian vector random process $\mathbf{X}(t)$, according to the theory of linear transformation, the vector process can be transformed into a standard Gaussian vector process with zero average and unit covariance matrix by means of orthogonal and standard transformations. That is to say, each sample $X(t)$ in the vector process $\mathbf{X}(t)$ at the arbitrary time t will be subjected to a standard Gaussian distribution. i.e.,

$$X(t) \sim \mathcal{N}[0, I] \quad (12\text{-}9)$$

Without compromising generality, suppose the derivative of the process $X(t)$ is

$$\dot{\mathbf{x}}(t) \sim \mathcal{N}[0, \mathrm{diag}(\sigma_{\dot{x}_i}^2)] \quad (12\text{-}10)$$

Then we get the following results:

(1) The safety boundary B with a single plane

It means that the number of limit state equations in the structural system is $m = 1$. In this case, the problem of solving the crossing rate v_B between vector process $\mathbf{X}(t)$ and boundary of plane B will degenerate into a one dimensional first-passage situation.

In addition,the plane B can be expressed in the form of a normal equation,

$$\sum_{i=1}^{n} n_i x_i = r \tag{12-11}$$

where, $n_i (i = 1, 2, \cdots, n)$ are the direction cosines of the unit normal line of plane B; r is the distance from coordinate origin to the plane B.

Let

$$Z = \sum_{i=1}^{n} n_i x_i \tag{12-12}$$

the safety region Ω_r of the structural system has,

$$\Omega_r = \{Z | Z \leqslant r\} \tag{12-13}$$

From formula (12-11), the expectation and the variance of variable Z and its derivative $\dot{Z}$ can be deduced as,

$$\begin{cases} \mu_Z = 0, & \sigma_Z^2 = \sum_{i=1}^{n} n_i = 1 \\ \mu_{[\dot{z}]} = 0, & \sigma_{\dot{Z}}^2 = \sum_{i=1}^{n} n_i^2 \sigma_{\dot{x}_i}^2 \end{cases} \tag{12-14}$$

Substituting the result from the above formula into the Rice formula (6-10), by which the crossing rate of structural system with the single degree-of-freedom could be found, yields the structural system crossing rate as,

$$\begin{aligned} v_B &= \int_0^{\infty} \dot{Z} f_{Z\dot{Z}}(r, \dot{Z}) d\dot{Z} \\ &= \int_0^{\infty} \dot{Z} \frac{1}{2\pi \sigma_Z \sigma_{\dot{Z}}} \exp\left\{ -\frac{1}{2} \left[\frac{r^2}{\sigma_Z^2} + \frac{\dot{Z}^2}{\sigma_{\dot{Z}}^2} \right] \right\} d \cdot Z \\ &= \frac{\sigma_{\dot{Z}}}{2\pi} \exp\left(-\frac{r^2}{2} \right) \end{aligned} \tag{12-15}$$

(2) The safety boundary B with q planes

Suppose the direction cosines of the unit normal line of the jth plane B_j among the boundary B are $n_{ji} (i = 1, 2, \ldots, n)$, the plane B_j can be expressed by its normal equation as follows,

$$\sum_{i=1}^{n} n_{ji} x_i = r_j \tag{12-16}$$

where, r_j is the distance from the coordinate origin to the plane B_j.

It is obvious that the crossing rate υ_B of structural system with respect to boundary B should be the sum of the crossing rates of all q planes.[17] That is,

$$\upsilon_B = \sum_{j=1}^{q} \upsilon_{B_j} = \frac{1}{\sqrt{2\pi}} \sum_{j=1}^{q} f(B_j)\sigma_j \tag{12-17}$$

in which

$$\sigma_j^2 = \sum_{i=1}^{n} n_{ji}^2 \sigma_{\dot{x}_i}^2 \tag{12-18}$$

$$\begin{aligned} f(B_j) &= \int_{B_j} f_{\mathbf{X}}(\mathbf{X}) d\mathbf{X} \\ &= \int_{B_j} \frac{1}{(2\pi)^{n/2}} \exp\left\{-\frac{1}{2}\sum_{i=1}^{n} x_i^2\right\} d\mathbf{X} \\ &= \varphi(r_j) \cdot \Phi_{n-1}(B_j) \end{aligned} \tag{12-19}$$

where, $\varphi(\cdot)$ is a 1-D standard Gaussian density function; $\Phi_{n-1}(\cdot)$ is a $(n-1) - D$ standard Gaussian distribution function.

The computational step of formula (12-19) can be stated as[18]: the projection of an $n - D$ joint probability density function $f_{\mathbf{X}}$ at the plane B_j is calculated first. The projected expression of $f_{\mathbf{X}}$ in the plane B_j can be obtained by eliminating one variable from $f_{\mathbf{X}}$ through the expression (12-16) of plane B_j (in the above deduction, the variable x_n is taken as the eliminated variable). Then the integration is done about this expression in plane B_j. Generally speaking, the analysis solution of this kind of integration is very difficult to be carried out.

5.13. THE UPPER AND THE LOWER BOUNDS OF DYNAMIC RELIABILITY FOR A MULTIDEGREE-OF-FREEDOM SYSTEM WITH A WEAKEST LINK

For static deterministic structures, the reliable logical model is a series model. For structures with local deterministic structures, they can also be represented as a series models from the viewpoint of macroscopic construction analysis or considerations of safety. In this kind of model, any one element's damage will lead to failure of the whole structure. The reliability of a structural system is determined by the reliability of the weakest element to a certain extent. Therefore, this kind of system is called the structural system with the weakest link, of which the dynamic reliability is discussed below.[13]

Under the action of a dynamic load, each link (or element) of the structure could be probably destroyed. If each link's damage is considered as one kind of dynamic destroyed model, the whole system with n links will have n kinds of destroyed models. Clearly, the failure probability $P_{fi}(t)$ of the ith link in $[0, T]$, i.e., the probability of the ith dynamic damage model of structural system in $[0, T]$, can be stated as,

$$P_{fi}(t) = 1 - P_{ri}(t)\ (0 \leqslant t \leqslant T)\ (i = 1, 2, \cdots, n) \tag{13-1}$$

$$P_{ri}(t) = P_r\{-\lambda_{li} \leqslant x_i(t) \leqslant \lambda_{ui}, 0 \leqslant t \leqslant T\} \tag{13-2}$$

where $P_{ri}(t)$ is the dynamic reliability function of the ith link; $x_i(t)$ is the dynamic response (displacement, stress or strain) of the ith link; λ_{li} and λ_{ui} are the predetermined upper and lower bounds of $x_i(t)$ respectively.

As for the structural system with multidegree-of-freedoms, it is impossible to calculate an accurate value of the structural dynamic reliability $P_r(t)$ due to the higher correlation among the responding elements. Therefore, only two extreme situations such as complete correlation (correlative coefficient $\rho_{ij} = 1$) and complete independence (correlative coefficient $\rho_{ij} = 0$) are considered here. According to the theory of the reliability[20], the upper and the lower bounds of dynamic reliability $P_r(t)$ for the structural system with a weakest link in the interval $[0, T]$ can be obtained as,

$$\prod_{i=1}^{n} P_{ri}(t) \leqslant P_r(t) \leqslant \min\{P_{ri}(t)\}\ (0 \leqslant t \leqslant T) \tag{13-3}$$

If $\rho_{ij} = 1$, the unequal symbol at the right hand side of formula (13-3) becomes equal; Conversely, if $\rho_{ij} = 0$, the unequal symbol at the left hand side of formula (13-3) will become equal; if $0 < \rho_{ij} < 1$, the unequal symbols of the above formula hold.

It can be seen theoretically from formula (13-3) that, in order to find the upper bound of $P_r(t)$, the dynamic reliability of the weakest link in a structural system should be calculated in advance. Further in order to find the lower bound of $P_r(t)$, the dynamic reliability of all links in structural system should be known in advance, and also the product at each time must be known immediately thereafter. However, what is done in practical computations for

$P_r(t)$'s lower bound is to find the dynamic reliability of those links with $P_{ri}(T) < 1$ in the interval $[0, T]$, without caring about the other links with $P_{ri}(T) = 1$. Based on the above consideration, the following strategy of finding the upper and the lower bounds of dynamic reliability for stationary system response has been determined[13] as follows.

Suppose there are $m(m \leqslant n)$ links, and their dynamic reliability are all less than 1 in the interval $[0, T]$. If the upper and the lower bounds of $P_r(t)$ are required to be known with the least computation, the first task is to determine the weakest and the next $m - 1$ weakest links among the dynamic reliability function set $\{P_{ri}(t)\}$. The detailed searching procedure is described as follows.

$\min_{i=1,n}\{P_{ri}(t)\}$ — to find the weakest link of the n links;

$\min_{i=1,n-1}\{P_{ri}(t)\}$ — to find the next weakest link of the $n - 1$ remaining links;

...

...

$\min_{i=1,n-(m-1)}\{P_{ri}(t)\}$ — to find the mth weakest link of the $n - (m - 1)$ remaining links.

If the Poisson formula (8-9) or Markov formula (9-1) is taken as the expression in the function set $\{P_{ri}(t)\}$, the dynamic reliability of all n links should be found and then be compared with each other. If it is done so, the computation is not decreased, but increased considerably. In order to avoid such a large computation in the task of searching for the weakest and the next $(m - 1)$ weakest links, the simplified Poisson formula (10-3) is addressed as the expression of the function set $\{P_{ri}(t)\}$. Consequently the problem of searching for the weakest link can be equivalent to the problem of searching for the maximum one in the following function set. i.e.,

$$\max_{i=1,n}\left\{\exp\left(-\frac{\lambda_{ui}^2}{2\sigma_i^2}\right) + \exp\left(-\frac{\lambda_{li}^2}{2\sigma_i^2}\right)\right\} \tag{13-4}$$

If the pregiven upper and lower limits are all symmetrical with respect to zero or have approximate symmetry, i.e., $\lambda_{ui} = \lambda_{li} = \lambda_i (i = 1, 2, \cdots, n)$, the problem of searching for the maximum one can be transformed into the problem of searching for the minimum one in the following function set. That is,

$$\min_{i=1,n}\left\{\frac{\lambda_i}{\sigma_i}\right\} \tag{13-5}$$

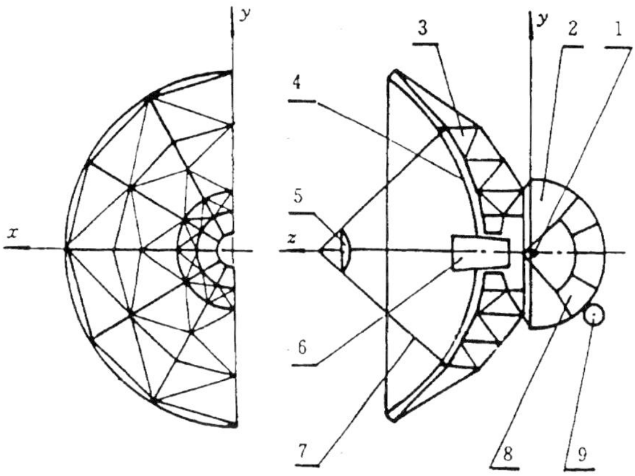

Figure 11. The large parabolic antenna structure. 1. Pitch, 2. Fan shaped gear, 3. Back-up structure, 4. Reflector, 5. Secondary reflector, 6. Feed horn, 7. Jackstays of the secondary reflector, 8. Balancing weight, 9. Driving gear.

By employing the formula (13-4) or (13-5) one by one, the weakest and the next $(m-1)$ weakest links can be easily found for the structural system. Then the accurate value can be obtained by applying the computation formula of computing dynamic reliability for the system with a single degree-of-freedom for these m links. To go a step further, the upper and the lower bounds of $P_r(t)$ for the structural system can be obtained by formula (13-3).

By contrast, in order to determine the strongest link of the system, it can be easily obtained by either searching for the minimum one from formula (13-4) or searching for the maximum one from formula (13-5). Once the dynamic reliability of the weakest and the strongest links of the system are obtained, important information can be provided for the reliability-based optimum design of structural systems.

By means of the above method Chen[13] has developed the computation of the upper and the lower bounds of systematic dynamic reliability for a large parabolic antenna structure as shown in Figure 11 under the action of the stationary Davenport wind power spectrum (the average wind speed is $\bar{v} = 70$ m/s). The computation result of the upper and the lower bounds of the system's reliability is described in Figure 12.

5.14. THE FINAL REMARKS

The study of structural dynamic reliability has been conducted over half a century since 1944 when Rice presented his probability computational

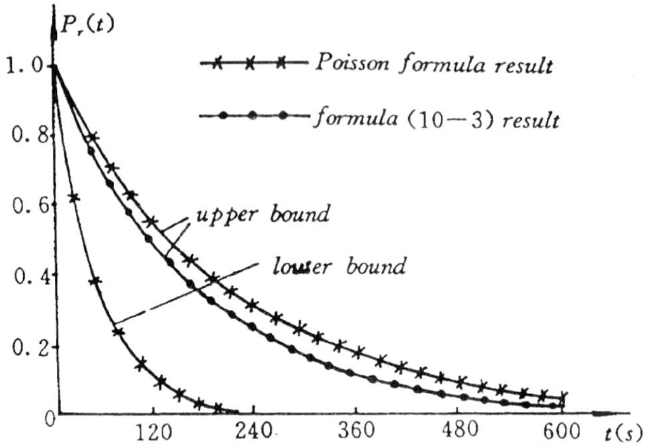

Figure 12. The computation result of dynamic reliability for the winding vibration response of a large parabolic antenna structure.

formula for the first-passage in a random process. During this period of time, some important researching achievements have been made by a number of researcher's, among whom are Rice, Siegert, Colenem, Crandall, Cramer, Roberts, Lin, Yang, Yan and so on. It is due to their research that the dynamic reliability of structures has become a developed research area.

However, its development is still relatively slow compared with some other subjects. The main constraint is due to the aspect of random mathematics, that is, from the random processes on which dynamic reliability is theoretically based. It can be anticipated that some new developments in dynamic reliability results will be made rapidly with the development of further techniques. We consider that the following interesting subjects about the dynamic reliability of structures are worthwhile for further development.

(1) More efficient numerical method for computing the dynamic reliability of structures;
(2) The determination of dynamic reliability failure models for structural systems;
(3) Dynamic reliability analysis methods for structural systems;
(4) Dynamic reliability computation for the response to nonstationary random processes;
(5) Dynamic reliability computations for nonlinear structural systems;
(6) Further study on the cumulative damaged criteria for cumulative damage;
(7) Dynamic reliability analysis for random structures;
(8) The theory and methods of fuzzy dynamic reliability of structures.

PART TWO: RELIABILITY ASPECTS IN STRUCTURAL OPTIMIZATION

5.15. INTRODUCTION

The reliability based optimum design of structures is a new design method developed recently, in which the analysis of structural reliability is combined with mathematical programming. That is to say, the requirement of structural reliability is taken as either the constraint or the objective, then the optimum design in the sense of reliability can be determined by means of optimization methods. Reliability based optimum design is the most reasonable design model. Consequently, it has become one of the important research and development directions in modern engineering structural design.

There are the following obvious advantages of reliability based optimum design compared with traditional optimum design.

(1) In traditional structural design, the deterministic method have been employed. That is to say, both the structure and the load to which the structure is subjected are all viewed as deterministic parameters. In this case, the influence of the random disturbances and the distribution of structural parameters, loads and so on are not considered. The usually employed method is either to maximize the load or to minimize the resistance, or to accept the safety coefficient method. It is obvious that the former method is an extreme one, and for the later method, a large safety coefficient is not equivalent to the higher reliability that can result by considering the problem from the view point of random processes.

(2) Further, in the design for minimum weight, it is quite often the case that the static undetermined structure tends to be a static determined structure, the static determined structure tends to be a geometrically unstable structure, and the optimum result is located at the boundary of the feasible region. Obviously, these optimum results are unacceptable in engineering. However, by applying the methods of structural reliability based on optimum design, the optimum design which satisfs the requirement of reliability can be obtained. it can also be stated in another way, i.e., under the conditions which satisfy the other requirements of a structure, the maximum reliability of the structure can be obtained by means of reliability based optimization methods.

(3) In traditional structural optimization, structural safety is guaranteed through the strength constraints of system elements. Even though the structural system is composed of those elements which satisfy the strength requirements, the system's reliability requirement may not be able to be

guaranteed if it is determined from the reliability point of view. However, since the structural system's reliability is usually taken as the constraint in the reliability based optimum design of structures, the optimum structure satisfying the system's reliability requirements can be obtained in the general sense.

In structural reliability optimization, the design variable $\mathbf{X} = (x_1, x_2, \ldots, x_n)^T$ may be the parameters to be determined parameters, random parameters or the combination of the both. As for the objective or the constraint functions, the cost $C(\mathbf{X})$, the weight $W(\mathbf{X})$ or the mass $M(\mathbf{X})$, and the reliability $P_r(\mathbf{X})$ or the failure probability $P_f(\mathbf{X}) = 1 - P_r(\mathbf{X})$ can all be objectives. Based on these considerations, the problem of structural reliability optimization can be classified into the following four kinds of models according to the different selections of the objective and the constraint.

I. $C(\mathbf{X}) \rightarrow \min,\ Pr(\mathbf{X}) \geqslant$ const.
II. $W(\mathbf{X}) \rightarrow \min,\ Pr(\mathbf{X}) \geqslant$ const.
III. $Pr(\mathbf{X}) \rightarrow \max,\ C(\mathbf{X}) \leqslant$ const.
IV. $Pr(\mathbf{X}) \rightarrow \max,\ W(\mathbf{X}) \leqslant$ const.

In the reliability optimization of structures, the form of reliability constraint is usually taken as,

$$P_r(\mathbf{X}) \geqslant P_r^*,\ or\ P_f(X) \leqslant P_f^*$$

where, P_r^* and P_f^* are the pregiven values of the reliability and the failure probability respectively.

If the time variable t is included in the constraint function, the reliability constraints can be categorized into two main parts as static and dynamic constraints.

(1) Static reliability constraints: the constraints are the functions of the space variables $\mathbf{X}$ only;
(2) Dynamic reliability constraints: the constraints are the function of both space variable $\mathbf{X}$ and time variable t. For instance, problems such as the dynamic reliability with the first-passage, the fatigue life and so on all belong to this kind of constraint.

To go a step further, according to the included factors in the constraint functions, both the static and the dynamic reliability constraints can be divided into three kinds of models.

(1) reliability constraints of element: strength, buckling, fatigue, size and so on;

(2) Reliability constraints of failure model: performance function;
(3) Reliability constraints of structural system: strength, stiffness, precision, stability, natural frequency and so on.

Since the reliability constraint of a structure is expressed in terms of the probability and is the inexplicit complex function of design variables, it will be much more difficult to deal with reliability constraints than general constraints.

Generally speaking, there are three main aspects to structural reliability optimization, these are structural analysis, reliability computation and optimum design. Among these the reliability computation is the basic and most difficult task. Due to the limit of structural reliability theory, the technique used to date, the approximate methods such as the method of the second order moments and so on are usually applied in engineering. The detailed discussion of these methods can be found in related treatises.[19,20]

5.16. STRUCTURAL ANALYSIS WITH THE RANDOM LOADING CASE

Without compromising generality, if the linear elastic structure is subjected to m random loads, the corresponding finite element equation can be written as follows,

$$[K](\mathbf{U}_1, \mathbf{U}_2, \cdots \mathbf{U}_m) = (\mathbf{L}_1, \mathbf{L}_2, \cdots, \mathbf{L}_m) \qquad ((2\text{-}1))$$

in which, $[K]$ is the structural stiffness matrix; $\mathbf{L}_k, \mathbf{U}_k (k = 1, 2, \ldots, m)$ are separately the kth random loading vector and the kth random displacement vector caused by $\mathbf{L}_k$.

Solving Equation (2-1) gives the displacement,

$$(\mathbf{U}_1, \mathbf{U}_2, \cdots, \mathbf{U}_m) = [K]^{-1}(\mathbf{L}_1, \mathbf{L}_2, \cdots, \mathbf{L}_m) \qquad (2\text{-}2)$$

According to the relationship between displacement and stress, the random stress vector of element (e) due to the kth random load can be expressed as

$$\{\mathbf{S}_k^{(e)}\} = [D^{(e)}][T^{(e)}]\{\mathbf{U}_k^{(e)}\}\ (e = 1, 2, \cdots n_e; k = 1, 2, \cdots, m) \qquad (2\text{-}3)$$

where, $\{\mathbf{S}_k^{(e)}\}$ is the random stress vector of element (e) due to the kth load; $[D^{(e)}]$ gives the mathematical relationship between stress and displacement; $[T^{(e)}]$ is the transformation matrix of the eth element; $\{\mathbf{U}_k^{(e)}\}$ is the nodal displacement vector of element (e) under the kth loading case; n_e is the total number of elements in the structure.

By means of the computational properties of random variable moments, the expectation and the variation of the structural random displacement and the element's random stress can be found by the following formulae, respectively.

$$E(\mathbf{U}_k) = [K]^{-1}E(\mathbf{L}_k)\ (k = 1, 2, \cdots, m) \tag{2-4}$$

$$\sqrt{(\mathbf{U}_k)} = [K]^{-1}\sqrt{(\mathbf{L}_k)}(k = 1, 2, \cdots, m) \tag{2-5}$$

$$E(\mathbf{S}_k^{(e)}) = [D^{(e)}][T^{(e)}]E(\mathbf{U}_k^{(e)})(e = 1, 2, \cdots, n_e; k = 1, 2, \cdots, m) \tag{2-6}$$

$$\sqrt{(\mathbf{S}_k^{(e)})} = [D^{(e)}][T^{(e)}]\sqrt{(\mathbf{U}_k^{(e)})}(e = 1, 2, \cdots, n_e; k = l, 2, \cdots, m) \tag{2-7}$$

where, $E(\cdot)$ and $D(\cdot)$ indicate the operators of expectation and variation separately.

According to superposition theory of linear structures, the ith random response component (displacement or stress) of a structure that is subjected to m random loads simultaneously can be easily obtained as,

$$Y_i = \sum_{k=1}^{m} Y_{ik} \tag{2-8}$$

where, Y_{ik} stands for the ith random response component (displacement or stress) caused by the kth loading.

Considering the moment property of the random variable, the expectation and the variation of Y_i can be separately written as

$$E(Y_i) = \sum_{k=1}^{m} \mu_{ik} \tag{2-9}$$

$$D(Y_i) = \sum_{k=1}^{m} \sigma_{ik}^2 + 2\sum_{1=k<r}^{m} \rho_{kr}\sigma_{ik}\sigma_{ir} \tag{2-10}$$

in which, μ_{ik} and σ_{ik}^2 are the expectation and the variation of the responding component Y_{ik} respectively; ρ_{kr} is the correlation coefficient between the kth and the rth random loads.

If m loads are independent of each other, i.e., $\rho_{kr} = 0(k \neq r; k, r = 1, 2, \cdots, m)$, $D(Y_i)$ can be expressed as,

$$D(Y_i) = \sum_{k=1}^{m} \sigma_{ik}^2 \tag{2-10a}$$

If the loads are correlated with each other completely, i.e., $\rho_{kr} = 1$ $(k \neq r; k, r = 1, 2, \cdots, m)$, $D(Y_i)$ will be,

$$D(Y_i) = (\sum_{k=1}^{m} \sigma_{ik})^2 \tag{2-10b}$$

If m loads are all subjected to normal distribution, i.e.,

$$\mathbf{L}_k \sim \mathcal{N}[E(\mathbf{L}_k), D(\mathbf{L}_k)], (k = 1, 2, \cdots, m) \tag{2-11}$$

according to this under the action of the normal loads the response distribution in a linear elastic structure will not be changed, both the random vectors of structural displacement and stress are also subjected to a normal distribution. That is to say,

$$\mathbf{U}_k \sim \mathcal{N}[E(\mathbf{U}_k), D(\mathbf{U}_k)] \ (k = 1, 2, \cdots, m) \tag{2-12}$$

$$\mathbf{S}_k^{(e)} \sim \mathcal{N}[E(\mathbf{S}_k^{(e)}), D(\mathbf{S}_k^{(e)})] \ \ (e = 1, 2, \cdots, n_e; k = 1, 2, \cdots, m) \tag{2-13}$$

Moreover, based on the reproducibilty of the normal variables under linear transformations, it is easily seen that the superposing element Y_i of the structure is also subjected to a normal distribution, i.e.,

$$Y_i \sim \mathcal{N}[E(Y_i), D(Y_i)] \tag{2-14}$$

The normal distribution is a popular probability model in engineering for the normal variables have many advantages but the abnormal variables have not. Furthermore, there exist some accurate and efficient methods in reliability analysis for the normal variables. In order that these methods can also be used for the abnormal variables, one method is to transform the abnormal variables into the equivalent normal variables at a point by means of the normal tail transformation.[20] It is generally, but not necessary always the case, that the normal distribution is quite applicable in practice.

5.17. THE ELEMENT RELIABILITY BASED OPTIMUM DESIGN OF TRUSS STRUCTURES

The truss structure is a basic engineering structures, in which each bar is subjected to axial load only. For the static deterministic truss, one element's failure will lead to the failure of the whole structure. For the statically undetermined truss with a lower order of static indetermination, even though the redundant bar's failure may not result in the whole structural failure, the whole system's reliability will be reduced to a certain extent. For the sake of simple and safe consideration, the element's reliability is taken as the constraint in the reliability based optimization of truss structures.

Suppose there are n bars in a truss structure, which is subjected to m random loads $L_k (k = 1, 2, \ldots, m)$ simultaneously, the problem now is to minimize the structural weight W with the determination of the cross-sectional areas $A_i (i = 1, 2, \cdots, n)$ of all elements whilst satisfying the requirement of strength reliability. This problem can be described in the following mathematical form,

$$\text{find} \quad \mathbf{A} = (A_1, A_2, \cdots, A_n)$$

$$\text{min.} \quad \mathbf{W}(\mathbf{A}) = \sum_{i=1}^{n} \rho_i l_i A_i \tag{3-1}$$

$$\text{s.t.} \quad P_{fi}(\mathbf{A}) \leqslant P_{fi}^* \qquad (i = 1, 2, \cdots, n) \tag{3-2}$$

$$-A_i \leqslant 0 \qquad (i = 1, 2, \cdots, n) \tag{3-3}$$

where, ρ_i, l_i are the density and the length of the ith element; P_{fi}^* is the pregiven value of the failure probability, the value of which can be determined by certain distributed criteria based on the requirement for the system's failure probability P_f^*; $P_{fi}(\mathbf{A})$ is the failure probability of the ith bar, which can be expressed by the performance function $Z_i(A)$ in the following form,

$$P_{fi}(\mathbf{A}) = P\{Z_i(\mathbf{A}) \leqslant 0\}, \quad (i = 1, 2, \cdots, n) \tag{3-4}$$

By implementing structural analysis, the general expression of the performance function $Z_i(A)$ of the ith bar can be determined as,

$$Z_i(\mathbf{A}) = A_i R_i - \sum_{k=1}^{m} b_{ik}(\mathbf{A}) L_k \quad (i = 1, 2, \cdots, n) \tag{3-5}$$

where R_i is the strength random variable of the ith bar, and $R_i = R_i^+$ (the strength of the antitension) if the bar is tensioned; $R_i = R_i^-$ (the strength of the anticompression) if the bar is compressed. The value of R_i^- can be determined from the condition of stability; $b_{ik}(\mathbf{A})$ is the loading coefficient of the ith bar under the kth load. For a static determined structure, b_{ik} is independent of the cross-sectional area $\mathbf{A}$.

It can be determined from formula (3-5) that if the variable $\mathbf{A}$ is known, $Z_i(\mathbf{A})$ will be the combination function of the random variables R_i and $L_k (k = 1, 2, \cdots, m)$. If the distributions of R_i and $L_k (k = 1, 2, \ldots, m)$ are known, the distribution of $Z_i(\mathbf{A})$ can be obtained as a random variable function, and furthermore, the failure probability $P_{fi}(\mathbf{A})$ can also be found from formula (3-4).

Murotsu[21] has presented a simple solution method based on the condition of both strength and load which are all subjected to normal distributions. The solution procedure can be stated as follows:

Since

$$R_i \sim \mathcal{N}(\mu_{Ri}, \sigma_{Ri}^2), \quad (i = 1, 2, \cdots, n) \tag{3-6}$$

and

$$L_k \sim \mathcal{N}(\mu_{Lk}, \sigma_{Lk}^2), \quad (k = 1, 2, \cdots, m) \tag{3-7}$$

considering formula (3-5) gives

$$Z_i \sim \mathcal{N}(\mu_{Z_i}, \sigma_{Z_i}^2), \quad (i = 1, 2, \cdots, n) \tag{3-8}$$

By applying the method of the second moment, the original reliability constraint (3-2) can be expressed by the following reliability index form.

$$\beta_i \geqslant \beta_i^* (i = 1, 2, \cdots, n) \tag{3-9}$$

where

$$\beta_i^* = -\Phi^{-1}(P_{fi}^*) \tag{3-10}$$

$$\beta_i = -\Phi^{-1}(P_{fi}) = \frac{\mu_{Z_i}}{\sigma_{Z_i}} \tag{3-11}$$

in which $\Phi^{-1}(\cdot)$ is the inverse function of the standard normal distribution.

From formula (3-5), the mean and the variation of function $Z_i(\mathbf{A})$ can be obtained by the following formulae,

$$\mu_{Z_i} = A_i \mu_{Ri} - \sum_{k=i}^{m} b_{ik}(\mathbf{A}) \mu_{Lk} \tag{3-12}$$

$$\sigma_{Z_i}^2 = A_i^2 \sigma_{Ri}^2 + \sum_{k=i}^{m} b_{ik}^2(\mathbf{A}) \sigma_{Lk}^2 + 2 \sum_{1=k<r}^{m} b_{ik}(\mathbf{A}) b_{ir}(\mathbf{A}) \rho_{kr} \sigma_{Lk} \sigma_{Lr} \tag{3-13}$$

where ρ_{kr} is the correlation coefficient between the loads L_k and L_r.

Noticing that at the optimum point which also satisfies the inequality constraints,

$$\beta_i = \beta_i^* \quad (i = 1, 2, \cdots, n) \tag{3-14}$$

and substituting formulae (3-11)–(3-13) into the above formula with some further organization yields the following algebraic equations by which the design variable A_i can be found,

$$a_{0i} A_i^2 - 2a_{1i} A_i + a_{2i} = 0 \quad (i = 1, 2, \cdots, n) \tag{3-15}$$

where,

$$a_{0i} = \mu_{Ri}^2 - \beta_i^{*2} \sigma_{Ri}^2 \tag{3-16}$$

$$a_{1i} = a_{1i}(\mathbf{A}) = \mu_{Ri} \sum_{k=1}^{m} b_{ik}(\mathbf{A})\mu_{Lk} \tag{3-17}$$

$$a_{2i} = a_{2i}(\mathbf{A}) = (\sum_{k=1}^{m} b_{ik}(\mathbf{A})\mu_{Lk})^2 - \beta_i^{*2}\left[\sum_{k=1}^{m} b_{ik}^2(\mathbf{A})\sigma_{Lk}^2 + 2\sum_{1=k<r}^{m} b_{ik}(\mathbf{A})b_{ir}(\mathbf{A})\rho_{kr}\sigma_{Lk}\sigma_{Lr}\right] \tag{3-18}$$

Since all the load coefficient b_{ik} are independent of the cross-sectional area A_i for static determined trusses, Equation (3-15) will become a quadratic equation of A_i, so that the closed form of the optimum solution A_i^* can be determined as,

$$A_i^* = \frac{1}{a_{0i}}(a_{1i} + \sqrt{a_{1i}^2 - a_{0i}a_{2i}}) \quad (i = 1, 2, \cdots, n) \tag{3-19}$$

As for the static undetermined trusses, Equation (3-15) will be a nonlinear inexplicit algebric equation with respect to the variable A_i $(i = 1, 2, \ldots, n)$ for $b_{ik}(\mathbf{A})$ is the function of $\mathbf{A}$. In this case, the following iteration method can be employed to get the solution.

STEP 1: takes the initial values of $P_{fi}^*(i = 1, 2, \ldots, n)$ and $\mathbf{A}^{(0)} = (A_1^{(0)}, A_2^{(0)}, \cdots, A_n^{(0)})$ and let $0 \Rightarrow k$;

STEP 2: calculate the load coefficients $b_{ik}(\mathbf{A})(i = 1, 2, \ldots, n; k = 1, 2, \ldots, m)$ which correspond to the values of $\mathbf{A}$;

STEP 3: calculate the failure probability $P_{fi}(\mathbf{A}^{(k)})$ for each element by the formula

$$P_{fi}(\mathbf{A}^{(k)}) = \Phi\left[-\frac{\mu_{Z_i}(\mathbf{A}^{(k)})}{\sigma_{Z_i}(\mathbf{A}^{(k)})}\right] \quad (i = 1, 2, \cdots, n)$$

STEP 4: If $|P_{fi}(\mathbf{A}^{(k)}) - P_{fi}^*|/P_{fi}^* \leqslant \varepsilon(i = 1, 2, \ldots, n)$, the iteration is stopped and $\mathbf{A}^{(k)}$ is taken as the optimum solution $\mathbf{A}^*$; otherwise, go to step 5;

STEP 5: carry out the $(k + 1)$ iteration for $\mathbf{A}$ according to the following formula,

$$\mathbf{A}^{(k+1)} = \frac{1}{a_{0i}}[a_{1i}(\mathbf{A}^{(k)}) + \sqrt{a_{1i}^2(\mathbf{A}^{(k)}) + a_{0i} \cdot a_{2i}(\mathbf{A}^{(k)})}] \quad (i = 1, 2, \cdots, n$$

then let $k + 1 \Rightarrow k$ and go to step 2.

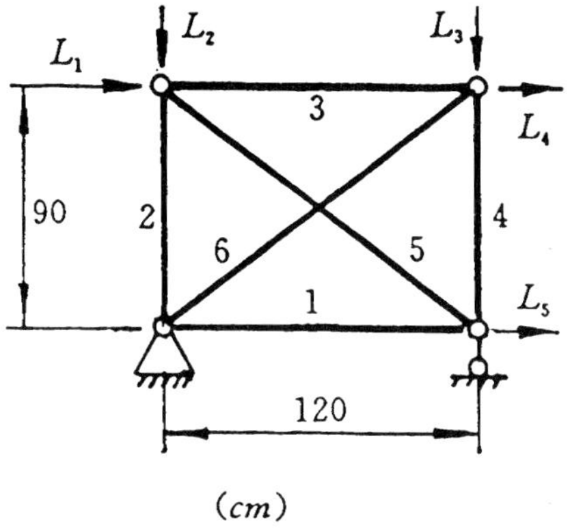

Figure 13. 6-Bar truss structure.

In order to demonstrate the method, the following example is provided here. Example:

This is a 6-bar truss structure (Figure 13). There are five groups of loads $L_k(k = 1, 2, \ldots, 5)$, which are independent of each other and subjected to normal distributions with the variation coefficient $V_{Lk} = 0.2(k = 1 \sim 5)$. The mean values are $\mu_{L1} = 50$ kN, $\mu_{L2} = \mu_{L3} = 30$ kN and $\mu_{L4} = \mu_{L5} = 20$ kN. The material strength $R_i(i = 1 \sim 6)$ of each bar (whether tensioned or compressed) are also subjected to the same normal distribution with the mean $\mu_{Ri} = 27.6$ kN/cm^2 $(i = 1 \sim 6)$, the corresponding variation coefficient $V_{Ri} = 0.05(i = 1 \sim 6)$. The density of the material is $\rho_i = 2.7 \times 10^{-3}$ kg/cm^3.

The failure probability P_{fi}^* of each bar can be determined by the pregiven value of system failure probability P_f^* according to the equivalent criteria, i.e., $P_{fi}^* = P_f^*/6$. The values of P_f^* are equal to 10^{-1}, 10^{-2}, 10^{-3}, 10^{-4} and 10^{-5} respectively. The initial cross-sectional areas of the elements are all the same as $A_i^{(0)} = 4.2$ cm^2 $(i = 1 \sim 6)$. The optimum result is given in Table 1.

It can be seen from the result that the cross-sectional areas of the bars are directly proportional to the requirement of the reliability.

5.18. THE FAILURE MODEL RELIABILITY BASED OPTIMUM DESIGN OF FRAME STRUCTURES

Frame structure is another kind of rod structure which are often applied in practical engineering. In this kind of structure, each beam can be subject to axial forces and also bending, shear and torsion loads.

Table 1. The optimum result of 6-bar truss

P_f^* bar No.	10^{-1} $A_i(\text{cm}^2)$	10^{-2} $A_i(\text{cm}^2)$	10^{-3} $A_i(\text{cm}^2)$	10^{-4} $A_i(\text{cm}^2)$	10^{-5} $A_i(\text{cm}^2)$
1	2.91	3.14	3.33	3.51	3.66
2	0.61	0.80	0.96	1.10	1.23
3	0.60	0.76	0.89	1.01	1.12
4	2.42	2.68	2.90	3.09	3.27
5	2.60	2.83	3.02	3.19	3.35
6	1.76	2.00	2.20	2.38	2.55
W(**A**) (kg)	3.64	4.07	4.42	4.74	5.03

According to the failure analysis of the frame structures, there may be several danger sections of each beam, i.e., there may exist several failure models for one beam. The method for trusses can not be employed to handle the frame structures simply. A new kind of method is necessary. In this section, a failure model reliability based optimum design for 2-D frame structures by Murotsu[22] is described.

Suppose the 2-D frame structure is composed of n beams and subjected to m random loads $L_k(k = 1, 2, \ldots, m)$ simultaneously. It is also assumed that the system is of m_f possible failure models (i.e., m_f danger sections). The task is to minimize the structural weight or mass by designing each bar's geometric parameters such as diameter, thickness and so on. This optimum design problem can be mathematically stated in the form of,

$$\text{find} \quad \mathbf{X} = (x_1, x_2. \cdots, x_n)$$

$$\text{min.} \quad \mathbf{W(X)} = \sum_{i=1}^{n} W_i = \sum_{i}^{n} \rho_i l_i A(x_i) \tag{4-1}$$

$$\text{s.t.} \quad P_{fj}(\mathbf{X}) \leqslant P_{fj}^* \qquad (j = 1, 2, \cdots, m_f) \tag{4-2}$$

$$-x_i < 0 \qquad (i = 1, 2, \cdots, n) \tag{4-3}$$

in which, x_i, ρ_i, l_i and $A(x_i)$ are the geometric characteristic parameter, the material density, the length and the cross-sectional area of the ith beam separately; $P_{fj}(\mathbf{X})$ and P_{fj}^* are the value of the real probability and the allowable value of the jth failure model of structure respectively.

It should be noted that the above optimization model is the same as that for truss structures described in formulae (3-1)–(3-3) in form. The solution procedure of the failure probability function $P_{fj}(\mathbf{X})$ is given below.

Now let us consider the failure model of each bar of the structure first. Because a 2-D frame structure is considered here,the bar is not subjected to twisting load. Suppose the ith bar is subjected to the bending moment and the axial force simultaneously (ignoring the shear force), and there are a total of m_i dange sections*in the bar, i.e., m_i failure models, the failure criterion of the jth dange section will be,

$$\frac{M_j}{M_{pj}} + \frac{N_j}{N_{pj}} \geqslant 1 \quad (j = 1, 2, \cdots, m_i) \tag{4-4}$$

$$\frac{M_j}{M_{pj}} + \frac{N_j}{N_{cj}} \geqslant 1 \quad (j = 1, 2, \cdots, m_i) \tag{4-5}$$

the above inequalities (4-4) and (4-5) are the failure criteria of tension-bending combination and the compression-bending combination of the jth dange section respectively. In the above formulae, M_j and N_j are the bending moment and the axial force at the jth dange section respectively; $M_{pj} = \mathrm{sign}(M_j) R_i A(x_i) J_{pj}(x_i)$, $N_{pj} = \mathrm{sign}(N_j) R_i A(x_i)$, and $N_{cj} = \mathrm{sign}(N_j) R_{ci} A(x_i)$ are the allowable bending moment, axial tension and axis compression at the jth dange section respectively; Here $\mathrm{sign}(\cdot)$ is the positive and the negative symbol of $(\cdot)$. In addition, the random variables R_i and R_{ci} are the material strength and the material yield strength of the ith beam respectively; $A(x_i)$ is the cross-sectional area of the ith bar, obviously it is the function of the design variable x_i; $J_{pj}(x_i)$ are the plastic section modules of the ith bar at the jth dange section, likewise, it is also the function of x_i.

Applying the matrix structure analysis method to the problem gives the bending moment and the axial force of the jth dange section under the m loads simultaneously,

$$M_j = \sum_{k=1}^{m} b_{jk}^{(M)}(\mathbf{X}) \cdot L_k \qquad (j = 1, 2, \cdots, m_i) \tag{4-6}$$

$$N_j = \sum_{k=1}^{m} b_{jk}^{(N)}(\mathbf{X}) \cdot L_k \qquad (j = 1, 2, \cdots, m_i) \tag{4-7}$$

in which, $b_{jk}^{(M)}(\mathbf{X})$ and $b_{jk}^{(N)}(\mathbf{X})$ are the affected coefficients of the bending and the axial forces at the jth dange section with respect to the load L_k respectively.

---

*The dange section means that at this section of the beam, the real stress is most likely over than the maximum allowable stress of the material.

Considering the failure criterion (4-4), (4-5) the above formulae yield the following performance function Z_j, which is corresponding to both the failure models of the tension-bending and the compression-bending at the jth dange section of the ith beam separately.

$$Z_j = R_i^2 A(x_i) J_{pj}(x_i) - \sum_{k=1}^{m} \{[\mathrm{sign}(N_j) b_{jk}^{(M)}(\mathbf{X}) A(x_i) + \mathrm{sign}(M_j) b_{jk}^{(N)}(\mathbf{X}) \cdot J_{pj}(x_i)] R_i L_k\} \quad (j = 1, 2, \cdots, m_i) \tag{4-8}$$

$$Z_{cj} = R_i R_{ci} A(x_i) J_{pj}(x_i) - \sum_{k=1}^{m} \{[sign(N_j) b_{jk}^{(M)}(\mathbf{X}) R_{ci} A(x_i) + \mathrm{sign}(M_j) b_{jk}^{(N)}(\mathbf{X}) \cdot R_i J_{pj}(x_i)] L_k\} (j = 1, 2, \cdot s, m_i) \tag{4-9}$$

As a result of this, the failure probability of the jth dange section in a frame structure, i.e., the probability of the jth failure model of the structure can be written as

$$\begin{aligned} P_{fj}(\mathbf{X}) &= P\{Z_j \leqslant 0 \cup Z_{cj} \leqslant 0\} \\ &= P\{Z_j \leqslant 0\} + P\{Z_{cj} \leqslant 0\} - P\{Z_j \leqslant 0 \cap Z_{cj} \leqslant 0\} \\ &\quad (j = 1, 2, \cdots, m_f) \end{aligned} \tag{4-10}$$

where

$$m_f = \sum_{i=1}^{n} m_i$$

Since the correlation between Z_j and Z_{cj}, the joint probability distribution function of Z_j and Z_{cj} must be known in advance to calculate the above formula precisely. However, this is very difficult. For the sake of simplicity and safety, the failure probability $P_{fj}(\mathbf{X})$ could take its upper limit in practical computation. That is to say,

$$P_{fj}(\mathbf{X}) = P\{Z_j \leqslant 0\} + P\{Z_{cj} \leqslant 0\} (j = 1, 2, \cdot s, m_f) \tag{4-11}$$

In practice, the approximation method (for instance, the method of the second order moment) is usually used to handle the formula (4-11).

Seen from the viewpoint of mathematics, the problem (4-1)–(4-3) is a nonlinear programming problem with constraints, for which many algorithms can be applied. Murotsu[22] has used the SLP method to solve this kind of problem with the following main procedures.

Suppose $\mathbf{X}^{(k)}$ is the design vector obtained in the kth iteration. To begin with, extending the objective and the constraint functions at the point $\mathbf{X}^{(k)}$ in the first order Taylor series, i.e.,

$$\mathbf{W}(\mathbf{X}) = \mathbf{W}(\mathbf{X}^{(k)}) + \triangledown \mathbf{W}(\mathbf{X}^{(k)})^T \cdot \triangle \mathbf{X} \tag{4-12}$$

$$P_{fj}(\mathbf{X}) = P_{fj}(\mathbf{X}^{(k)}) + \triangledown P_{fj}(\mathbf{X}^{(k)})^T \cdot \triangle \mathbf{X} (j = 1, 2, \cdots, m_f) \tag{4-13}$$

then transforming the original programming problem into a sequential LP problems, which can be solved by the simplex method.

$$\begin{aligned} &\text{find} && \triangle \mathbf{X} \\ &\text{min.} && \triangledown \mathbf{W}(\mathbf{X}^{(k)})^T \cdot \triangle \mathbf{X} && (4\text{-}14) \\ &\text{s.t.} && \triangledown P_{fj}(\mathbf{X}^{(k)})^T \cdot \triangle \mathbf{X} \leqslant P_{fj}^* - P_{fj}(\mathbf{X}^{(k)}) (j = 1, 2, \cdots, m_f) && (4\text{-}15) \end{aligned}$$

Once the optimum design $\triangle \mathbf{X}^*$ of the above LP problem is obtained, the new design point of the next iteration can be determined by means of,

$$\mathbf{X}^{(k+1)} = \mathbf{X}^{(k)} + \triangle \mathbf{X}^* \tag{4-16}$$

next, replacing $\mathbf{X}^{(k)}$ by $\mathbf{X}^{(k+1)}$ and going to the next iteration. The procedure of this method is continued repeatedly till convergence is reached. Now the following example, the door frame structure (Figure 14), is given to demonstrate the efficiency of the method.

The structure is subjected to two random loads L_1 and L_2 with the same variation coefficient simultaneously. There are eight dange sections in the structure. The pillar (1) and the beam (2) are all selected as the pipe cross-section with the constant ratio of the thickness to the outer diameter. The radius of the pillar (1) and the beam (2) are considered as the design variables x_1 and x_2 respectively. The material strength $R_i (i = 1, 2)$ of the two elements are the random variables with the same variation coefficient.

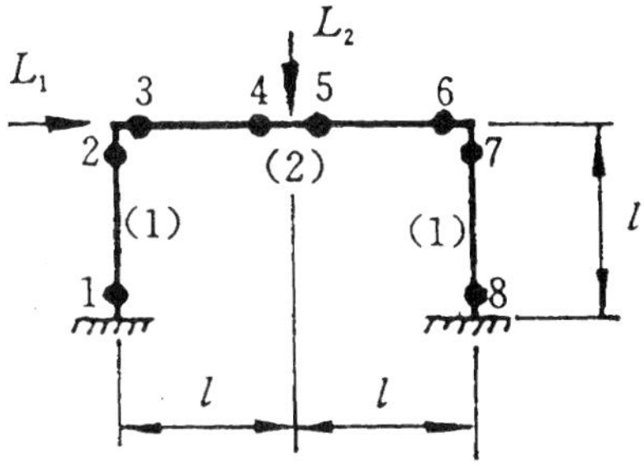

Figure 14. The door frame structure.

Table 2. The optimum result with $\rho_{L_1L_2} = 0.0$

v_{Ri}/v_{Lk}	0.05			0.1		
	$X_1^*(m)$	$X_2^*(m)$	$\mathbf{W}^*(kg)$	$X_1^*(m)$	$X_2^*(m)$	$\mathbf{W}^*(kg)$
0.15	0.143	0.134	931	0.151	0.143	1040
0.3	0.156	0.144	1083	0.161	0.151	1171
0.6	0.176	0.161	1368	0.181	0.166	1450

Table 3. The optimum result with $\rho_{L_1L_2} = 0.5$ and 1.0 and $v_{Ri} = 0.05$

$\rho_{L_1L_2}/v_{Lk}$	0.5			1.0		
	$X_1^*(m)$	$X_2^*(m)$	$\mathbf{W}^*(kg)$	$X_1^*(m)$	$X_2^*(m)$	$\mathbf{W}^*(kg)$
0.15	0.145	0.137	955	0.146	0.139	976
0.3	0.158	0.148	1130	0.159	0.152	1166
0.6	0.179	0.168	1455	0.182	0.174	1527

The fundamental parameters are $\mu_{L1} = 50$ kN, $\mu_{L2} = 40$ kN; $\mu_{R1} = \mu_{R2} = 274M$ pa; $l = 5$ m. The modules of elasticity is $E = 210$ Gpa. The density of material is $\rho_i = 7.85 \times 10^{-3}$ kg/cm^3 $(i = 1, 2)$. The ratio between the thickness and the outer diameter is 0.05. The limit values of the failure probability of all the danger sections are the same as 0.001, i.e., $P_{fj}^* = 0.001(j = 1 \sim 8)$.

Under the condition of different loading correlation coefficient and different combination of the loading and the strength variation coefficients, the practical computation has been carried out.The optimum results are shown in the Tables 2 and 3.

It can be seen from the results that the final structural weight is directly proportional to the variation coefficient of the random variables and the correlation coefficient of the loads respectively.

5.19. SYSTEM RELIABILITY BASED STRUCTURAL OPTIMIZATION

Studying a structure from the viewpoint of system reliability, the structure that satisf's the reliability requirement for each element does not mean that the demand for structural system reliability is also satisfied. In addition, because it is impossible to consider the correlations among the failure models in the constraints of the failure model's reliability, it can not be certain that system reliability is achieved although the requirements of the reliability for each failure model are satisfied. Therefore it can be said that the optimum design model with system reliability constraints will be the most reasonable and the most rational design model for the reliability based optimum design. Taking the minimum weight design as an example, the general mathematic model can be described in the following form.

$$\text{find} \quad \mathbf{X} = (x_1, x_2, \cdots, x_n)$$

$$\text{min.} \quad W(\mathbf{X}) = \sum_{i}^{n} W_i(x_i) \tag{5-1}$$

$$\text{s.t.} \quad P_f(\mathbf{X}) \leqslant P_f^* \tag{5-2}$$

where $P_f(\mathbf{X})$ and P_f^* are the real and the allowable values of the failure probability of the structural system separately.

If the above model is compared with both the element reliability based model and the failure reliability based model, it seems to be simpler than the other two models in the form because so many reliability constraints in formulae (3-2) and (4-2) are replaced by the one constraint of formula (5-2). As a matter of fact, however, it becomes more difficult to be treated, particularly the computation of the failure probability $P_f(\mathbf{X})$ of the structural system.[23]

It is easy to show from the theory of structural reliability that there are many possible failure models in the complex or the statically undetermined structural system under random loading condition. Nevertheless, the system's failure probability will depend on all of these failure models. What should be pointed out here is that two main difficult problems in structural reliability analysis are still research topics. One is how to determine all the failure models (particularly the main failure model) of a structural system. The other is how to calculate the failure probability of a structure precisely according to those failure models.

Suppose that the structure is subjected to m random loads L_k $(k = 1, 2, \cdots, m)$ and that there are a total of m_f known possible failure models, for which the corresponding performance functions are denoted by $Z_j(\mathbf{X})(j = 1, 2, \cdots, m_f)$. Obviously, any one of the m_f failure models can lead to the failure of whole system. Based on these considerations, the failure probability $P_f(\mathbf{X})$ of the structural system will be equal to the probability of summing the events of m_f failure models. i.e.,

$$P_f(\mathbf{X}) = P\{\cup_{j=1}^{m_f}(Z_j(\mathbf{X}) \leqslant 0)\} \tag{5-3}$$

Since all the failure models are usually correlated to each other, it is almost impossible to calculate the above formula exactly. In structural reliability based optimum design, one typical way is to replace $P_f(\mathbf{X})$ by the upper bound of the first or the second order of system's failure probability which is both simple and which provides safety of the structure[24], is as follows,

$$\text{Let}\quad P_f(\mathbf{X}) = P_f^{(1)}(\mathbf{X}) = \sum_{j=1}^{m_f} P\{Z_j(\mathbf{X}) \leqslant 0)\} = \sum_{j=1}^{m_f} P_{fj}(\mathbf{X}) \qquad (5\text{-}4)$$

$$\text{or let}\quad P_f(\mathbf{X}) = P_f^{(2)}(\mathbf{X}) = \max\nolimits_{1\leqslant j\leqslant m_f}(P\{Z_j(\mathbf{X}) \leqslant 0\} + \sum_{j=2}^{m_f} \min\nolimits_{1\leqslant i\leqslant j-1}[P\{(Z_i(\mathbf{X}) \leqslant 0) \cap (Z_j(\mathbf{X}) > 0)\}] \qquad (5\text{-}5)$$

where $P_f^{(1)}(\mathbf{X})$ and $P_f^{(2)}(X)$ are separately the upper bound of the first and the second order failure probability of structural system.

In addition, by means of the reliability analysis method of structures such as the loading increasement method, the response surface method and so on, the general expression of the performance function $Z_j(\mathbf{X})$ can be obtained as follows,

$$Z_j(\mathbf{X}) = \sum_{q=1}^{n_j} a_{jq} R_q f_q(x_q) - \sum_{k=1}^{m} b_{jk}(\mathbf{X}) L_k \qquad (5\text{-}6)$$

in which a_{jq} is the strength coefficient of the jth failure model, which is determined by the element q; R_q is the random variable of material's strength of element q; $f_q(x_q)$ is the geometric characteristic function or the related function of the design variable x_q of element q; $b_{jk}(\mathbf{X})$ is the loading coefficient corresponding to the jth failure model under the kth loading condition, and it is independent of the design variable $\mathbf{X}$ for a static determined structure; n_j is the total element number involved in the jth failure model.

In the practical computation of formulae (5-4) and (5-5) about $P\{Z_j(\mathbf{X}) \leqslant 0\}$, the second moment method is usually employed. As for solving the joint probability $P\{Z_i(\mathbf{X}) \leqslant 0 \cap Z_j(\mathbf{X}) > 0\}$, another method by Ditleven[25] can be employed. Up to this point, it can be noticed that the problem (5-1) and (5-2) is a nonlinear programming problem with constraints, which can be easily solved by any NLP algorithm such as the sequential unconstrained minimizing technique (SUMT) or some other indirect method.

Example:

Figure 15 shows a statistically determined 5-bar truss sstructure with the parameters $l_1 = 0.9$ m, $l_2 = 1.2$ m, the density of material $\rho_i = 2.7 \times 10^{-3}$ kg/cm^3 ($i = 1 \sim 5$); The loads $L_k (k = 1 \sim 5)$ are independent of

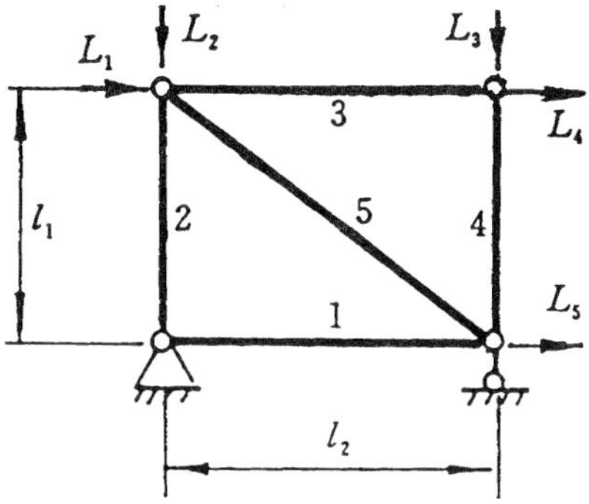

Figure 15. 5-Bar truss structure.

each other and are characterized by a normal distribution. The mean values of the loads are $\mu_{L1} = 50$ kN, $\mu_{L2} = \mu_{L3} = 30$ kN, and $\mu_{L4} = \mu_{L5} = 20$ kN. The variation coefficients are all the same, namely, $v_{Lk} = 0.1(k = 1 \sim 5)$; The tension and the compression strength of each bar are normal variables with the same mean and the same variation coefficient, namely, $\mu_{Ri} = 2.76 \times 10^5$ kN/m^2 and $v_{Ri} = 0.05(i = 1 \sim 5)$. The limited failure probability of the structure is $P_f^* = 1.0 \times 10^{-4}$. The cross-sectional areas of the five bars with the initial failure value of $A_i^{(0)} = 18.12$ cm$^2(i = 1 \sim 5)$ are selected as design variables. The structural weight is $W = \Sigma_{i=1}^5 \rho_i A_i l_i$. The failure probability of the system $P_f(\mathbf{X})$ is approximated by the upper limit of $P_f^{(1)}(\mathbf{X})$, the first order failure probability of structural system.

Since this is a statistically determined structure, the failure of each element will be one failure model of the system. By means of structural analysis, the performance function corresponding to the damage of every bar can be obtained as,

$$Z_1 = A_1 R_1 - L_1 - L_4 - L_5 \qquad \text{(tension)}$$
$$Z_2 = A_2 R_2 - (l_2/l_1)L_1 + L_2 - (l_2/l_1)L_4 \qquad \text{(tension)}$$
$$Z_3 = A_3 R_3 - L_4 \qquad \text{(tension)}$$
$$Z_4 = A_4 R_4 - L_3 \qquad \text{(compression)}$$
$$Z_5 = A_5 R_5 - (\sqrt{l_1^2 + l_2^2}/l_1)L_1 - (\sqrt{l_1^2 + l_2^2}/l_1)L_4 \qquad \text{(compression)}$$

The mean μ_{Zj} and the variation σ_{Zj}^2 of the function Z_j can be deduced from the above formulae. As a result of this, the probability of the jth failure model of the structural system becomes,

$$P_{fj}(\mathbf{A}) = \Phi(-\beta_j) \qquad (j = 1, 2, \cdots, 5)$$

where

$$\beta_j = \frac{\mu_{Zj}}{\sigma_{Zj}} \qquad (j = 1, 2, \cdots, 5)$$

The corresponding reliability constraint of the system can be expressed as,

$$P_f(\mathbf{A}) = P_f^{(1)}(\mathbf{A}) = \sum_{j=1}^{5} P_{fj}(\mathbf{A}) = \sum_{j=1}^{5} \Phi(-\beta_j) \leqslant P_f^* = 1.0 \times 10^{-4}$$

Murotsu[26] has solved the above problem on the computer Burrough-6700 by applying SUMT and the indirect method respectively with the result shown in Table 4.

It can be observed from the results that the optimum solutions by the two methods are quite close, but the CPU time required by SUMT is much more than that required by the indirect method due to the many computations of the derivatives for the probability function in the SUMT method. In addition, the most desirable design means that each element of the structure is of the same reliability. This is easy to be found from the result and is in accordance with the result of other optimization problem methods.

What should be pointed out about this method is that even though the design model of the systematic reliability based optimum design is one of the most rational models, it is only appropriate to the circumstance that the failure model and the performance function are easily solved. For complex structures or highly static undetermined structures, the implementation of this kind of optimization is considerably difficult owing to the difficulty of computing the structural reliability.

Table 4. The optimum result of 5 bar truss

method / bar number	SUMT		indirect method	
	A_i^* (cm^2)	$P_{fj}(\times 10^{-4})$	A_i^* (cm^2)	$P_{fj}(\times 10^{-4})$
1	4.49	0.2288	4.52	0.2000
2	1.66	0.1261	1.64	0.2000
3	1.12	0.0826	1.09	0.2000
4	1.68	0.0947	1.65	0.2000
5	4.45	0.4071	4.54	0.2000
$P_f = \sum P_{fj}$		0.9993×10^{-4}		1.0000×10^{-4}
W (kg)	4.44		4.46	
CPU(s)	84.15		0.37	

5.20. PRECISION RELIABILITY BASED OPTIMIZATION OF ANTENNA STRUCTURES

The parabolic reflector antenna is an often used antenna in engineering. For this kind of antenna, the antenna's performance depends upon the root mean square (r.m.s) value of the reflector's deformation. Thus, the reflector's surface precision (r.m.s) is usually taken as the design criterion when the parabolic antenna, particular large and high performance antennas (such as radio astronomy, satellite earth station) are designed. Since the wind load is the main load for the antenna structure, the reflector's precision based optimum design of an antenna structure under the loading case of random wind is discussed in the paper[27], which is summarized as follows.

5.20.1. The Precision Reliability Analysis of the Reflector

Utilizing statistical methods to describe the wind to which the antenna is subjected, the wind pressure can be approximately considered to be a normal distribution with complete correlation. Consequently the response (displacement and stress) of the antenna structure to the wind is also a normal distribution with correlation with each other completely. The reliability P_r of the antenna reflector's precision is defined as the probability that the real value of r.m.s. $U^2_{\rm rms}$ (random variable) of the reflector in the axial or the normal direction is less than or equal to its allowable value $U^{*2}_{\rm rms}$ (determined value). That is to say,

$$P_r = P_{\rm rob}\{U^2_{\rm rms} \leqslant U^{*2}_{\rm rms}\} \tag{6-1}$$

$$U^{*2}_{\rm rms} = \delta^2_{\rm rms} - S^2_{\rm rms} \tag{6-2}$$

where, $\delta^2_{\rm rms}$ is the value of the reflector's precision, which is determined by the Ruze formula[28]; $S^2_{\rm rms}$ is the r.m.s. value of the reflector due to the system's load such as self-weight.

Because the distribution, to which the variable $U^2_{\rm rms}$ is submitted, is not known, it is impossible to calculate the reliability P_r from the formula (6-1). For the sake of finding P_r, the following equivalent event "$U^2_{\rm rms} \leqslant U^{*2}_{\rm rms}$" is determined.

Suppose that there are n_r nodes on the reflector, and the direction of electronic axis of the antenna coincide with the axis z of the global coordinate system. Then $U^2_{\rm rms}$ can be expressed by means of the displacement u_{zi} $(i = 1, 2, \cdots, n_r)$ along the direction of axis z in the following form,

$$U^2_{\rm rms} = \frac{1}{n_r}\sum_{i=1}^{n_r} u^2_{zi} \tag{6-3}$$

As a result of this, the event "$U_{\rm rms}^2 \leqslant U_{\rm rms}^{*2}$" can be described as,

$$\sum_{i=1}^{n_r} u_{zi}^2 \leqslant n_r U_{\rm rms}^{*2} = \sum_{i=1}^{n_r} U_{\rm rms}^{*2} \tag{6-4}$$

Because

$$u_{zi} \sim \mathcal{N}(\mu_{zi}, \sigma_{zi}^2) \qquad (i = 1, 2, \cdots, n_r) \tag{6-5}$$

and

$$\rho(u_{zi}, u_{zj}) = 1 \qquad (i \neq j; i, j = 1, 2, \cdots, n_r) \tag{6-6}$$

the following expression can be obtained by normalizing both sides of the formula (6-4) with μ_{zj} and σ_{zj} as,

$$\sum_{i=1}^{n_r} \left(\frac{u_{zi} - \mu_{zi}}{\sigma_{zi}}\right)^2 \leqslant \sum_{i=1}^{n_r} \left(\frac{U_{\rm rms}^{*} - \mu_{zi}}{\sigma_{zi}}\right)^2 \tag{6-7}$$

in which, μ_{zi} and σ_{zi} are the mean and the variation of the random variable u_{zi} respectively; $\rho(u_{zi}, u_{zj})$ is the correlative coefficient of the variables u_{zi} and u_{zj}.

Considering that all the random variables $\left(\frac{u_{zi}-\mu_{zi}}{\sigma_{zi}}\right)^2$ $(i = 1, 2, \cdots, n_r)$ are submitted to χ^2 distribution with one degree of freedom and that the variables $u_{zi} (i = 1, 2, \cdots, n_r)$ are correlated with each other completely yields the following equation,

$$\sum_{i=1}^{n_r} \left(\frac{u_{zi} - \mu_{zi}}{\sigma_{zi}}\right)^2 = n_r \cdot \chi^2(1) \tag{6-8}$$

where, $\chi^2(1)$ is the χ^2 distribution with one degree of freedom.

By applying the above formula, the inequality (6-7) can be written as,

$$\chi^2(1) \leqslant \frac{1}{n_r} \sum_{i=1}^{n_r} \left(\frac{U_{\rm rms}^{*} - \mu_{zi}}{\sigma_{zi}}\right)^2 = (\delta^*)^2 \tag{6-9}$$

clearly, the above formula is equivalent to the event "$U_{\rm rms}^2 \leqslant U_{\rm rms}^{*2}$". Therefore the precision reliability of the reflector (6-1) can be expressed by this equivalent event as,

$$P_r = P_{\rm rob}\{\chi^2(1) \leqslant (\delta^*)^2\} = \int_0^{(\delta^*)^2} f(x)dx \tag{6-10}$$

where, $f(x)$ is the probability density function of variable $\chi^2(1)$.

6.20.2. Statement of the Optimum Design

In order to highlight the electronic performance of a large antenna, the requirement for reflector's precision under certain loading conditions becomes important. In addition, it is also necessary to make sure that the antenna as a large structure should be strong enough to withstand a large loading environment. As a result of these requirements, the reliability based optimum design of the antenna can be mathematically stated as that: searching for the design variable of all bar cross-sectional areas of the antenna's backup structure and minimizing the structural weight whilst satisfying the following three requirements, such as (1) the reflector's precision reliability; (2) the reliability of each element's strength including the yield; (3) the bound constraint of cross-sectional area. That is to say,

$$\text{find} \quad \mathbf{A} = (A_1, A_2, \cdots, A_n$$

$$\text{min.} \quad \mathbf{W}(\mathbf{A}) = \sum_{i=1}^{n} W_i = \sum_{i=1}^{n} \rho_i l_i A_i \tag{6-11}$$

$$\text{s.t.} \quad P_r^* - P_r(\mathbf{A}) \leqslant 0 \quad \text{under the working load} \tag{6-12}$$

$$P_{rj}^* - P_{rj}(\mathbf{A}) \leqslant 0 \quad (j = 1, 2, \cdots, n_e) \tag{6-13}$$

under the worst load environment

$$A_{\min}^* - A_i \leqslant 0 \quad (i = 1, 2, \cdots, n) \tag{6-14}$$

where $P_{rj}(\mathbf{A})$ and P_{rj}^* are the actual reliability and the allowable value of the reliability of the jth element; n_e is the total number of the elements in structure.

According to the experience of practical design, the main problem of a large antenna structure is to satisfy the precision requirement. In other words, the strength constraints can be satisfied automatically as long as the constraints of precision can be achieved. For the sake of this, the precision constraints of formula (6-12) is usually considered as the active constraint in the optimum design of antenna structures. As a result of this, the optimality criteria method, which will be discussed in the next section, can be applied to solve the above problem.

By noticing that the precision reliability $P_r(\mathbf{A})$ is a quadratic function of the design variable $\mathbf{A}$ and that $P_r(\mathbf{A})$ is directly proportional to the variable $A_i (i = 1, 2, \cdots, n)$, Chen[27] has presented the following iteration formula to find the design variable A_i,

$$\begin{cases} A_i^{(k+1)} = A_i^{(k)} \left\{ \dfrac{\dfrac{\partial P_r^{(k)}(\mathbf{A})}{\partial A_i^{(k)}}}{\left[\dfrac{1}{n}\sum_{i=1}^{n}\left(\dfrac{\partial P_r^{(k)}(\mathbf{A})}{\partial A_i}\right)^2\right]^{1/2}} \right\}^{r_1(k)} \cdot \left(\dfrac{P_r^*}{P_r^{(k)}(\mathbf{A})}\right)^{r_2(k)} & (i = 1, 2, \cdots, n) \\ A_i^{(k+1)} \geqslant A_{\min} \end{cases} \tag{6-15}$$

Once the iteration is terminated, the strength reliability constraint (6-13) for each bar must be checked completely. If some of them are not met, the related design variable will be adjusted so that the satisfaction of the constraints is achieved.

5.20.3. Numerical Example of an 8-m Antenna

This is an often used parabolic antenna with the diameter of eight meters. The backup structure is composed of 12 radial beams, 4 ring beams and some other supporting elements. Only one fourth of the structure is considered due to symmetry and it consists of 97 bars shown in Figure 16.

The structural parameters are: modules of elasticity $E = 2.1 \times 10^6$ kg/cm^2, material density $\rho = 7.85 \times 10^{-3}$ kg/cm^3. The strength of each bar is a specific value, the allowable tension strength is $R_j^+ = 2100$ kg/cm^2 $(j = 1, 2, \cdots, n_e)$. The permissible compression strength R_j^- is determined by the stable condition of each bar. The pregiven r.m.s. value of reflector precision is $U_{\rm rms}^* = 0.85$ mm.

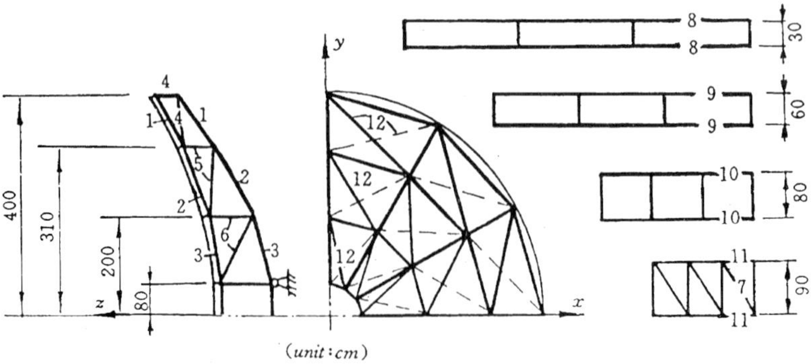

Figure 16. The 8-m parabolic antenna structure.

The cross-sectional area of the bar is taken as the design variable. For this example, the 97 elements are linked to 12 independent design variable groups by the consideration of symmetry and so on. In addition, each group is a pipe element with different thickness but the same inner diameter, 20 mm.

The loading condition is that the horizontal-pointed antenna is subjected to structural self-weight and the random wind pressure load horizontally with the wind variation $V_L = 0.2$.

The constraint conditions:

(1) The precision reliability of the antenna is greater than or equal to 0.95 under the random wind with the mean $E(V) = 20$ m/s, i.e.,

$$P_r(\mathbf{A}) = P\{U_{\text{rms}} \leqslant 0.85 \text{ mm}\} \geqslant P_r^* = 0.95$$

(2) The strength reliability of each bar is greater than or equal to 0.99 under the random wind load with mean $E(\mathbf{V}) = 50$ m/s, i.e.,

$$P_{rj}(\mathbf{A}) \geqslant P_{rj}^* = 0.99 (j = 1, 2, \cdots, n_e)$$

(3) The bound constraints

$$A_i \geqslant A_{\min} = 0.5 \text{ cm}^2 (i = 1, 2, \cdots, n)$$

The iteration converges after six iterations and the optimum result is shown in Figure 17 and Table 5.

5.21. OPTIMALITY CRITERIA METHOD FOR STRUCTURAL RELIABILITY OPTIMIZATION

In general, two main approaches have been used to solve structural synthesis problems, namely mathematical programming (MP) and optimality criteria (OP) methods. These approaches have often been portrayed as opposing alternatives. The argument against MP is that its efficiency depends on the number of design variables, while the argument against OP is that the problem is not always solved in an exact mathematical way, and in some cases yields divergence of the iteration process. Also, the identification of active and passive design variables can be tedious in OP.

Table 5. The numerical optimum result of 8-m antenna structure

variables (cm^2)	A_1	A_2	A_3	A_4	A_5	A_6	A_7	A_8	A_9	A_{10}	A_{11}	A_{12}
initial value	1.5	2.0	3.0	1.0	1.5	1.5	3.0	1.0	1.5	2.0	3.0	1.0
optimum value	0.580	1.745	2.602	0.5	0.5	1.590	0.643	1.493	1.867	0.5	0.5	0.5

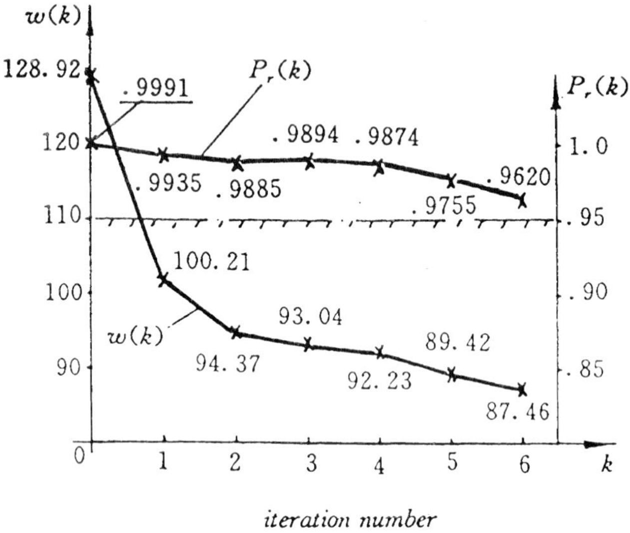

	W (kg)	P_r	$\min\{P_{rj}\}$	iteration number
initial value	128.92	0.99914	1.0000	0
optimum value	87.46	0.96199	0.9913	6

Figure 17. The optimum result of the 8-m antenna structure.

Considering both methods with their respective problems, a number of researchers have developed mixed methods (MM) based on both MP and OP. Theoretically speaking, the mixed method utilizes the advantages of both MP and OP methods. Therefore, it has been applied in engineering design frequently.

Nevertheless, three kinds of methods such as MP, OP and MM are still employed to solve reliability based optimum design of structures according to the different characters of specific problems. This is because one of the three methods may be especially efficient for some specific problems.

So far in this chapter, the MP has been the principle method employed to solve the optimum design problem. In section nine, the mixed method is used to solve a problem there. In this section, the OP method is specifically discussed. The iteration design expression in the OP method will change according to different problems. That is to say, concrete optimality criteria must be deduced for different kinds of problems, there is no general type of OP as for the case of MP. Consequently, it is impossible to present all OP methods in this limited section. Here, only one type of OP methods[29] is given to show what the OP method is and how optimum criteria are deduced.

Take the truss structure as an example, the problem is to minimize the structural weight by finding the sectional areas of n bars whilst satisfying the reliability requirement of the structural system. The problem can be mathematically described in the following form,

$$\text{find} \quad \mathbf{A} = (A_1, A_2, \cdots, A_n)$$

$$\text{min.} \quad \mathbf{W}(\mathbf{A}) = \sum_i^n w_i = \sum_i^n \rho_i l_i A_i \tag{7-1}$$

$$\text{s.t.} \quad P_r^* - P_r(\mathbf{A}) \leqslant 0 \tag{7-2}$$

in which, $P_r(\mathbf{A})$ and P_r^* are the real and the allowable values of the structural reliability respectively.

The above problem is a nonlinear programming (NLP) problem, which is solved by the OP method here. To begin with, the Lagrange function of the problem is constructed as,

$$\begin{aligned} L(\mathbf{A}, \lambda) &= \mathbf{W}(\mathbf{A}) + \lambda[P_r^* - P_r(\mathbf{A})] \\ &= \sum_{i=1}^n \rho_i l_i A_i + \lambda[P_r^* - P_r(\mathbf{A})] \end{aligned} \tag{7-3}$$

where, λ is the largrange multiplier.

By applying the Kuhn-Tucher condition, we obtain the following equations from (7-3),

$$\frac{\partial L}{\partial A_i} = \rho_i l_i - \lambda \frac{\partial P_r(\mathbf{A})}{\partial A_i} = 0 \quad (i = 1, 2, \cdots, n) \tag{7-4}$$

$$\frac{\partial L}{\partial \lambda} = P_r^* - P_r(\mathbf{A}) = 0 \tag{7-5}$$

Theoretically, the optimum design $A_i^*(i = 1, 2, \cdots, n)$ and λ^* (a vector with $n+1$ elements) can be obtained by solving the above Equations (7-4) and (7-5). $A_i^*(i = 1, 2, \cdots, n)$ are the optimum design of the original problem. However, it is considerably difficult to solve Equations (7-4) and (7-5) directly for the reliability $P_r(\mathbf{A})$ which is a complex implicit function of the variable $\mathbf{A}$. To overcome this difficulty, the iteration method is utilized here.

From formula (7-4), the following optimality criteria can be deduced as,

$$\frac{1}{\rho_i A_i} \frac{\partial P_r(\mathbf{A})}{\partial A_i} = \frac{\partial P_r(\mathbf{A})}{\partial w_i} = \frac{1}{\lambda} (i = 1, 2, \cdots, n) \tag{7-6}$$

It can be observed from the above formula that, when the reliability constraint is satisfied and the optimum design is reached, the varying rates of systematic reliability will converge to a constant $1/\lambda$ owing to the shift of design variables. This is an optimality criterion of the system reliability based optimum design. According to this criterion and other relative physical aspects, the new design point at the $(k+1)$th iteration can be obtained as,

$$A_i^{(k+1)} = A_i^{(k)} + \Delta A_i^{(k)} \quad (i = 1, 2, \cdots, n) \tag{7-7}$$

where, $\Delta A_i^{(k)}$ is the increment at the kth step of the ith variable.

It is easy to show from the structural reliability that if $\Delta A_i^{(k)} > 0$, $\frac{\partial P_r(\mathbf{A})}{\partial A_i}$ will be proportional to $\Delta A_i^{(k)}$ directly, and further $\frac{|P_r^* - P_r^{(k)}(\mathbf{A})|}{P_r^*}$ is also directly proportional to $\Delta A_i^{(k)} (i = 1, 2, \cdots, n)$. Here, $P_r^{(k)}(\mathbf{A})$ is the systematic reliability at the kth step. As a final result, the iteration formula of $\Delta A_i^{(k)}$ can be set up as,

$$\Delta A_i^{(k)} = \text{sign}[P_r^* - P_r^{(k)}(\mathbf{A})] \cdot C^{(k)} \left[\frac{\dfrac{\partial P_r^{(k)}(\mathbf{A})}{\partial A_i^{(k)}}}{\dfrac{1}{n} \sum\limits_{i=1}^{n} \dfrac{\partial P_r^{(k)}(\mathbf{A})}{\partial A_i^{(k)}}} \right]^{r_1(k)} \cdot \left[J \frac{|P_r^* - P_r^{(k)}(\mathbf{A})|}{P_r^*} \right]^{r_2(k)} \quad (i = 1, 2, \cdots, n) \tag{7-8}$$

when $P_r^* - P_r^{(k)}(\mathbf{A}) \geqslant 0$, take $C^{(k)} > 0$; otherwise, $C^{(k)} < 0$.

In the above formula, $\left[\frac{\frac{\partial P_r^{(k)}(\mathbf{A})}{\partial A_i^{(k)}}}{\frac{1}{n} \sum_{i=1}^{n} \frac{\partial P_r^{(k)}(\mathbf{A})}{\partial A_i^{(k)}}} \right]$ is the ratio between the sensitivity of the ith variable at the kth iteration and the mean value of all the variables, this term shows how to adjust the direction of the gradient to increase ΔA_i at the kth iteration; $\frac{|P_r^* - P_r^{(k)}(\mathbf{A})|}{P_r^*}$ is the relative difference between the structural reliability at the kth iteration and the pregiven value, which presents the step length of the variable's increment at the kth iteration; $C^{(k)}$, $r_1^{(k)}$ and $r_2^{(k)}$ are the convergent exponents at the kth iteration, whose values are determined by the following expressions.

If the iteration converges rapidly, $C^{(k)} = 1$; If the iteration does not converge, $C^{(k)} = \left(\frac{1}{2}\right)^m$ (m is the number of the nonconverged iterations). The exponent $r_1^{(k)}$ is a constant 1/2. The exponent $r_2^{(k)}$ is determined by,

$$r_2^{(k)} = \frac{1}{2}, \quad \text{if } 10^{-6} \leqslant |P_r^* - P_r^{(k)}(\mathbf{A})|/P_r^* < 10^{-5}$$
$$r_2^{(k)} = \frac{1}{2} \sim 1, \quad \text{if } 10^{-5} \leqslant |P_r^* - P_r^{(k)}(\mathbf{A})|/P_r^* < 10^{-3}$$
$$r_2^{(k)} = 1, \quad \text{if } 10^{-3} \leqslant |P_r^* - P_r^{(k)}(\mathbf{A})|/P_r^* < 10^{-1}$$

According to the optimality criteria, the convergent condition is taken as,

$$\begin{cases} \left\{ \sum_{i=1}^{n} \left[\frac{\partial P_r^{(k)}(\mathbf{A})}{\partial A_i^{(k)}} - \frac{1}{n} \sum_{i=1}^{n} \frac{\partial P_r^{(k)}(\mathbf{A})}{\partial A_i^{(k)}} \right]^2 \right\}^{1/2} \leqslant \eta \\ |P_r^* - P_r^{(k)}(\mathbf{A})|/P_r^* \leqslant \varepsilon \end{cases} \tag{7-9}$$

where, η and ε are the parameters which indicate satisfactory convergence.

In practical computation, sensitivity $\frac{\partial P_r^{(k)}(\mathbf{A})}{\partial A_i^{(k)}}$ can be obtained by numerical methods such as the finite difference method. That is to say, replacing the differentiation approximately by the finite difference.

$$\frac{\partial P_r^{(k)}(\mathbf{A})}{\partial A_i^{(k)}} \simeq \frac{\Delta P_r(\Delta A_i^{(k)})}{\Delta A_i^{(k)}}, \qquad (i = 1, 2, \cdots, n) \tag{7-10}$$

where, $\Delta A_i^{(k)}$ is usually taken as $0.01 \sim 0.05 A_i^{(k-1)}$.

Example:

This is a 5-bar truss as shown in Figure 18. The random loads L_1 and L_2 are submitted to the same logarithm-normal distribution with complete correlation each other. The mean and the variation coefficient are separately $\mu_{Lk} = 10$ and $V_{Lk} = 0.2 (k = 1, 2)$. The material tension and compression strength $R_i (i = 1 \sim 5)$ are the same and submitted to the same normal distribution, the mean and the variation coefficient are $\mu_{Ri} = 40$ and $v_{Ri} = 0.1$ $(i = 1 \sim 5)$ respectively. The correlation coefficient among the strengths are $\rho_{R_i R_j} = 0.5$ $(i \neq j; i, j = 1 \sim 5)$. The material density is $\rho_i = 1$ $(i = 1 \sim 5)$. The required system reliability is $P_r^* = 0.9999$. By using the above OP method, the optimum result is obtained after 3 iterations, which is shown in Table 6.

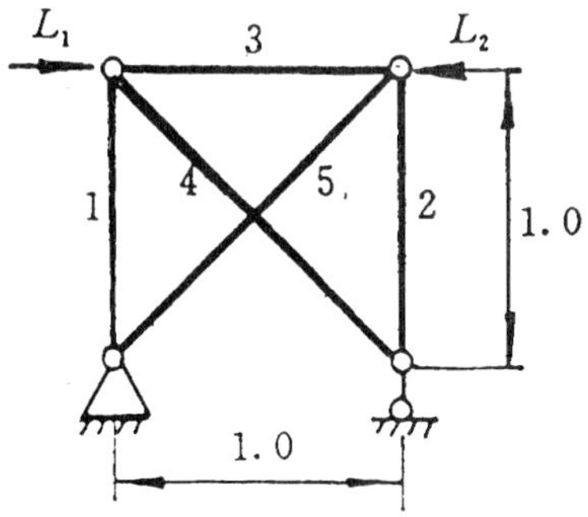

Figure 18. The 5-bar truss structure.

5.22. THE DYNAMIC RELIABILITY BASED OPTIMUM DESIGN OF ENGINEERING STRUCTURES

So far, most work about structural reliability optimization have concentrated on the static problem. The research achievements about structural dynamic reliability based optimization are relatively few due to its difficulty. But anyway, some methods about this kind of problem have been presented by some researchers. Nigam[30] has presented a general description and a solving method for structural dynamic reliability based optimization with determined design variables. Rao[31] has treated the problem with random design variables. In this section, the general model of structural dynamic reliability based optimization and the corresponding solution method for the case of determined design variable are discussed.

Table 6. The optimum result of 5-bar truss by the OP method

structural parameters	iteration number k			
	0	1	2	3
$A_1^{(k)} = A_2^{(k)} = A_3^{(k)}$	0.5	0.5018	0.50333	0.50296
$A_4^{(k)} = A_5^{(k)}$	$0.5\sqrt{2}$	$0.5025\sqrt{2}$	$0.5035\sqrt{2}$	$0.5032\sqrt{2}$
$P_r(\mathbf{A})$	0.9998843	0.9998940	0.9999045	0.9999001
$\frac{P_r^* - P_r^{(k)}(\mathbf{A})}{P_r^*}$	1.574×10^{-5}	5.99×10^{-6}	-4.5×10^{-7}	-8.0×10^{-8}
$\mathbf{W}^{(k)}$	3.5	3.5154	3.5240	3.5204

5.22.1. Dynamic Reliability Optimum Design Based on the Failure Mechanism of the First-Passage

Without compromising generality, assume that the structural system has m possible dynamic failure models (strength, displacement, velocity, acceleration and so on) under random dynamic loads. Moreover, these m failure models all belong to the failure category of the first passage.*In addition, consider constraint conditions in structural systems. The general mathematical model of structural dynamic reliability based optimization can be described as follows.

find $D = (d_1, d_2, \cdots, d_n)$

min. $F(\mathbf{D})$ (8-1)

s.t.: (1) constraint of system's dynamic reliability

$$P_f(\mathbf{D}, t) = P\left\{\cup_{j=1}^{m} g_j[Y(\mathbf{D}, \mathbf{X}, t)] \geqslant \lambda_j\right\} \leqslant P_f^* \quad (0 \leqslant t \leqslant T^*, \mathbf{X} \in R^n) \qquad (8\text{-}2)$$

(2) dynamic constraint

$$G_i[Y(\mathbf{D}, \mathbf{X}, t)] \leqslant \alpha_i^* \quad (0 \leqslant t \leqslant T^*, \mathbf{X} \in R^n) \quad (i = 1, 2, \cdots, n_1) \qquad (8\text{-}3)$$

(3) constraint of moment

$$E\{G_i[Y(\mathbf{D}, \mathbf{X}, t)]\} \leqslant \beta_i^* \quad (0 \leqslant t \leqslant T^*, \mathbf{X} \in R^n) \quad (i = 1, 2, \cdots, n_2) \qquad (8\text{-}4)$$

(4) static constraint

$$f_i(\mathbf{D}) \leqslant \gamma_i^* \quad (i = 1, 2, \cdots, n_3) \qquad (8\text{-}5)$$

(5) constraint of the eigenvalue

$$\omega_i^{(l)} \leqslant \omega_i(\mathbf{D}) \leqslant \omega_i^{(u)} \quad (i = 1, 2, \cdots, n_4) \qquad (8\text{-}6)$$

in which, t is the time variable; T^* is the service life of a structure; $\mathbf{X}$ is the coordinate vector; R^n is the structural related coordinate space; $Y(\cdot)$ is the dynamic response of the structure; $g_j(\cdot)$ is the performance function corresponding to the jth dynamic failure model; λ_j is the bound value corresponding to the function $g_j(\cdot)$; P_f^* is the permissible value of the failure probability of structural system; $G_i(\cdot)$ is the dynamic constraint function; $f_i(\cdot)$ is the static constraint function; α_i^*, β_i^* and γ_i^* are the allowable values of the constraint functions respectively; $\omega_i^{(u)}$ and $\omega_i^{(l)}$ are the pregiven upper

*Recall again, that the term first passage refers to exceeding the structural design stress limit.

and lower limits of the ith eigenvalue of the structure. n_1, n_2, n_3 and n_4 are the total number of each type of constraint, respectively.

It should be noted that the constraints (8-2), (8-3) and (8-4) are all functions of the space and time simultaneously. They should be satisfied at each point in space R^n within the time interval $[0, T^*]$. Constraint (8-5) means the added conditions of size, weight and so on as indicated by practical implementation of the structure. The structural eigenvalue is constrained in the form of constraint (8-6) in order to avoid the phenomenon of resonance.

The main difficulty in solving the (8-1)–(8-6) results from dealing with the constraints (8-2)–(8-4), particularly the dynamic reliability constraint (8-2). One basic method for dealing with time related constraint functions is to eliminate the time factor so that it could become equivalent to a static constraint. Based on this consideration, Zhu[32] has presented the following equivalent method for dealing with constraints (8-2)–(8-4).

In constraint (8-2), denote the occurrence of the jth dynamic failure model by the random event E_j, that is to say,

$$E_j = \left\{g_j[Y(\mathbf{D}, \mathbf{X}, t)] \geqslant \lambda_j\right\} (0 \leqslant t \leqslant T^*, \mathbf{X} \in R^n) \quad (j = 1, 2, \cdots, m) \tag{8-7}$$

the constraint (8-2) can be expressed by the event $E_j (j = 1, 2, \cdots, m)$ as,

$$P_f(\mathbf{D}, t) = P\left\{\cup_{j=1}^m E_j\right\} \leqslant P_f^* \tag{8-8}$$

If there is only one failure point in each failure model, the coordinate $\mathbf{X}$ in formula (8-2) can be taken as the coordinate value at the failure point. In addition, according to the property that the dynamic failure probability is a nondecreasing function of the time t, and the maximum failure probability occurs at time $t = T^*$, the resultant probability of the jth failure model can be expressed by its maximum value $q_j(\mathbf{D}, T^*)$ in the following form.

$$P\{E_j\} = q_j(\mathbf{D}, T^*)(j = 1, 2, \cdots, m) \tag{8-9}$$

Since the correlation among all the failure models is very difficult to determine, the probability of the event on the left hand side of formula (8-8) is generally difficult to determine exactly. As a result of this, only two extreme situations such as complete correlation and complete independence among all the events $E_j (j = 1, 2, \cdots, m)$ are considered in a practical computation,respectively.

If $E_j (j = 1, 2, \cdots, m)$ are correlated each other completely,

$$P\left\{\cup_{j=1}^m E_j\right\} = \max(P\{E_j\}) \tag{8-10}$$

If $E_j(j = 1, 2, \cdots, m)$ are independent of each other,

$$P\left\{\cup_{j=1}^{m} E_j\right\} = \sum_{j=1}^{m}(P\{E_j\}) \tag{8-11}$$

But, actually, the value of $P\left\{\cup_{j=1}^{m} E_j\right\}$ will be certainly located between the above two extreme values, i.e.,

$$\max(P\{E_j\}) \leqslant P\left\{\cup_{j=1}^{m} E_j\right\} \leqslant \sum_{j=1}^{m} P\{E_j\} \tag{8-12}$$

Substituting formula (8-9) into the above formula gives

$$\max\{q_j(\mathbf{D}, T^*)\} \leqslant P\left\{\cup_{j=1}^{m} E_j\right\} \leqslant \sum_{j=1}^{m} q_j(\mathbf{D}, T^*) \tag{8-13}$$

According to the above expression and based on the consideration of safety, the original dynamic reliability constraint (8-2) can be described by the equivalent static active constraint,

$$\sum_{j=1}^{m} q_j(\mathbf{D}, T^*) \leqslant P_f^* \tag{8-14}$$

A simple method of eliminating the dependence of the left hand side in formulae (8-3) and (8-4) upon both the space coordinate $\mathbf{X}$ and the time t is to replace the original function by the supremum of the function in the space and the time interval. Likewise, the constraints (8-3) and (8-4) can also be expressed by the equivalent static active constraint separately in the following form,

$$\sup_{(\mathbf{X},t)} \{G_i[Y(\mathbf{D}, \mathbf{X}, t)]\} \leqslant \alpha_i^*, (i = 1, 2, \cdots, n_1) \tag{8-15}$$

$$\sup_{(\mathbf{X},t)}\{E\{G_i[Y(\mathbf{D}, \mathbf{X}, t)]\} \leqslant \beta_i^*(i = 1, 2, \cdots, n_2) \tag{8-16}$$

Up to this point, the constraints of the original problem have been transformed into static constraints. Therefore, the problem of dynamic reliability based optimum design can be transformed into a problem of static reliability based on optimum design of structures.

Example:

This example comes from the reference.[33] Suppose a vehicle is driving with a constant speed along a road, the mechanical model can be simplified

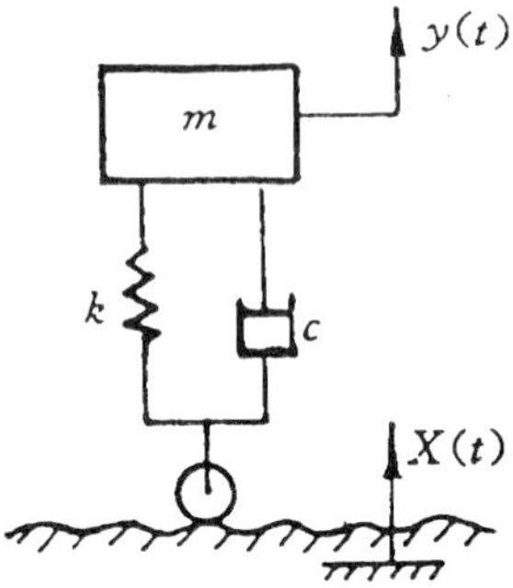

Figure 19. The vibration system with a single degree-of-freedom.

as a vibration system with a single degree-of-freedom (in Figure 19). The excitation $\mathbf{X}(t)$ caused by the road surface roughness is a stationary process with zero mean value, and $y(t)$ is the response process of the absolute displacement of the system. The problem is to find the design vector $\mathbf{D} = (\omega_o, \xi)$ and to minimize the objective function (the expected value $E[\ddot{y}_{max}(\mathbf{D})]$ of the absolute acceleration peak) by satisfying the following constraint conditions.

(1) dynamic reliability constraint of the relative displacement

$$P\{|y(t) - X(t)| \geqslant \lambda_0\} \leqslant P_f^*, \qquad (0 \leqslant t \leqslant T^*) \tag{a}$$

(2) constraint of the moment of relative displacement

$$E[|y(t) - X(t)|]^2 \leqslant 0.3^2\lambda_0^2, \qquad (0 \leqslant t \leqslant T^*) \tag{b}$$

(3) damped constraint

$$\xi^{(l)} \leqslant \xi \leqslant \xi^{(u)} \tag{c}$$

(4) constraint of eigenvalue

$$\omega^{(l)} \leqslant \omega_0 \leqslant \omega^{(u)} \tag{d}$$

where, ω_0 and ξ are the natural frequency and the damped ratio of the system, respectively.

The power spectrum density of $X(t)$ is $S_x(\omega) = S_0\frac{\omega_r^2}{\omega^2}(\omega^{(l)} \leqslant \omega \leqslant \omega^{(u)})$, in which, S_0 is the ratio factor and ω_r is the reference frequency. The parameter's values are:

$$\omega^{(l)} = 0.1\omega_r, \quad \omega^{(u)} = 1000\omega_r, \quad T^* = 600/\omega_r\ txtand\ P_f^* = 0.001$$

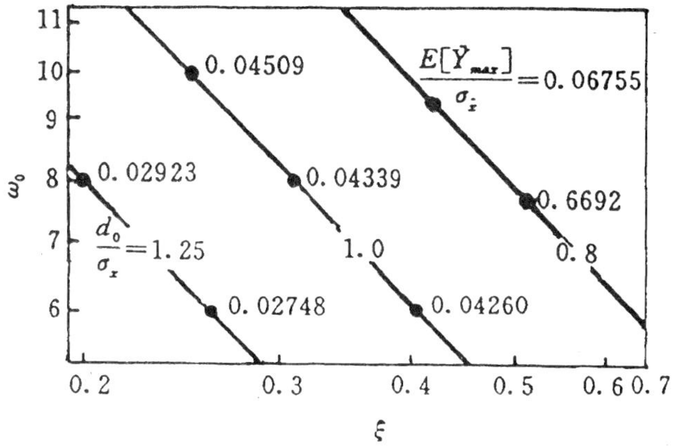

Figure 20. The optimum results for different limits of λ_0, ω_0 and ξ.

It can be determined from the constraints (a) and (b) that the left hand side of the dynamic reliability constraint (a) should be computed by means of a Poisson process model, because the limit λ_0 of the relative displacement is more than three times as much as the value of the mean square deviation. In addition, as the system tends to be a stationary response, the t in the left hand side of constraint (a) can be taken as T^* and the left hand side of constraint (b) will be independent of the time, t.

To solve this programming problem, the mixed sequential unconstrained minimizing technique (MSUMT) is employed. The optimum results for different limit λ_0, ω_0 and ξ are shown in Figure 20. For instance, if $\lambda_0 = \sigma_x$, the optimum design results are: $\omega_0^* = 10$ rad/s and $\xi^* = 0.25$. In this case, the value of objective function is $E[\ddot{y}_{\max}] = 0.04509\sigma_{\ddot{X}}$.

5.23. RELIABILITY OPTIMUM DESIGN BASED ON THE FAILURE MECHANISM OF THE REPETITIVE STRESS

It is easy to determine from the knowledge of structural dynamic mechanics that the main failure mechanism of a structure is failure resulting from repetitive stress if the peak value of the stress response random process $S(t)$ is lower than the maximum allowable stress. For this case, the dynamic reliability constraint of the structure should be replaced by the fatigue life constraint of the element. That is to say,

$$T_e(\mathbf{D}) \geqslant T^* \qquad (e = 1, 2, \cdots, n_e) \tag{8-17}$$

in the above formula, T^* is the useful design life of a structure; $T_e(\mathbf{D})$ is the fatigue life of the eth element, which is usually an implicit complex function of design variables $\mathbf{D}$; n_e is the total number of elements of the structure.

From formula (11-10) in part one, the computational formula of $T_e(\mathbf{D})$ is given as,

$$T_e(\mathbf{D}) = \frac{C_e}{\delta_e \nu_o^+(e)(\sqrt{2}\sigma_{se})^{\beta_e} \cdot \Gamma(1 + \frac{\beta_e}{2})} \tag{8-18}$$

where: C_e and β_e are the material constants of element (e); σ_{se} is the r.m.s. value of the stress response $S_e(t)$ of the eth element; $\nu_0^+(e)$ is the crossing rate of the stress response $S_e(t)$ of the eth element with respect to zero line; δ_e is the adjustment factor of the eth element's fatigue life, and its value is discussed in the eleventh section of part one.

Therefore the expressions in the model of structural reliability optimization based on the repetitive stress failure mechanism will be the same as usual except for replacing the dynamic reliability constraint (8-2) of the system by the fatigue life constraint (8-17) of the element.

Jha[34] has presented techniques for weight minimizing design for a satellite antenna structure based upon the fatigue life constraint.

5.24. THE DISPLAYING OF THE RELIABILITY CONSTRAINTS

In structural reliability based optimization, the reliability constraint is usually expressed in the form of probability, and, is therefore a complex implicit function of the design variables. This leads to the traditional optimization method, particularly those sensitivity information based algorithms can not be employed. If the reliability constraints can be expressed in terms of traditional constraint, not only the traditional methods can be utilized directly but also tedious reliability computations can be avoided in each step of the optimization. Recently some work regarding the reliability constraints has been developed by a number of researchers.[35,36]

The results presented here are by Chen and Duan.[35] To begin with, the original reliability constraint is expressed in the form of a reliability index through the theory of second order moments. Then, the original reliability constraint is equivalently displayed in the same form as the traditional safety design criteria by introducing the reliability safety coefficient.

The general expression of the element reliability constraint of the structure can be stated as,

$$P_r \geqslant P_r^* \tag{9-1}$$

$$P_r = P_{\text{rob}}\{Z(R, S) > 0\} \tag{9-2}$$

in which, $Z(R, S)$ is the performance function of the element, which can always be described as the function of both the element strength random variable R and the working stress random variable S.

If the distribution of function Z is known, the constraint (9-1) can be expressed in the form of the probability density function $f_Z(\cdot)$ or distribution function $F_Z(\cdot)$ of Z as follows,

$$\int_{o}^{\infty} f_Z(z)dz = 1 - F_Z(0) \geqslant P_r^* \tag{9-3}$$

If the distribution of the function Z is unknown, the constraint (9-1) can be expressed by a reliability index by the method of the second order moment in the following form.

$$\beta \geqslant \beta^* \tag{9-4}$$

$$\beta^* = \Phi^{-1}(P_r^*) \tag{9-5}$$

$$\beta = \Phi^{-1}(P_r) = \frac{\mu_Z}{\sigma_Z} \tag{9-6}$$

If $Z(R, S) = R - S$, i.e., the performance function expresses the strength redundancy of the element, the active expression of the inequality constraint (9-4) can be written as,

$$(\mu_R - \mu_S)(\sigma_R^2 + \sigma_S^2)^{-1/2} = \beta^* \tag{9-7}$$

Substituting the variation coefficients ν_R and ν_S of variables R and S into the above formula gives the following quadratic equation,

$$(-\beta^{*2}\nu_R^2)K_\beta^2 - 2K_\beta + (1 - \beta^{*2}\nu_S^2) = 0 \tag{9-8}$$

where,

$$K_\beta = \frac{\mu_R}{\mu_S} \tag{9-9}$$

is called the reliability safety coefficient, whose mean is no longer the same as that in traditional safety coefficient. The value of K_β can be found from,

$$K_\beta = \frac{1 + \beta^*(\nu_R^2 + \nu_S^2 - \beta^{*2}\nu_R\nu_S)^{1/2}}{1 - \beta^{*2}\nu_R^2} \tag{9-10}$$

If $Z(R, S) = R/S - 1$, the active expression of the constraint (9-4) and the corresponding reliability safety coefficient can be written as,

$$\left(1 - \frac{\mu_S}{\mu_R}\right)\left(v_R^2 + v_S^2\right)^{-1/2} = \beta^* \tag{9-11}$$

$$K_\beta = \frac{1}{1 - \beta^*(\nu_R^2 + \nu_S^2)^{1/2}} \tag{9-12}$$

If $Z(R, S) = \ln(R/S)$, the critical expression of the constraint (9-4) and the corresponding reliability safety coefficient become,

$$\ln\left(\frac{\mu_R}{\mu_S}\right)(\nu_R^2 + \nu_S^2)^{-1/2} = \beta^* \tag{9-13}$$

and

$$K_\beta = \exp[\beta^*(\nu_R^2 + \nu_S^2)^{1/2}] \tag{9-14}$$

It can be observed from the above formula that when the variation coefficient v_R and v_S of the variables R and S are fixed, the reliability safety coefficient K_β can only be determined by the pregiven reliability index. From formula (9-9), the original reliability constraint (9-1) can be expressed by the reliability safety coefficient as follows,

$$\mu_R \geqslant K_\beta \mu_S \tag{9-15}$$

where, the strength mean μ_R can be obtained from the statistical information of the variable R directly, while the stress mean $\mu_S = \mu_S[E(\mathbf{L}), \mathbf{X}]$ is a function of both means of the random loading vector $\mathbf{L}$ and the design vector $\mathbf{X}$, the value of μ_S can be determined by structural analysis.

The above formula (9-15) is the expression of the reliability constraint of the original problem. Its form is the same as the traditional safety design criteria. It shows that the problem of structural elements which satisfy reliability constraints is equivalent to the problem that the strength mean of the element is greater than or equal to the product of element's stress mean and the reliability safety coefficient. When variables R and S are given by the normal distribution, formula (9-15) will be an exact expression.

Example:

This is a 10 bar static undetermined truss structure (as shown in Figure 21), about which the paper[35] has shown some details. The loading case: two loads L_1 and L_2 are all subjected to the same normal distribution with both the mean $\mu_{Lk} = 100$ K and the variation coefficient $\nu_{Lk} = 0.1(k = 1, 2)$. Ten bars have the same material with density $\rho = 0.1$ lb/in^3 and elasticity $E = 10^7$ psi. The tension and the compression strength for all bars are the same. The design variables are taken as both the cross-sectional areas of ten bars and the coordinates of joint 1, 3 as well as the y coordinate of joint 5. i.e.,

$$(\mathbf{A}, \mathbf{X}) = (A_1, \cdots, A_{10}, x_1, y_1, x_3, y_3, y_5)^T \tag{a}$$

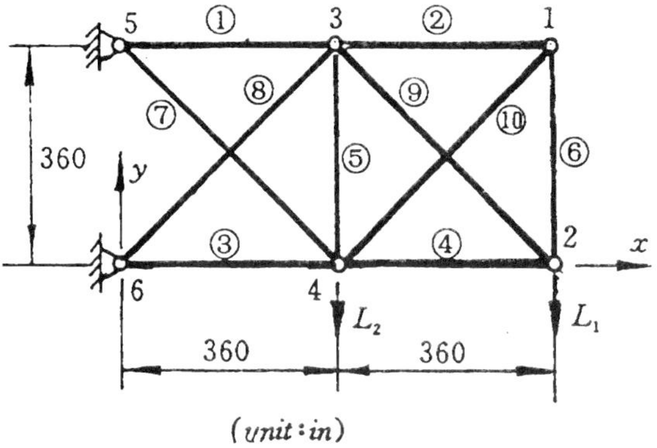

Figure 21. The 10-bar truss structure.

The problem is to minimize the structural weight

$$\mathbf{W}(\mathbf{A}, \mathbf{X}) = \sum_{i=1}^{10} \rho_i l_i A_i \tag{b}$$

whilst satisfying the following constraints:
(1) displacement reliability

$$\min_{i=1,4} \{P_{\delta_i}(\mathbf{A}, \mathbf{X})\} = \min_{i=1,4} (P\{[\delta] - \delta_{iy} \geqslant 0\}) \geqslant P_\delta^* \tag{c}$$

(2) element reliability

$$\min_{j=1,10} \{P_{\sigma_j}(\mathbf{A}, \mathbf{X})\} = \min_{j=1,10} (P\{[\sigma] - \sigma_j \geqslant 0\}) \geqslant P_\sigma^* \tag{d}$$

(3) bound

$$A_j \geqslant A_{j\,\min}^* (j = 1, 2, \cdots, 10) \tag{e}$$

in which, δ_{iy} is the displacement random variable at the ith joint in the y-direction; $[\delta]$ is the allowable displacement value (determinant); σ_j is the working stress random variable of the jth bar; $[\sigma]$ is the tension and the compression strength of the element (determinant). In this example, the predetermined parameters include: $[\delta] = 2.0$ in, $[\sigma] = 25000$ psi, $P_\delta^* = 0.985$ and $P_\sigma^* = 0.9999$.

By using the equations (9-10) and (9-15) of the reliability constraints, the equivalent expressions for the constraints (c) and (d) become, the equivalent displacement constraint

$$[\delta] \geqslant K_\beta \cdot \min_{i=1,4} \{\mu_{\delta iy}\} \tag{c'}$$

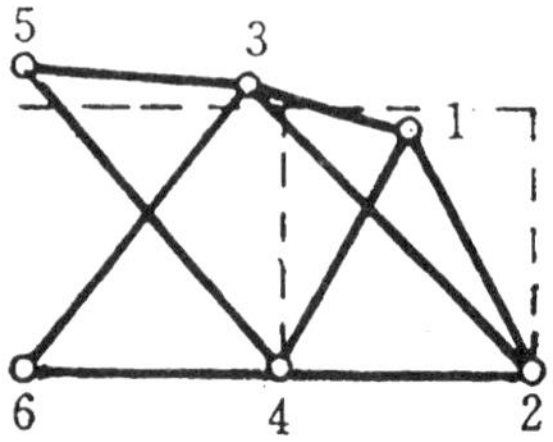

Figure 22. The optimum result of 10-bar truss.

where

$$K_\beta = 1 + \Phi^{-1}(P_\delta^*) \cdot \nu_{Lk} = 1 + \Phi^{-1}(0.985) \times 0.1 = 1.217$$

and the equivalent stress constraint

$$[\sigma] \geqslant K_\beta \cdot \min_{j=1,10} \{\mu_{\sigma j}\} \qquad \text{(d')}$$

where

$$K_\beta = 1 + \Phi^{-1}(P_\sigma^*) \cdot \nu_{Lk} = 1 + \Phi^{-1}(0.9999) = 1.371$$

Up to this stage, the original reliability optimization problem has already been transformed into a traditional NLP Problem, which is a kind of mixed method presented by Duan and Ye[38] to solve this problem. In order to check the influence of the loading correlation upon the optimum result, both the load correlating coefficients of $\rho_{L_1L_2} = 0$ (completely independent of each other) and of $\rho_{L_1L_2} = 1.0$ (completely correlated of each other) are taken, respectively, in practical computation. The optimum results are given in Figure 22 and Table 7.

From the optimum result, it can be seen that the load's correlation has a strong influence upon the optimum design. Further, the higher the correlation is, the worse the loading conditions. Consequently, the optimum solution value of variables will become bigger. This conclusion is in accordance with the results in Section 4.

Table 7. The optimum result of 10-bar truss

Variables	A_1	A_2	A_3	A_4	A_5	A_6	A_7	A_8	A_9	A_{10}
Initial value	35	35	35	35	35	35	35	35	35	35
$\rho = 0.0$ design results	25.19	0.1	20.96	20.60	0.1	0.38	17.10	20.82	22.32	0.1
$\rho = 1.0$ design results	27.82	0.1	21.96	18.61	0.1	0.18	19.70	20.04	25.37	0.1

Table 7. (continued)

Variables	x_1	y_1	x_3	y_3	y_5	W	$\min\{P_{\delta_i}\}$	$\min\{P_{\sigma_j}\}$
initial value	720	360	360	360	360	14687.6	1.0000	1.0000
$\rho = 0.0$ design results	542.62	346.69	290.24	400.00	428.34	5565.8	0.9890	1.0000
$\rho = 1.0$ design results	540.00	337.83	340.50	400.00	415.00	5959.5	0.9902	1.0000

5.25. THE EQUIVALENT DETERMINATION OF THE RANDOM PROGRAMMING PROBLEM

Random programming (RP) is a newly emerged method stemming from mathematical programming. In mathematical programming, if some or all the design variables are viewed as the random parameters, both the objective and the constrained functions are all the functions of the random variables, this programming problem is called the random programming problem. Seen from this point of view, the reliability optimization problem with the specified design variables presented in previous sections is just one special case of the RP problem. As for solving the RP problem, there is no doubt about that solving it is more difficult than the traditional NLP problem. Even for the linear random programming, its solution technique are not matured yet.[37] In order to overcome this difficulty, some researchers have presented alternative treatment techniques — transforming the RP problem into a traditional programming problem. Recently, Chen etc.[36] have developed a technique such that the RP problem can be transformed into an equivalent traditional programming problem for the case of normal distributed design variables. The procedure is described as follows.

Suppose the random programming structure is:

$$\begin{aligned} &\text{find random variables} && \mathbf{X} = (x_1, x_2, \cdots, x_n)^T \\ &\text{minimize} && F(\mathbf{X}) && (10\text{-}1) \\ &\text{subject to} && P_{fj}(\mathbf{X}) = P\{G_j(\mathbf{X}) \leqslant 0\} \leqslant P_{fj}^* \\ & && (j = 1, 2, \cdots, m) && (10\text{-}2) \end{aligned}$$

where, $x_i (i = 1 \sim n)$ are independent of each other and have a normal distribution, i.e.,

$$x_i \sim \mathcal{N}(\mu_i, \sigma_i^2) \qquad (i = 1, 2, \cdots, n) \tag{10-3}$$

In this model, $F(\mathbf{X})$ and $G_j(\mathbf{X})(j = 1, 2, \cdots, m)$ are the nonlinear functions of $\mathbf{X}$, and $G_j(\mathbf{X})$ is the performance function corresponding to the jth failure model of structure. $P_{fj}(\mathbf{X})$ and P_{fj}^* are the system probability and the allowable limit of the jth failure model, respectively. m is the total number of the failure models of the structure.

The above problem (10-1) and (10-2) is an RP problem. In order to transform it into a deterministic programming problem, the functions $F(\mathbf{X})$ and $G_j(\mathbf{X})(j = 1, 2, \cdots, m)$ are expanded in the Taylor series near the mean point $\boldsymbol{\mu} = (\mu_1, \cdots, \mu_n)^T$ of variables $\mathbf{X}$. And the first three terms of $F(\mathbf{X})$ and the first two terms of $G_j(\mathbf{X})$ are retained, i.e.,

$$(\mathbf{X}) = F(\boldsymbol{\mu}) + \sum_{i=1}^{n} \left(\frac{\partial F}{\partial x_i}\right)_{\mathbf{X}=\boldsymbol{\mu}} \cdot (x_i - \mu_i) + \sum_{i=1}^{n} \sum_{k=1}^{n} \left(\frac{\partial^2 F}{\partial x_i \partial x_k}\right)_{\mathbf{X}=\boldsymbol{\mu}} \cdot (x_i - \mu_i)(x_k - \mu_k) \tag{10-4}$$

$$G_j(\mathbf{X}) = G_j(\boldsymbol{\mu}) + \sum_{i=1}^{n} \left(\frac{\partial G_j}{\partial x_i}\right)_{\mathbf{X}=\boldsymbol{\mu}} \cdot (x_i - \mu_i)(j = 1, 2, \cdots, m) \tag{10-5}$$

Taking the expectation on both sides of the above formulae and considering the independence among the variables $x_i (i = 1 \sim n)$ yields

$$E[F(\mathbf{X})] = F(\boldsymbol{\mu}) + \sum_{i=1}^{n} \left(\frac{\partial^2 F}{\partial x_i^2}\right)_{\mathbf{X}=\boldsymbol{\mu}} \cdot \sigma_i^2 \tag{10-6}$$

$$E[G_j(\mathbf{X})] = G_j(\boldsymbol{\mu}) \quad (j = 1, 2, \cdots, m) \tag{10-7}$$

Taking the variation about both sides of formula (10-5) and noticing the independence of variables $x_i (i = 1 \sim n)$ leads to

$$D[G_j(\mathbf{X})] = \sum_{i=1}^{n} \left(\frac{\partial^2 G_j}{\partial x_i^2}\right)_{\mathbf{X}=\boldsymbol{\mu}} \cdot \sigma_i^2, \quad (j = 1, 2, \cdots, m) \tag{10-8}$$

It can be seen from formula (10-5) that, according to the retention of normality for linear combinations of normal variables, the function $G_j(\mathbf{X})$ can also be considered to be a normal distribution, approximately. Consequently, the original reliability constraint (10-2) is equivalent to an inequality constraint by employing the theory of the second moment of reliability in the following form.

$$\beta_j(\mathbf{X}) \geqslant \beta_j^* \quad (j = 1, 2, \cdots, m) \tag{10-9}$$

where

$$\beta_j(\mathbf{X}) = \frac{E[G_j(\mathbf{X})]}{\sqrt{D[G_j(\mathbf{X})]}} \quad (j = 1, 2, \cdots, m) \tag{10-10}$$

$$\beta_j^* = -\Phi^{-1}(P_{fj}^*) \quad (j = 1, 2, \cdots, m) \tag{10-11}$$

Substituting formulae (10-7) and (10-8) into formula (10-9) gives the following equivalent deterministic expression of the original constraint (10-3),

$$G_j(\boldsymbol{\mu}) - \beta_j^* \left[\sum_{i=1}^{n} \left(\frac{\partial^2 G_j}{\partial x_i^2} \right)_{\mathbf{X}=\boldsymbol{\mu}} \cdot \sigma_i^2 \right]^{1/2} \geqslant 0 \quad (j = 1, 2, \cdots, m) \tag{10-12}$$

Here, the original RP problem (10-1) and (10-2) can be transformed into the following determined programming problem.

$$\begin{aligned}
&\text{find} && (\boldsymbol{\mu}, \boldsymbol{\sigma}) = (\mu_1, \mu_2, \cdots, \mu_n, \sigma_1, \sigma_2, \cdots, \sigma_n) \\
&\text{min. (or max)} && E[F(\mathbf{X})] = F(\boldsymbol{\mu}) + \sum_{i=1}^{n} \left(\frac{\partial^2 F}{\partial x_i^2} \right)_{\mathbf{X}=\boldsymbol{\mu}} \cdot \sigma_i^2 && (10\text{-}13) \\
&\text{s.t.} && G_j(\boldsymbol{\mu}) - \beta_j^* \left[\sum_{i=1}^{n} \left(\frac{\partial^2 G_j}{\partial x_i^2} \right)_{\mathbf{X}=\boldsymbol{\mu}} \cdot \sigma_i^2 \right]^{1/2} \geqslant 0 \\
& && (j = 1, 2, \cdots, m) && (10\text{-}14) \\
& && \sigma_j \geqslant 0 \quad (j = 1, 2, \cdots, n) && (10\text{-}15)
\end{aligned}$$

Obviously, this is a general NLP problem with $2n$ design variables and $(m + n)$ constraints, which can be solved by the mature algorithms in mathematics. Solving this problem can find not only the optimum means of the variables but also the optimum variations of the variable.

Example 1:

The structural resistance force $R \sim \mathcal{N}(\mu_R, \sigma_R^2)$, the load $S \sim \mathcal{N}(\mu_S, \sigma_S^2)$. The total cost function of the structure is supposed to be

$$F(\mu_R, \mu_S, \sigma_R, \sigma_S) = 0.2\mu_R^{1.5} + 100\sigma_R^{-1.2} + 100\mu_S^{-1.6} + 50\sigma_S^{-0.7}$$

The failure probability of the structure is limited as $P_f^* = 0.01$. Therefore the specific problem can be stated in the words that to find the optimum mean and variation of the variable (R, S) and also to minimize the total cost of the structure.

The problem can be shown in mathematical language as follows.

$$\begin{aligned} &\text{find} && \mathbf{X} = (\mu_R, \mu_S, \sigma_R, \sigma_S) \\ &\text{min.} && F(\mu_R, \mu_S, \sigma_R, \sigma_S) && \text{(a)} \\ &\text{s.t.} && P_f(\mathbf{X}) = P\{R - S \leqslant 0\} \leqslant P_f^* = 0.01 && \text{(b)} \end{aligned}$$

In this model, since the objective F itself is a function of the variable's mean and variation, it is unnecessary to expand it in a Taylor series. As the performance function $G(R, S) = R - S$ is the linear function of design variables, it is also unnecessary to make a Taylor series expansion. According to the theory of the second moment, the equivalent deterministic expression of the constraint (b) can be written as,

$$\mu_R - \mu_S - \beta^*(\sigma_R^2 + \sigma_S^2)^{-1/2} \geqslant 0 \qquad \text{(b}'\text{)}$$

in which

$$\beta^* = -\Phi^{-1}(P_f^*) = -2.33$$

So the original RP problem can be transformed into a deterministic nonlinear programming problem described by formulae (a) and (b′). By the Lagrange multiplier method, the optimum design of the problem can be obtained as follows.

$$\mu_R^* = 27.1250, \quad \sigma_R^* = 5.6124, \quad \mu_S^* = 9.7720, \quad \sigma_S^* = 4.8929, \quad \text{and}$$
$$F^* = 8278.83$$

Example 2:

This is the dual problem of the example 1, whose mathematical model is,

$$\begin{aligned} &\text{find} && \mathbf{X} = (\mu_R, \mu_S, \sigma_R, \sigma_S) \\ &\text{max.} && P_r(\mathbf{X}) = P\{R - S > 0\} && \text{(c)} \\ &\text{s.t.} && F(\mathbf{X}) = 0.2\mu_R^{1.5} + 100\sigma_R^{-1.2} + 100\mu_S^{-0.6} + 50\sigma_S^{-0.7} = 7500 && \text{(d)} \end{aligned}$$

where, $P_r(\mathbf{X})$ is the reliability function of structure.

By the method of the second moment, the objective function (c) can be expressed as an equivalent deterministic formation,

$$\beta = (\mu_R - \mu_S)(\sigma_R^2 + \sigma_S^2)^{-1/2} \to \text{max.} \qquad \text{(c}'\text{)}$$

Then, the original RP problem is transformed into a deterministic NLP problem as described in formulae (c′) and (d).The optimum solution by the Largrange multiplier method is:

$$\mu_R^* = 24.9380,\quad \sigma_R^* = 6.5254,\quad \mu_S^* = 10.0375,\quad \sigma_S^* = 5.8513,$$
$$\beta^* = 1.7001 \text{ and } P_r^* = \Phi(\beta^*) = 0.95543.$$

5.26. CONCLUSIONS

The reliability based optimization of engineering structures has become an important research subject as that the mathematical model becomes more rational, that the optimum results are more practical, and so on. Some interesting work has been presented since the paper about the structural reliability optimization, by Hillon and Feigen, was published three decades ago. A number of researchers such as Moses, Stevenson, Susitgky, Murthy, Murotsu, Thoft-Christensen and so on have made some worthwhile contributions to this area. Particularly, Moses is not only the person who open up the researching area of structural reliability based optimization but also the person who has made an important contribution to the researching area of structural system reliability. With the development of the structural reliability theory and method, the reliability based optimization of structures has shown its rapidly development. However, lots of work about the theory, method and application of the structural reliability optimization need to be done further. Seeing from this point of view, it is necessary to give more deeply research to the following research subjects:

(1) sensitivity analysis of the random function;
(2) optimum distribution of the reliability for structural system or each subsystem;
(3) shape and topological optimum design of structures by considering the reliability;
(4) structural optimum design based on the dynamic reliability;
(5) fuzzy reliability based optimum design of structures;
(6) random programming method for structural optimum design.

References

1. Lu, Z.Z. and Y.S. Feng, 1993, A preliminary analysis method for reliability of structural vibration, *Mechnical Strength*, **15**(2), 1–6 (in Chinese).
2. Nigam, N.C., 1983, *Introduction to Random Vibrations*, The MIT Press.

3. Rice, S.O., Mathematical analysis of random noise, *Bell System Technical Journal*, **23**, 282–332, 1994, and **24**, 52–162, 1995.
4. Middleton, D., 1960, *An Introduction to Statistical Communication Theory*, McGraw-Hill.
5. Huston, W.B. and T.H. Skopinski, *Probability and Frequency Characteristics of Some Flight Buffer Load*, NACA TN3733.
6. Shinoguka, M. and J.H. Yang, 1971, Peak structural response to nonstationary random excitation, *J. Sound Vibration*, **14**(4), 505–517.
7. Yang, J.N. and S.C. Liu, 1980, Statistical Interpretation and Application of Response Spectrum, *Proc. 7th WCEE6*, pp. 657–664.
8. Cramer, H., 1966, On the intersection between the trajectories of a normal stationary stochastic process and a high level, *Arkiv. Mat.*, **6**(2), 337–349.
9. Lin, Y.K., 1967, *Probabilistic Theory of Structural Dynamics*, McGraw-Hill.
10. Vanmarcke, E.H., *First passage and other failure criteria in the narrow-band random vibration: A discrete state approach*, Ph.D. Dissertation, Dept. of Civil Engineering, MIT.
11. Crandall, S.H., 1970, First-crossing probabilities of the linear oscillator, *Journal of Sound and Vibration*, **12**(3), 285–299.
12. Corotis, R.B., E.H. Vanmarcke and C.A. Cornell, 1972, First passage of nonstationary random process, *J. Eng. Mech. Div., Proc. ASCE 98(EMI)*, pp. 401–414.
13. Chen, J.J., B.Y. Duan and Y.G. Zen, 1977, Study on dynamic reliability analysis of the structures with multidegree-of-freedom system, *Computers & Structures*, **62**(5), 877–881.
14. Miner, M.A., 1945, Cumulative fatigue damage, *Trans. ASME, J. Appl. Mech.*, **12**(3), A159–A169.
15. Wirshing, P.H. and C.L. Mark, 1980, Fatigue under the wide band random stress, *J. Structure Devision ASCE.*
16. Lin, S.S., 1988, Application of Mechanical Reliability in Strength Design and Fatigue Life Prediction, *Astronautic Press*, Beijing (in Chinese).
17. Bolotin, V.V., 1969, *Statistical method in structural mechanics*, San Francisco, Holder-Day, (English translation).
18. Li, G.Q., H. Cao, Q. S.Li and D. Huo, 1993, Theory of structural dynamic reliability and its application, *Earthquake Press*, Beijing (in Chinese).
19. Augusti, C., A. Baratta and F. Casciati, 1984, *Probabilistic method in structural engineering*, Chapman and Hall.
20. Thoft-Christensen, P. and Y. Murotsu, 1986, *Application of structural systems reliability theory*, Springer-Verlag.
21. Murotsu, Y., S. Miwa, K. Niwa and K. Taguchi, 1981, Design of truss structure with specified values of reliability (in Japanese), *Transactions of Japan Society of Mechanical Engineers*, **47**(416), 211–220.
22. Murotsu, Y., M. Kishi, M. Yonezawa and K. Taguchi, 1984, Probabilistically Optimum design of frame structure, in: P. Thoft-Christensen (ed), *System Modelling and Optimization*, Springer-Verlag, pp. 545–554.
23. Moses, F. and J.D. Stevenson, 1970, Reliability based structural design, *J. of the Structural Division, ASCE*, **96**(ST2), 221–244.
24. Murotsu, Y., M. Yonezawa, F. Oba and K. Niwa, 1979, Method for reliability analysis of structures, in: Burns, J.J., Jr. (ed), *Advances in Reliability and Stress Analysis, ASME Publication*, **H00119**, 3–21.
25. Ditlevsen, O., 1979, Narrow reliability bounds for structural system, *J. Struct. Mech.*, **7**, 453–472.
26. Murotsu, Y., M. Yonzawa, F. Oba and K. Niwa, 1979, *Optimum structural design under the constraint on failure probability*. ASME Paper, 79-DET-114.
27. Chen, J.J. and B.H. Dai, 1989, Optimum design of antenna structure based on reliability, *Acta Electronic Sinica*, Supplement, pp. 44–54.
28. Ruty, J., 1952, The effect of aperture errors on the antenna radiation pattern, *Supplual Nuovo Cimento*, **9**, 364–380.
29. Feng, Y.S. and F. Moses, 1986, A method of structural optimization based on structural system reliability, *J. Struct, Mech.*, **14**, 437–453.

30. Nigam, N.C., 1986, Optimum design of systems operating in random vibration environment, *Random Vibration-status and recent development*, Elsevier.
31. Rao, S.S., 1981, Reliability-based optimization under the random vibration environment, *Computers and Structures*, **14**, 345–355.
32. Zhu, W.Q., 1992, *Random Vibration*, Sicience and Technology Press, Beijing (in Chinese).
33. Dahlberg, T., 1977, Parametric optimization of a 1-DOF vehicle travelling on a randomly profiled road, *J. Sound Vib.*, **55**, 245–253.
34. Jha, V.K., T.S. Sankar and R.B. Bht, 1986, Optimization of aerospace structures subjected to random vibration and fatigue constraints, *The Shock and Vibration Bulletin*, **56**, Part 2.
35. Chen, J.J. and B.Y. Duan, 1994, Structural optimization by displaying the reliability constraints, *Computers and Structures*, **50**(6), 777–783.
36. Chen, S.H., D.T. Song and W.Z. Han, 1994, Stochastic optimization for vibration systems with the mutildegree of freedoms, *Acta Mechanica Sinica*, **26**(4), 432–439 (in Chinese).
37. Voll, P., 1976, *Stochastic Linear Programming*, Spring-Verlag.
38. Duan, B.Y. and S.H. Ye, 1987, A mixed method for shape optimization of skeletal structures, *Engineering Optimization*, **21**(3), 183–197.

COLOR PLATE I. *See* H. Saadatmanesh, Figure 17, page 163.

COLOR PLATE II. *See* H. Saadatmanesh, Figure 36, page 180.

COLOR PLATE III. *See* H. Saadatmanesh, Figure 37, page 180.

COLOR PLATE IV. *See* H. Saadatmanesh, Figure 39, page 182.

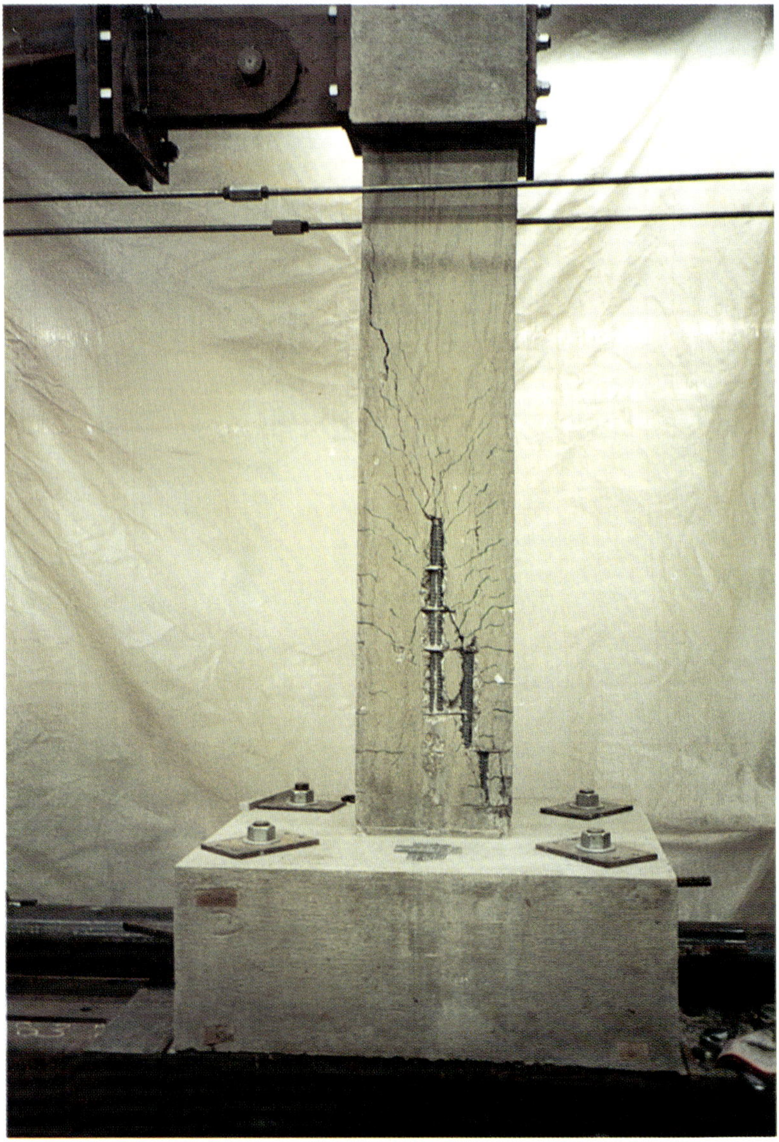

COLOR PLATE V. *See* H. Saadatmanesh, Figure 47, page 189.

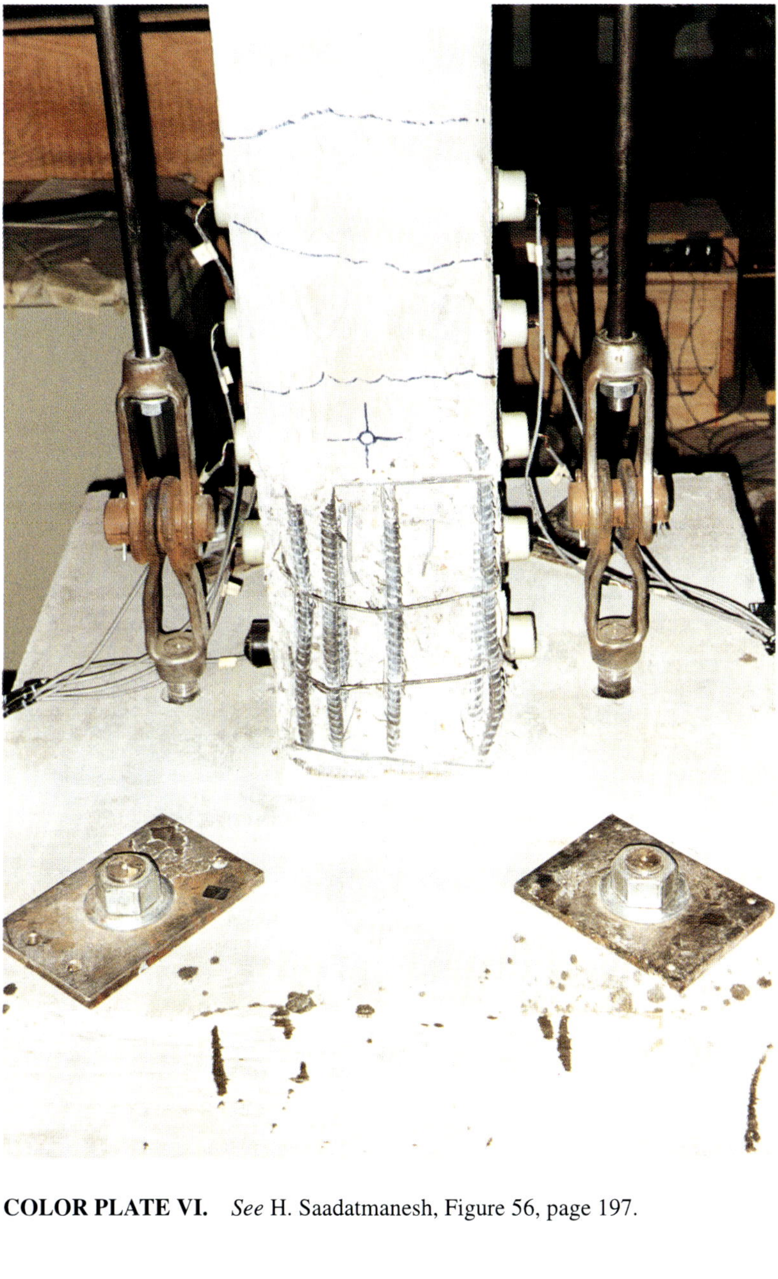

COLOR PLATE VI. *See* H. Saadatmanesh, Figure 56, page 197.